CURRENT CONCEPTS IN PLANT TAXONOMY

Editor-in-Chief
D. L. Hawksworth PhD DSc FLS FIBiol
Commonwealth Mycological Institute, Kew

Proceedings of an International
Conference held in Reading

THE SYSTEMATICS ASSOCIATION
SPECIAL VOLUME No. 25

CURRENT CONCEPTS IN PLANT TAXONOMY

Edited by

V. H. HEYWOOD

and

D. M. MOORE

Department of Botany, University of Reading, England

1984

Published for the

SYSTEMATICS ASSOCIATION

by

ACADEMIC PRESS

(Harcourt Brace Jovanovich, Publishers)

LONDON ORLANDO

SAN DIEGO SAN FRANCISCO NEW YORK TORONTO

MONTREAL SYDNEY TOKYO SÃO PAULO

ACADEMIC PRESS INC. (LONDON) LTD.
24–28 Oval Road
London NW1 7DX

US Edition published by
ACADEMIC PRESS INC.
(Harcourt Brace Jovanovich, Inc.)
Orlando, Florida 32887

British Library Cataloguing in Publication Data
Current concepts in plant taxonomy. – (The Systematics Association special volume, ISSN 0309–2593; no. 25)
1. Botany – Classification
I. Heywood, V. H. II. Moore, D. M.
III. Series
581′.012 QK95

ISBN 0–12–347060–9
LCCN 83–73145

TYPESET BY GLOUCESTER TYPESETTING SERVICES
PRINTED IN GREAT BRITAIN BY
ST EDMUNDSBURY PRESS, BURY ST EDMUNDS, SUFFOLK

Contributors

Ashton, P. S., *Arnold Arboretum of Harvard University, 22 Divinity Avenue, Cambridge, MA 02138, USA*
Ayensu, E. S., *Office of Biological Conservation, Smithsonian Institution, Washington, D.C. 20560, USA*
Barthlott, W., *Institut für Systematische Botanik und Pflanzengeographie, Freie Universität Berlin, Altensteinstrasse 6, D-1000 Berlin 33, Federal Republic of Germany*
Bisby, F. A., *Department of Biology, Building 44, The University, Southampton SO9 5NH, England*
Blackmore, S., *Department of Botany, British Museum (Natural History), Cromwell Road, London SW7 5BD, England*
Cullen, J., *Royal Botanic Garden, Inverleith Row, Edinburgh EH3 5LR, Scotland*
Cutler, D. F., *Royal Botanic Gardens, Kew, Richmond, Surrey TW9 3AB, England*
Funk, V. A., *National Museum of Natural History, Smithsonian Institution, Washington, D.C. 20560, USA*
Greilhuber, J., *Institut für Botanik und Botanischer Garten der Universität Wien, Rennweg 14, A-1030 Wien, Austria*
Harborne, J. B., *Department of Botany, The University, Whiteknights, Reading RG6 2AS, England*
Heywood, V. H., *Department of Botany, The University, Whiteknights, Reading RG6 2AS, England*
Humphries, C. J., *Department of Botany, British Museum (Natural History), Cromwell Road, London SW7 5BD, England*
Kay, Q. O. N., *Department of Botany and Microbiology, University College of Swansea, Swansea SA2 8PP, Wales*
Kubitzki, K., *Institut für Allgemeine Botanik und Botanischer Garten Hamburg, Ohnhorststrasse 18, D-2000 Hamburg 52, Federal Republic of Germany*
McNeill, J., *Department of Biology, Faculty of Science and Engineering, University of Ottawa, 30 Somerset E., Ontario, Canada K1N 6N5*
Moore, D. M., *Department of Botany, The University, Whiteknights, Reading RG6 2AS, England*
Prance, G. T., *New York Botanical Garden, Bronx, New York, NY 10458, USA*
Snaydon, R. W., *Department of Agricultural Botany, The University, Whiteknights, Reading RG6 2AS, England*

Stace, C. A., *Department of Botany, The University, University Road, Leicester LE1 7RH, England*
Tomlinson, P. B., *Harvard Forest, Harvard University, Petersham, MA 01366, USA*

Preface

This volume is based on papers presented at an International Conference on Current Concepts in Plant Taxonomy, organized under the auspices of the Systematics Association and held in the University of Reading on 7–9 July 1982. The Conference was attended by some 117 delegates from 11 countries.

Fifteen years ago a comparable conference assessed the then state of plant taxonomy. The ensuing proceedings ("Modern Methods in Plant Taxonomy", V. H. Heywood (ed.), Academic Press, 1968), proved remarkablys uccessful and were reprinted three times, almost unprecedented for a symposium volume. After one and a half decades it became apparent that a contemporary overview would again be profitable and a number of leading and active systematists were invited to present a critical but balanced assessment of their special fields, identifying challenges, problems and misconceptions as well as highlighting achievements and advances. The contributors accepted the challenge admirably, and some of them cast a very critical eye on the current procedures of taxonomy and even questioned basic assumptions. The results therefore constitute a valuable updating of taxonomists' views of taxonomy and of themselves.

A comparison of this volume with that published 15 years ago reveals the rise to prominence of chemosystematics and micromorphological studies, then still in their infancy, the continuing importance of biosystematics, though no longer holding the centre of the stage as it did in the 1960s, and major advances in the establishment and use of taxonomically orientated data-bases. A chapter on cladistics acknowledges a currently contentious subject that has affected plant taxonomy only marginally hitherto. Whether it will come to occupy the role that its proponents assert or whether it will be viewed as a divine afflatus is for the next generation to decide. The increased preoccupation with conservation in the past decade or so has led to serious reappraisals of the contributions made so far by taxonomists to inventories of the world's resources and these, in turn, have led taxonomists to look again at some of their more traditional practices.

The organization of the conference on which this book is based was sponsored by the Systematics Association, which also provided some funds. We are grateful to the Association, and to The Royal Society of London, which provided travel funds for some overseas speakers. We acknowledge the support of the University of Reading in arranging accommodation and lecture facilities, as well as for providing a much-enjoyed sherry reception in St Patrick's Hall. We are indebted to Miss Alice Hezlett and Miss Valerie Norris, our secretaries, for looking after so many of the arrangements before and during the conference. Finally, we must accord our appreciation of the assistance given by the editorial staff of Academic Press in seeing this volume through to publication.

Reading, *November 1983*

V. H. Heywood
D. M. Moore

Contents

Recent Approaches in Morphology and Anatomy

Chemistry, Taxonomy and Systematics

Data Processing and Taxonomy

Introduction

1 | The Current Scene in Plant Taxonomy

V. H. HEYWOOD

Department of Botany, University of Reading, England

Abstract: In the 15 years that have intervened since the 1967 symposium on "Modern Methods in Plant Taxonomy" (Heywood 1968), plant taxonomy has undergone major developments and some of the subdisciplines which in 1967 were only beginning to emerge have now not only flourished but can be assessed in a wider context. Biochemical systematics has come of age and is no longer the subject of exaggerated claims; numerical taxonomy has developed a diverse and highly sophisticated set of procedures available for classificatory and discriminatory purposes but is less frequently used than one might have predicted; electronic data processing has been applied to the information processing side of taxonomy, a hitherto neglected field; scanning electron microscopy has reached very high levels of refinement and is virtually an everyday tool for the study of micro-characters of surface features and has, along with ultrasonic fracturing or sectioning, revolutionized systematic palynology; again new techniques, such as banding, have opened up new vistas for karyology in taxonomy; biosystematics, experimental taxonomy, genecology and contributory disciplines have been the subject of considerable reappraisal and the future pattern has not yet clearly developed; the problem of endangered species and communities, especially in tropical regions, has focussed attention on the global tasks and priorities for taxonomy.

INTRODUCTION

In attempting a review of the current scene in plant taxonomy and systematics it is useful to have a base-line against which to make comparisons. I have chosen 1967 (i.e. 15 years ago) when a conference on "Modern Methods in Plant Taxonomy" was held at Liverpool and the proceedings published in 1968

Systematics Association Special Volume No. 25, "Current Concepts in Plant Taxonomy", edited by V. H. Heywood and D. M. Moore, 1984. Academic Press, London and Orlando.
ISBN 0 12 347060 9

(Heywood 1968). Fifteen years may seem a short time but as we shall see they were very fertile ones as regards developments and activity in nearly all areas of taxonomy and systematics.

In the mid-1960s and the succeeding ten years or so, plant systematics and taxonomy, in common with many other branches of science and learning, were living through a golden age. This coincided with the heyday of liberal budgets and expansion, the founding of new institutions, and the launching of all manner of ambitious projects. Alas, by the mid-1970s the beginnings of what was to be a long period of recession were already affecting science budgets. Looking back over the period under review a number of major events or developments can be highlighted.

THE RISE OF NUMERICAL TAXONOMY

The emergence of numerical taxonomy or taximetrics as a major subdiscipline after the initial landmark of Sokal and Sneath's "Principles of Numerical Taxonomy" (1963) was a major feature that had many subsequent repercussions on other branches of taxonomy. As Shetler, writing in 1974 noted, "Ten years ago, with the appearance of Sokal and Sneath's (1963) book, numerical taxonomy burst onto the biological systematics scene, triggering a revolution of sorts that would not have been possible before the advent of the electronic digital computer". Already in 1969 Sokal, in the Foreword to the Symposium volume "Numerical Taxonomy" (Cole 1969), was writing, "numerical taxonomy seems to have come of age. The literature has grown so that no-one can seriously hope to keep up with it. The controversy accompanying its early development has largely died down". Other textbooks soon followed such as Jardine and Sibson's "Mathematical Taxonomy" (1971), and in 1973 Sneath and Sokal's new edition of their key work under the title "Numerical Taxonomy". A valuable addition to the literature is Dunn and Everitt's "An Introduction to Numerical Taxonomy" (1982). In addition an Annual Conference on Numerical Taxonomy was established in 1967 and the literature grew at such a pace as to make Sokal's plaintive comments just quoted seem almost naïve.

Initially the interests of taxonomists in the computer revolution were focussed on the techniques involved in numerical classification, a field often referred to as numerical phenetics. This has been reviewed by Sneath (1976) and by Duncan and Baum (1981). Today it is almost standard practice to use some form of numerical classification in taxonomic revisionary work and phenograms are often presented as a basis against which to measure more conventionally produced results. Much of the earlier enthusiasm for numerical phenetics has been dissipated with the realization that it does not provide taxonomists with an unambiguous methodology for constructing "correct" classifications. On the contrary, it is probably true that progress in this area is now largely dependent on taxonomists themselves formulating more precisely taxonomic theory and being able to indicate much more clearly what it is they wish a classification to show and how they wish it to be measured (cf. Sneath 1976). The computer has been successfully used in the generation of keys, initially for bacteria but later applied to virtually all groups of organisms, and a valuable collection of papers on biological identification with computers will be found in the Systematics Association Symposium of that title edited by Pankhurst (1975). A useful discussion of more conventional keys in systematics is given by Nelson and Platnick (1981, chapter 2).

Almost more important than the actual classificatory or discriminatory roles of numerical taxonomy were the effects it had on the basic concepts and principles of taxonomy, leading to a reappraisal of many aspects such as the concept of characters and character-states, weighting, homology, the distinction between phenetic and phylogenetic classification, the analysis of phylogeny into its cladistic, patristic and chronistic components, relative and absolute ranking, the clearer definition of and distinction between the different types of tree diagram (dendrograms)* – many of these topics are still unresolved and the subject of renewed debate today by the proponents of cladistics as a form of systematics or phylogenetic reconstruction. (It may be noted in passing that the English translation of Hennig's 1950 "Grundzüge einer Theorie der phylogenetischen Systematik" appeared in 1966 under the title "Phylogenetic Systematics" but was largely unremarked by the botanical community.)

Another major consequence of the computer revolution in systematics was its effect on the information processing scene. This was slow to come about despite the pioneering efforts of people such as Sydney Gould in the late 1940s and 1950s (Shetler 1974; Heywood 1974), although by 1973 most of the contributors to the ICSEB-I symposium "Computer Revolution in Systematics" (published in *Taxon* **23**(1) 21–100, 1974) dealt with non-classificatory aspects of the subject. As Shetler (1974) wryly comments, "this lag is the more surprising when one considers the colossal information retrieval problem that the taxonomist always has faced, but perhaps the lag reflects in part the constitutional aversion of the scientist to slowing up the pace of primary discovery for the sake of organizing and synthesizing what is already discovered"; it is probably

* The latest term to be introduced is that of synapomorphogram to refer to cladograms *sensu* Nelson and Platnick (1981) by Sneath (1982).

also a consequence of the traditional disregard that the professional taxonomist has had for the consumer (assuming that a consumer has even been identified) since the post-Linnaean period.

Important landmarks in this field of information–processing in systematics were more the development of the TAXIR (taxonomic information retrieval system) in 1968 (Estabrook and Brill 1969) and the establishment of the Flora North America programme in 1967 which was described as "the most ambitious data-banking effort ever undertaken in systematic biology on an integrated basis" (Shetler 1974). FNA was a casualty of what its executive secretary and later programme director was later to call the myths of data banking and also of a failure to heed repeated warnings of the need for short-term concrete products. The "Flora de Veracruz" project is an example of a Flora using electronic data processing methods for the processing of herbarium data (Gómez-Pompa and Nevling 1973). One of the first major reviews of the subject was a symposium volume published by the Systematics Association – "Data Processing in Biology and Geology" (Cutbill 1971).

Progress in recent years has been steady but unspectacular, despite the quite remarkable advances in computer technology and microprocessors which have had a widespread impact on all aspects of life (even penetrating the home with the personal mini/microcomputer concept). Following an international conference aimed at exploring the scope for the use of electronic data processing in major European plant taxonomic collections in 1973, a working party was established to pursue various problems in detail such as systems and software, descriptors and the production of a type register (Brenan *et al.* 1975). A subsequent volume of proceedings appeared (Brenan *et al.* 1979) under the title "Computers in Botanical Collections" but sadly there are few substantial results that we can point to in the following years.

NEW SOURCES OF TAXONOMIC CHARACTERS

In addition to an increased sophistication in the study and handling of the more traditional kinds of morphological character, two main sources of data – cytology and chemistry – showed a dramatic expansion, the latter consequential to a large degree on the developments in analytical procedures and techniques. A third source of data stemmed entirely from the application of a further technological development, the scanning electron microscope (and to a lesser degree the transmission electron microscope), which revolutionized the use of microcharacters in taxonomy.

(1) Chromosome information – cytotaxonomy

Cytological, especially karyological, data became much more abundant and not only did an increasing percentage of species become subjected to some degree of chromosome study but the intensity of analysis/sampling improved. Chromosome counts have continued to be compiled in a series of handbooks – the most comprehensive is that of Bolkhovskikh *et al.* (1969) "Chromosome Numbers of Flowering Plants" – while important regional compilations include Löve and Löve's Chromosome Atlas of the Slovenian Flora (1974) and the Arctic Flora (1975). A major difficulty is that the pace of publication has accelerated so much in the period under review that such chromosome handbooks soon become substantially out of date. The annual "Index to Plant Chromosome Numbers" published in *Regnum Vegetabile* (1970–74) and the IOPB Chromosome Number Reports regularly published in *Taxon* are useful additional sources but still leave the retrieval of chromosome information about plants a difficult and tortuous process, a point highlighted by Favarger (1978) in an essay on the philosophy of chromosome counting.

There have been quite remarkable advances in the knowledge of certain areas. In the Mediterranean region, for example, the number of papers containing chromosome counts, published per annum, increased more than tenfold between 1965 and 1980. Variation of chromosome number within species became widely appreciated and accepted despite last-ditch stands by traditionalists such as Löve, and intensive sampling of populations within species and of species within genera became more common, as in such classic examples as *Crocus* (Brighton 1976, 1978) and *Claytonia* (Lewis 1970; Lewis and Oliver 1971).

Conversely, uniformity of chromosome number in large genera such as the amphiatlantic *Hymenaea* (Lee and Langenheim 1975) and the surprising total absence of polyploidy in the large genus *Eucalyptus* (Rye 1979) has been revealed. An extensive review of chromosome numbers in the angiosperms presented according to the classificatory system of Cronquist is given by Raven (1975). This reveals that no cytological information had been reported by 1975 for 44 of the 354 families recognized by Cronquist nor for 49 of the additional families recognized by Thorne or by Takhtajan in their respective systems. Moreover, as Moore (1978) notes in a general discussion of chromosomes and plant taxonomy, for even well-researched families such as the Compositae only about 30% of the estimated 20 000 species have chromosome counts available and the figure for the Umbelliferae is 30% while for the Euphorbiaceae counts are known for only about 5% of species. Thus, despite the great increase in our knowledge of chromosome numbers over the past 15 years there still remain vast lacunae in our knowledge. For temperate groups the situation is becoming more satisfactory although there is still a great need for more detailed sampling of most species and genera, while for Mediterranean groups the position has, as we have noted, improved dramatically over the past 15 years. The state of our knowledge of the cytology of tropical groups remains alarmingly poor and no substantial improvement can be foreseen.

Quantitative estimation of DNA in plant chromosomes and specialized staining and banding techniques have led to a fuller understanding of the structure of eukaryotic chromosomes and of karyological variation and its taxonomic and evolutionary significance (Greilhuber, Chapter 9).

(2) Chemical information – chemosystematics

In the sixteen years since 1967, chemosystematics may be said to have come of age in that the vast mass of data published has led to a sober appraisal of the principles, practice and achievements of this young discipline, thus moderating the excessive pioneering claims which we sometimes heard earlier. Not surprisingly attention has recently been diverted away from solving taxonomic problems (or attempting to solve them) or from confirming relationships already elucidated from various other lines of evidence, towards the study of various ecological and geographical relationships and coevolutionary plant–animal interactions, notably the role of secondary compounds as defence substances against animal herbivores (cf. Rosenthal and Janzen 1979; Swann 1979; Harborne 1977, 1982).

During the period under review many reviews of chemosystematics and its applications have appeared, covering the field as a whole (e.g. Fairbrothers *et al.* 1975; Bendz and Santesson 1974; Hunziker 1969; Mabry *et al.* 1968; Swain 1973; Harborne 1968) or certain aspects or classes of compounds such as flavonoids (e.g. Harborne 1975), amino acid sequence data (e.g. Boulter 1973, 1976; Boulter *et al.* 1979), serology (e.g. Fairbrothers 1968), and electrophoresis (Hurka 1980). One of the latest books in this field is the proceedings of a Systematics Association Symposium published under the title "Chemosystematics: Principles and Practice" (Bisby *et al.* 1980). In the Preface of that volume it is noted that the Association played a pioneer role in organizing a symposium on chemotaxonomy in 1967 (Hawkes 1968) and that "To a large extent the more reasonable claims of the 1967 symposium are now accepted: a range of techniques is capable of producing chemically reliable comparative data on micromolecules and macromolecules for many groups of organisms, and these are of great systematic interest" (Bisby 1980). If I can echo some of the comments of the two authors entrusted with summing up the symposium (P. H. A. Sneath and K. A. Joysey), it is now accepted that good quality data are needed, well sampled and of known origin, that meet the requirements of both the chemists and the taxonomists; and also that much more effort will have to be applied to devising new and better methods of data analysis and handling, both chemical and mathematical, for taxonomic and evolutionary interpretation. But this, as in other fields, depends to a large extent on agreeing what we want to do with our data – in other words decisions have to be made on fundamental principles of taxonomy and systematics, just as in the instance of numerical taxonomy mentioned above.

(3) Scanning and transmission electron microscopy – microcharacters

It was about sixteen years ago that the first scanning electron microscopes became commercially available. My own first tentative studies were made on a demonstration model in 1967 and the first instrument (a Jeol JSM–2) which we installed in the Botany Department at Reading in 1968 was apparently the first production model and is going to be placed on display in the Royal Scottish Museum (via the Royal Botanic Garden, Edinburgh). The impact of the SEM on the study of surface features – seeds, spores, fruits, pollen grains, leaf surfaces was immediate and, as I ventured to predict in 1971, the scanning electron microscope has become a routine tool in biological research. From a barely attainable resolution of 50 nm (500 Å) on the first instruments, a resolution of down to 2 nm (20Å) is guaranteed on modern machines. Hundreds of thousands of SE micrographs are produced each year and several annual conferences are held and specialized bibliographies issued. A consequence has been that the study of microcharacters is now an important field (assessments are given by Cutler and Barthlott in Chapters 6 and 7 of this volume) and in particular systematic palynology has been revolutionized by the rich information on ektexine surface features, almost to the neglect of the important characters provided by light microscopy. From the point of view of systematics and taxonomy, the transmission electron microscope has proved of much less value although it plays an important role in the study of pollen grains, following the pioneer work of Skvarla and Turner (1966), still being pursued today (cf. Skvarla and Nowicke 1982; Nowicke and Skvarla 1979). The role of ultrastructural studies using the transmission electron microscope, such as those by Behnke (1975, 1977) on sieve-tube plastids, is of limited application in the systematics of flowering plants although a more important role is played in other groups such as the algae whose ultrastructural features of the flagellar apparatus for example, have some evolutionary significance (Melkonian 1982). A useful review of ultrastructural data for practical taxonomy is given by Stuessy (1979).

BIOSYSTEMATICS AT THE CROSSROADS

"Biosystematics at the Crossroads" was the title of a symposium introduced by Ehrendorfer, during the eleventh International Botanical Congress at Seattle in 1969. It reflected the growing concern amongst systematic biologists that the details of the patterns and processes involved in the evolution of populations at and below the species level did not easily fit into the models that had been erected in the 1940s and 1950s, partly derived from zoological studies, nor did they often help much in the solution of practical taxonomic problems. The position is well summarized by Raven (1974):

> Perhaps the most important discovery of the 25 years (1947–1972) is that biosystematic studies do *not* lead to an unequivocal definition of the taxonomic units in most cases: they contribute to an understanding of the populations and the processes by which they have changed and are changed but they do not dictate the taxonomic decisions that must be made in the light of this information.

In particular, attempts to replace the taxonomists' species definition by objective criteria led to the gradual adoption of the so-called biological species concept, at least in theory, by both zoologists and botanists. This curious episode in plant and animal systematics has been discussed by Raven (1977a, 1979), Sokal and Crovello (1970), Heywood (1976b). Some of us (e.g. Davis and Heywood 1963) have argued for the last 16 years (or even longer, e.g. Gilmour 1960) that a distinction has to be made between the evolutionary dynamics of populations and formal classifications of variation (but see also Snaydon, Chapter 11). The biological species concept was neither acceptable on theoretical or practical grounds. Its demise, now widely agreed, is largely the result of detailed studies on the nature and structure of populations and of gene flow within and between them (Raven 1974, 1977b), (see below).

One unfortunate effect of the biological species concept was to give an even more spurious sense of reality to such units than did the taxonomic species concept. As Raven (1977a) says, the dangerous delusion (fomented by the biological species concept) that those series of populations classified as species have a number of biological properties in common must be discarded "because it leads to false assumptions as we try to assess the properties of populations, their evolution and functioning as members of complex eco-systems we are rapidly destroying".

As I have noted previously (Heywood 1980):

When one considers that the theory of biosystematics has been in the main, based on a limited knowledge of temperate plants, while our understanding of reproductive biology and population structure of tropical species (by far the majority) is woefully inadequate, the need for a reorientation of our investigations into the nature of species on a more representative basis is obvious.

Moreover, we are "lamentably ignorant of the reproductive systems of all but a tiny fraction of the angiosperms, and for the tropics, where the problems are most urgent, the proportion is the least" (Heslop-Harrison 1976).

On a more positive note, major conceptual and practical advances have been made in the areas of population biology as related to systematics. Raven (1980) lists four in particular.

(*i*) A growing realization of the extreme local nature of plant populations and the consequent limited evolutionary significance of "gene flow". The subject has been reviewed by Levin and Kerster (1974, 1975) and Raven (1977b) and the debate still continues.

(*ii*) A better understanding of the nature of linkage and homeostasis and the conditions under which single genes or gene complexes can be substituted in populations (see Raven 1978).

(*iii*) Increased knowledge of molecular genetics, allowing us in particular to understand better the nature of local adaptation in plant populations (see Solbrig, Jain, Johnson and Raven, "The Population Biology of Plants", 1979).

(*iv*) An increased appreciation of the role of hybridization in the adaptation and evolution in plant populations (see reviews by Stace 1975; Raven 1980).

In addition, important advances have been made in areas of reproductive biology such as cost-benefit analysis on reproduction (Lloyd 1979; Solbrig 1979) pollen-ovule ratios, sex distribution and resource costs (Bawa 1980; Lloyd 1979; Llovet-Doust 1980). Poppendieck (1981), in a review of the field in *Progress in Botany*, remarks somewhat cynically "It is amusing to see now that the more fashionable point of view expects evolution to operate as a manager of a multi-national enterprise would direct his business; parsimoniously pursuing his strategy, cost efficient, and with an optimal outcome".

FLORISTICS

The last 16 years can be regarded as one of the great ages of floristics. It is a period that has seen the completion of the "Flora Europaea" project (Tutin *et al.* 1964–1980). Merxmuller's "Prodromus einer Flora von Südwest Afrika" (1966–72), Standley *et al.* "Flora of Guatemala" (1958–76), "Flora of West Tropical Africa" (1954–72); and the initiation of the "Flora of Thailand, Flora de Moçambique" (1969), "Flora of Australia" (1981–), "Flora of Ecuador" (1973), "Flora of Venezuela" (1968), "Flora Iranica" (1967), the "Iconographia Cormophytorum Sinicorum" (1972) (which will illustrate 10 000 of the estimated 30 000 Chinese species of Cormophyta), "Flora de Veracruz" (1978), to mention some of the more significant. It also saw the rise and fall of "Flora North America" as noted earlier. Floristic works are reviewed by Heywood (this volume, Chapter 19).

TAXONOMIC INDEXES AND BIBLIOGRAPHIES

A whole series of new indexes, abstracts, journals and bibliographical works were initiated or published during this period, including: "Kew Record of Taxonomic Literature", initiated in 1971 (1974), with the hope "that each volume will appear about 12 months after the end of the year served" – alas the gap ranges from 4–5 years! "Asher's Guide to Botanical Periodicals" which began in 1973 and expired in a fit of optimism. In 1968 the BPH or "Botanico-Periodicum-Huntianum" was published by the then Hunt Botanical Library, listing 12 000 periodical journals and their recommended and other abbreviations, containing plant science papers published between 1646 and 1946. In the same year the "Gray Herbarium Index" was published in ten volumes, listing about 250 000 plant names for the Western Hemisphere. In 1979, the "Index Nominum Genericorum" comprising 2000 pages in 3 volumes was published by IAPT. Three volumes of Stafleu and Cowan's monumental "Taxonomic Literature" have appeared, and an "International Register of Specialists and Current Research in Plant Systematics" was published by the Hunt Institute for Botanical Documentation, Pittsburgh.

Also worth noting is a whole series of newsletters covering topics such as Umbelliferae, Compositae, the Bean Bag, IOPB, Chemical Plant Taxonomy, etc. and it is pleasing to record that the *Flora Malesiana Bulletin*, founded in 1947, continues from strength to strength.

SYMPOSIA AND CONFERENCE VOLUMES

Symposia and conferences leading to a staggering array of volumes on a wide range of topics have been published:

– The outstanding series published by Missouri Botanical Garden such as "The

Bases of Angiosperm Phylogeny" (1975), "Perspectives in Tropical Botany" (1977), "Phytogeography of Africa" (1978).
– The remarkable series sponsored by the Systematics Association – such as "Scanning Electron Microscopy" (Heywood 1971a), "Taxonomy and Ecology" (Heywood 1973), "Biological Identification with Computers" (Pankhurst 1975), "Modern Approaches to the Taxonomy of Red and Brown Algae" (Irvine and Price 1978), "Bryophyte Systematics" (Clarke and Duckett 1979).
– The "Biology and Chemistry" series which was started in 1971 with the Umbelliferae (Heywood 1971b), followed by the Cruciferae (Vaughan *et al.* 1976), the Compositae (Heywood *et al.* 1977), the Solanaceae (Hawkes *et al.* 1979), the Leguminosae (Polhill and Raven 1981), each successive volume emulating the one before and setting apparently impossible standards.
– A random sample of others includes "Tropical Botany" (Larsen and Holm-Nielsen 1979), "Tropical Botany" (Gardens' Bulletin Singapore, 1977), "Systematic Botany and Biosphere Conservation" (Hedberg 1979), "Plants and Islands" (Bramwell 1979).
– "The Annual Review of Ecology and Systematics" (1970–) provides a forum for high-level discussion of various aspects of taxonomic theory, methodology and related topics.

Nor must we forget the very large number of monographs and revisions published during the period despite the concentration by taxonomists on floristic projects.

EARLY EVOLUTION OF THE ANGIOSPERMS

This is an area where, in Kubitzki's words, "it is sometimes difficult to separate facts from fiction", but during the past 16 years, especially the last five or so, there has been a significant increase in the corpus of factual data regarding pollen and leaf fossil material. In a recent review Doyle (1978) comments that the "past 20 years have seen unprecedented progress towards the solution of Darwin's 'abominable mystery' of the origin of the angiosperms". In 1970 Muller provided a review of palynological evidence showing the first appearance in the fossil record of various angiosperm families and in 1981 he published a major update of the fossil pollen records of extant angiosperms. Other reviews have been provided in Beck's "The Origin and Early Evolution of the Angiosperm" (1976) which includes papers by Doyle and Hickey, Walker and others, and by Wolfe *et al.* (1975), Crepet (1979), Dilcher (1979) and others. Dilcher makes the important point that the various phylogenetic schemes that have been presented have been based mainly on information derived from living plants while the fossil history of the flowering plants has had little influence upon current concepts of the primitive angiosperm/flower. There is, he says, "a concept of the 'primitive angiosperm' so well established in the literature that it is difficult to approach the topic of the origin and evolution of angiosperm reproduction system without being biased by this preconception. This is what I have termed elsewhere 'canalization of conceptual thinking' " (Heywood 1977). Doyle (1978) and most others still believe that the pattern of evolution in the early Cretaceous pollen and leaf records, and such distinctive features of modern angiosperms as their leaf and ovule morphology, are better explained by a monophyletic origin near the Jurassic-Cretaceous boundary, rather than by any other hypothesis.

Hughes (1976, 1977), Meeuse (1965, 1972, 1975, 1979, etc.) are amongst those who on various grounds have challenged the monophyletic viewpoint, in favour of a pleiophyletic one, and Dilcher's (1979) review of early angiosperm reproduction would seem to favour both unisexual anemophilous and bisexual entomophilous flowers as ancient conditions in the early angiosperm and it is not possible to determine which is more primitive. These are serious challenges to the Russo-American credo (Guédès 1979). The recent discovery of well-preserved flower structures from the Upper Cretaceous (Tiffney 1977), which cannot be identified with any contemporary group, has reinforced the difficulties of extrapolating from our present day knowledge in our attempts to reconstruct the angiosperm ancestors, a difficulty stressed by Stebbins (1975) and others. Cladistic methods have been applied by Hill and Crane (1982) to the problem of the origin of angiosperms, with inconclusive results.

SYSTEMS OF CLASSIFICATION OF HIGHER TAXA

The past 16 years have seen unprecedented activity in the area of higher systematics or system building. New systems for the angiosperms have been proposed by Dahlgren (1975), Thorne (1976) (both of them using the transection of an imaginary tree as a diagrammatic basis), Stebbins (1974) (an amalgam of the systems of Thorne, Takhtajan and Cronquist) and Takhtajan (1980a) (an outline of which and index has been published in *Taxon* by Bedell and Reveal (1982)); while Dahlgren and Clifford (1982) have published a major comparative systematic study of the monocotyledons. In 1981 Cronquist produced a detailed and revised account of his system and it is summarized in the vast two-volume "Synopsis and Classification of Living Organisms" (ed. Parker 1981). In addition, Heywood (1978) and Takhtajan (1980b, 1981) have both edited large illustrated accounts of the angiosperm families.

The resumption of "Die natürlichen Pflanzenfamilien" with a multi-authored treatment of the Loganiaceae (Leeuwenberg 1980) is greatly to be welcomed. Thonner's Analytical Key to the Families of Flowering Plants has been issued in a revised version by Geesink *et al.* (1981) and a second edition of "The Identification of Flowering Plant Families" has been prepared by Davis and Cullen (1979). Textbooks of taxonomy published during the period include Heywood (1968, 1976a), Bell (1969), Hawksworth (1974), Stace (1980), Jeffrey (1982), and Radford *et al.* (1974).

CONCLUSIONS

Looking back over the past fifteen years, the enormous richness and diversity of activity in plant taxonomy and systematics, the highlights of which has been sketched in above, leads to the inexorable conclusion that an enormous effort of digestion is needed – and a period of sober and mature reflection.

The absence of any major new development in plant systematics and taxonomy in the past few years (corresponding to cytotaxonomy, chemotaxonomy, etc.) has given taxonomy such a breathing space.

It is also a period of soul-searching and deciding on priorities, for we are now living in the shadow of what the poet Valéry called "*le monde fini*" – time is limited, plant life is at risk all over the world and in many areas if it is not sampled and studied now, there will not be another opportunity later.

Limited budgets, recession, and the threat of the bulldozer should cause us to focus more clearly on our priorities.

REFERENCES

Bawa, K. S. (1980). Evolution of dioecy in flowering plants. *Ann. Rev. Ecol. Syst.* **11,** 15–39.

Beck, C. B. (ed.) (1976). "The Origin and Early Evolution of the Angiosperms". Columbia University Press, New York, London.

Bedell, H. G. and Reveal, J. L. (1982). An outline and index to Takhtajan's 1980 classification of Flowering Plants. *Taxon* **31,** 211–232.

Behnke, H.-D. (1975). The bases of angiosperm phylogeny ultrastructure. *Ann. Missouri Bot. Gard.* **62,** 647–663.

Behnke, H.-D. (1977). Transmission electron microscopy and systematics of flowering plants. *In* "Flowering Plants: Evolution and Classification of Higher Categories" (K. Kubitzki, ed.) pp. 155–178. Springer Verlag, Vienna, New York.

Bell, C. R. (1969). "Plant Variation and Classification". Macmillan, London

Bendz, G. and Santesson, J. (eds) (1974). "Chemistry in Botanical Classification". *Proc. 25th Nobel Symposium.* Academic Press, London, New York

Bisby, F. A. (1980). Preface. *In* "Chemosystematics: Principles and Practice" (F. A. Bisby, J. G. Vaughan and C. A. Wright, eds) pp. vii–viii. Academic Press, London, New York.

Bolkhovskikh, Z., Grif, T., Matvejeva, T. and Zakharyeva, O. (1969). "Chromosome Numbers of Flowering Plants". Komarov Botanical Institute, Leningrad.

Boulter, D. (1973). The use of comparative amino acid sequence data in evolutionary studies of higher plants. *In* "Progress in Phytochemistry" (L. Reinhold and Y. Liwschitz, eds) Vol. 3, 199–229. Interscience, London.

Boulter, D. (1976). The evolution of plant proteins with special reference to higher plant cytochromes *c*. *In* "Commentaries in Plant Science" (H. Smith, ed.) pp. 79–91. Pergamon, Oxford.

Boulter, D., Peacock, D., Guise, D., Gleaves, J. T. and Estabrook, G. (1979). Relationships between the partial amino acid sequences of plastocyanin from members of ten families of flowering plants. *Phytochemistry* **18,** 603–608.

Bramwell, D. (ed.) (1979). "Plants and Islands". Academic Press, London, New York.

Brenan, J. P. M., Franks, J. W., Raynal, J. and Cullen, J. (1975). Report of a working party on electronic data processing in major European plant taxonomic collections, *Adansonia* 2, **15,** 7–24.

Brenan, J. P. M., Ross, R. and Williams, J. T. (eds) (1979). "Computers in Botanical Collections". Plenum Press, London, New York.

Brighton, C. A. (1976). Cytological problems in the genus *Crocus* (Iridaceae): I. Crocus *Crocus vernus* aggregate. *Kew Bull.* **31,** 33–46.

Brighton, C. A. (1978). Telocentric chromosomes in Corsican *Crocus* L. (Iridaceae) *Pl. Syst. Evol.* **129,** 299–314.

Clarke, G. C. S. and Duckett, J. G. (1979). "Bryophyte Systematics". Academic Press, London, New York.

Cole, A. J. (ed.) (1969). "Numerical Taxonomy". Academic Press, London, New York.

Crepet, W. L. (1979). Some aspects of the pollination biology of Middle Eocene angiosperms. *Rev. Palaeobot. Palynol.* **27,** 213–328.

Cutbill, J. L. (1971). "Data Processing in Biology and Geology". Academic Press, London, New York.

Dahlgren, R. (1975). A system of classification of the angiosperms to be used to demonstrate the distribution of characters. *Bot. Not.* **128,** 119–147.

Dahlgren, R. and Clifford, H. T. (1982). "The Monocotyledons. A Comparative Study". Academic Press, London, New York.

Davis, P. H. and Cullen, J. (1979). "The Identification of Flowering Plant Families". Cambridge University Press, Cambridge.

Davis, P. H. and Heywood, V. H. (1963). "Principles of Angiosperm Taxonomy". Oliver and Boyd, Edinburgh.

Dilcher, D. L. (1979). Early angiosperm reproduction: an introductory report. *Rev. Palaeobot. Palynol.* **27,** 291–328.

Doyle, J. A. (1978). Origin of angiosperms. *Ann. Rev. Ecol. Syst.* **9,** 365–392.

Duncan, T. and Baum, B. R. (1981). Numerical phenetics: its uses in botanical systematics. *Ann. Rev. Ecol. Syst.* **12,** 307–404.

Dunn, G. and Everitt, B. S. (1982). "An Introduction to Numerical Taxonomy". Cambridge University Press, Cambridge.

Estabrook, G. F. and Brill, R. C. (1969). The theory of the TAXIR accessioner. *Math. BioSci.* **5,** 327–340.

Fairbrothers, D. E. (1968). Chemosystematics with emphasis on systematic serology. *In* "Modern Methods in Plant Taxonomy" (V. H. Heywood, ed.) pp. 141–174. Academic Press, London, New York.

Fairbrothers, D. E., Mabry, T. J., Scogin, R. L. and Turner, B. L. (1975). The bases of angiosperm phylogeny: chemotaxonomy. *Ann. Missouri Bot. Gard.* **62,** 765–800.

Favarger, C. (1978). Philosophie des comptages de chromosomes. *Taxon* **27,** 441–448.

Geesink, R., Leewenberg, A. J. M., Ridsdale, C. E. and Veldkamp, J. F. (1981). "Thonner's Analytical Key to the Families of Flowering Plants". Leiden University Press, The Hague.

Gilmour, J. S. L. (1960). Taxonomy. *In* "Contemporary Botanical Thought" (A. M. McLeod and L. S. Cobley, eds) pp. 27–45. Edinburgh, Botanical Society.

Gómez-Pompa, A. and Nevling, L. I. (1973). The use of electronic data processing methods in the flora of Veracruz programme. *Contr. Gray Herb.* **203,** 49–64.

Guédès, M. (1979). Magnolioid island plants and angiosperm evolution. *In* "Plants and Islands" (D. Bramwell, ed.) pp. 307–328. Academic Press, London, New York.

Harborne, J. B. (1968). Biochemical systematics: the use of chemistry in plant classification. *In* "Progress in Phytochemistry" (L. Reinhold and Y. Liwschitz, eds) Vol. 1, pp. 545–588. Interscience, London.

Harborne, J. B. (1975). Biochemical aspects of flavonoids. *In* "The Flavonoids" (J. B. Harborne, T. J. Mabry and H. Mabry, eds) pp. 31–72. Academic Press, London, New York.

Harborne, J. B. (1977). "Introduction to Ecological Biochemistry". Adcademic Press, London, New York.

Harborne, J. B. (1982). "Introduction to Ecological Biochemistry" (2nd edn). Academic Press, London, New York.

Hawkes, J. G. (ed.) (1968). "Chemotaxonomy and Serotaxonomy". Academic Press, London, New York.

Hawkes, J. G., Lester, R. N. and Skelding, A. D. (1979). "The Biology and Taxonomy of the Solanaceae". *Linn. Soc. Symp.* Series 7. Academic Press, London, New York.

Hawksworth, D. L. (1974). "Mycologist's Handbook. An Introduction to the Principles of Taxonomy and Nomenclature in the Fungi and Lichens". Commonwealth Mycological Institute, Kew.

Hedberg, I. (ed.) (1979). "Systematic Botany, Plant Utilization and Biosphere Conservation". Almquist and Wiksell, Stockholm.

Hennig, W. (1950). "Grundzüge einer Theorie der Phylogenetischen Systematik". Deutscher Zentralverlag, Berlin.

Heslop-Harrison, J. (1976). Reproductive physiology. *In* "Conservation of Threatened Plants" (J. B. Simmons *et al.*, eds) pp. 199–205. NATO Conference, Series 1, Ecology Vol. 1. Plenum Press, New York, London.

Heywood, V. H. (ed.) (1968). "Modern Methods in Plant Taxonomy". Edward Arnold, London.

Heywood, V. H. (ed.) (1971a). "Scanning Electron Microscopy". Academic Press, London, New York.

Heywood, V. H. (ed.) (1971b). "The Biology and Chemistry of the Umbelliferae". Academic Press, London, New York.

Heywood, V. H. (ed.) (1973). "Taxonomy and Ecology". Academic Press, London, New York.

Heywood, V. H. (1974). Systematics – the Stone of Sisyphus. *Biol. J. Linn. Soc.* **6,** 169–178.

Heywood, V. H. (1976a). "Plant Taxonomy" (2nd edn). Edward Arnold, London.

Heywood, V. H. (1976b). Contemporary objectives in systematics. *In Proc. 8th Int. Conf. Numerical Taxonomy* (G. F. Estabrook, ed.) pp. 258–283. W. H. Freeman, San Francisco.

Heywood, V. H. (1977). Principles and concepts in the classification of higher taxa. *In* "Flowering Plants: Evolution and Classification of Higher Categories" (K. Kubitzki, ed.) pp. 1–12. Springer Verlag, Vienna, New York.

Heywood, V. H. (ed.) (1978). "Flowering Plants of the World". Oxford University Press, Oxford.

Heywood, V. H. (1980). The impact of Linnaeus on botanical taxonomy – past, present and future. *Veröff. Joachim Jungius-Ges. Wiss. Hamburg* **43,** 97–115.

Heywood, V. H., Harborne, J. B. and Turner, B. L. (1977). "The Biology and Chemistry of the Compositae", Vols 1 and 2. Academic Press, London, New York.

Hill, C. R. and Crane, P. R. (1982). Evolutionary cladistics and the origin of angiosperms. *In* "Problems of Phylogenetic Reconstruction" (K. A. Joysey and A. E. Triday, eds) pp. 269–361. Academic Press, London, New York.

Hughes, N. F. (1976). "Palaeobiology of Angiosperm Origins". Cambridge University Press, Cambridge.

Hughes, N. F. (1977). Palaeo-succession of earliest angiosperm evolution. *Bot. Rev.* **43,** 105–127.

Hunziker, J. H. (1969). Molecular data in systematics. *In* "Systematic Biology", Publ. 1962, pp. 280–312. National Academy of Science, Washington, D.C.

Hurka, H. (1980). Enzymes as a taxonomic tool: a botanist's view. *In* "Chemosystematics: Principles and Practice" (F. A. Bisby, J. G. Vaughan and C. A. Wright, eds) pp. 103–121. Academic Press, London, New York.

Irvine, D. E. G. and Price, J. H. (1978). "Modern Approaches to the Taxonomy of Red and Brown Algae". Academic Press, London, New York.

Jardine, N. and Sibson, R. (1971). "Mathematical Taxonomy". Wiley, London.

Jeffrey, C. (1982). "An Introduction to Plant Taxonomy". Cambridge University Press, Cambridge.

Larsen, K. and Holm-Nielsen, L. B. (1979). "Tropical Botany". Academic Press, London, New York.

Lee, Y.-T. and Langenheim, J. H. (1975). *Univ. Calif. Publ. Bot.* **69,** 1–109.

Leeuwenberg, A. J. M. (ed.) (1980). Loganiaceae. *In* A. Engler and K. Prantl, "Die natürlichen Pflanzenfamilien" (2nd edn) **28b** I, pp. 1–255.

Levin, D. A. and Kerster, H. W. (1974). Gene flow in seed plants. *Evol. Biol.* **7,** 139–220.

Levin, D. and Kerster, H. (1975). The effect of gene dispersal on the dynamics and status of gene substitution in plants. *Heredity* **35,** 317–336.

Lewis, W. H. (1970). Chromosomal drift, a new phenomenon in plants. *Science* **168,** 1115–1116.

Lewis, W. H. and Oliver, R. L. (1971). Meiotic chromosomal variation in *Claytonia virginica*. *J. Hered.* **62,** 379–380.
Lloyd, D. G. (1979). Parental strategies of angiosperms. *New Zealand J. Bot.* **17,** 595–606.
Lovett-Doust, J. (1980). Floral sexual ratios in andromonoecious Umbelliferae. *New Phytol.* **85,** 265–273.
Löve, Á. and Löve, D. (1974). "Cytotaxonomical Atlas of the Slovenian Flora". J. Cramer, Lehre.
Löve, Á. and Löve, D. (1975). "Cytotaxonomic Atlas of the Arctic Flora". J. Cramer, Vaduz.
Mabry, T. J., Alston, R. E. and Runeckles, V. C. (eds) (1968). "Recent Advances in Phytochemistry", Vol. 1. Appleton-Century-Crofts, New York.
Meeuse, A. D. J. (1965). "Angiosperms – Past and Present". Inst. Adv. Sci. Culture, New Delhi.
Meeuse, A. D. J. (1972). Facts and fiction in floral morphology with special reference to the Polycarpicae. *Acta Bot. Neerl.* **21,** 113–127; 235–252; 351–365.
Meeuse, A. D. J. (1975). Origin of the angiosperms – problem or inaptitude? *Phytomorphology* **25,** 373–379.
Meeuse, A. D. J. (1979). Why were the early angiosperms so successful? A morphological, ecological and phylogenetic approach (Parts I and II). *Proc. K. ned. Akad. Weta.* **82,** 343–369.
Melkonian, M. (1982). Structural and evolutionary aspects of the flagellar apparatus in green algae and land plants. *Taxon* **31,** 255–265.
Moore, D. M. (1978). The chromosomes and plant taxonomy. *In* "Essays in Plant Taxonomy" (H. E. Street, ed.) pp. 39–56. Academic Press, London, New York.
Muller, J. (1970). Palynological evidence on early differentiation of angiosperms. *Biol. Rev. Cambridge Philos. Soc.* **45,** 417–450.
Muller, J. (1981). Fossil pollen records of extant angiosperms. *Bot. Rev.* **47,** 1–142.
Nelson, G. and Platnick, N. (1981). "Systematics and Biogeography: Cladistics and Vicariance". Columbia University Press, New York.
Nowicke, J. W. and Skvarla, J. W. (1979). Pollen morphology: the potential influence in higher order systematics. *Ann. Missouri Bot. Gard.* **66,** 633–700.
Pankhurst, R. J. (ed.) (1975). "Biological Identification with Computers". Academic Press, London, New York.
Parker, S. P. (1981). "Synopsis and Classification of Living Organisms", Vols 1 and 2. McGraw-Hill, New York.
Polhill, R. M. and Raven, P. H. (1981). "Advances in Legume Systematics", Parts 1 and 2. Royal Botanic Gardens, Kew.
Poppendieck, H.-H. (1981). Taxonomy. I. Systematics and evolution of seed plants. *Progress in Botany* **93,** 188–235.
Radford, A. E., Dickinson, W. C., Massey, J. R. and Bell, C. R. (1974). "Vascular Plant Systematics". Harper and Row, New York, London.
Raven, P. H. (1974). Plant systematics 1947–1972. *Ann. Missouri Bot. Gard.* **61,** 166–178.
Raven, P. H. (1975). The bases of angiosperm phylogeny: cytology. *Ann. Missouri Bot. Gard.* **62,** 724–764.
Raven, P. H. (1977a). The systematics and evolution of higher plants. *In* "Changing

Scenes in Natural Sciences, 1776–1976", pp. 59–83. Academy of Natural Sciences, Philadelphia, Spec. Publ. 12. Lancester, Penn.

Raven, P. H. (1977b). Systematics and plant population biology. *Syst. Bot.* **1,** 284–316.

Raven, P. H. (1979). Future directions in plant population biology. *In* "Topics in Plant Population Biology" (O. T. Solbrig, S. Jain, G. B. Johnson and P. H. Raven, eds) pp. 461–487. Columbia University Press, New York.

Raven, P. H. (1980). Hybridization and the nature of species in higher plants. *Bull. Canada Bot. Assoc.* **13(1),** suppl. 3–10.

Rosenthal, G. A. and Janzen, D. H. (eds) (1979). "Herbivores – Their Interaction with Secondary Plant Metabolites". Academic Press, London, New York.

Rye, B. L. (1979). Chromosome number in the Myrtaceae and its taxonomic implications. *Austral. J. Bot.* **27,** 547–573.

Shetler, S. G. (1974). Demythologizing biological data banking. *Taxon* **23,** 71–100.

Skvarla, J. and Turner, B. L. (1966). Systematic implications from electron microscope studies of Compositae pollen – a review. *Ann. Missouri Bot. Gard.* **53,** 220–256.

Skvarla, J. and Nowicke, J. W. (1982). Pollen fine structure and relationships of *Achatocarpus* Triana and *Phaulothamnus* A. Gray. *Taxon* **31,** 244–249.

Sneath, P. H. A. (1976). Phenetic taxonomy at the species level and above. *Taxon* **25,** 437–450.

Sneath, P. H. A. (1982). Review of G. Nelson and N. Platnick's "Cladistics and Biogeography: Cladistics and Vicariance". *Syst. Zool.* **31,** 208–217.

Sneath, P. H. A. and Sokal, R. R. (1973). "Numerical Taxonomy". W. H. Freeman, San Francisco.

Sokal, R. R. and Crovello, T. J. (1970). The biological species concept: a critical evaluation. *Amer. Naturalist* **104,** 127–153.

Sokal, R. R. and Sneath, P. H. A. (1963). "Principles of Numerical Taxonomy". W. H. Freeman, San Francisco.

Solbrig, O. T. (1979). A cost-benefit analysis of recombination in plants. *In* "Topics in Plant Population Biology" (O. T. Solbrig, S. Jain, G. B. Johnson and P. H. Raven, eds) pp. 114–130. Macmillan, London.

Stace, C. A. (1975). "Hybridization and the Flora of the British Isles". Academic Press, London, New York.

Stace, C. A. (1980). "Plant Taxonomy and Biosystematics". Edward Arnold, London.

Stebbins, G. L. (1974). "Flowering Plants: Evolution above the Species Level". Harvard University Press, Cambridge Mass.

Stebbins, G. L. (1975). Deduction about trans-specific evolution through extrapolation from processes at the population and species level. *Ann. Missouri Bot. Gard.* **62,** 825–834.

Stuessy, T. F. (1979). Ultrastructural data for the practising plant systematist. *Amer. Zool.* **19,** 621–635.

Swain, T. (ed.) (1973). "Chemistry in Evolution and Systematics". Butterworth, London.

Swain, T. (1979). Phenolics in the environment. *In* "Biochemistry of Plant Phenolics" (T. Swain, J. B. Harborne and C. van Sumere, eds) pp. 617–640. Plenum Press, New York, London.

Takhtajan, A. L. (1980a). Outline of the classification of flowering plants (Magnoliophyta). *Bot. Rev.* **46,** 225–359.

Takhtajan, A. L. (1980b). "The Life of Plants", Vol. 5(1). "Flowering Plants". Culture, Moscow.

Takhtajan, A. L. (1981). "The Life of Plants", Vol. 5(2). "Flowering Plants". Culture, Moscow.

Thorne, R. F. (1976). A phylogenetic classification of the Angiospermae. *Evol. Biol.* **9,** 35–106.

Tiffney, B. H. (1977). Dicotyledonous angiosperm flowers from the Upper Cretaceous of Martha's Vineyard, Massachusetts. *Nature* **265,** 136–137.

Vaughan, J., MacLeod, A. J. and Jones, B. M. G. (eds) (1976). "The Biology and Chemistry of the Cruciferae". Academic Press, London, New York.

Wolfe, J. A., Doyle, J. A. and Page, V. M. (1975). The basis of angiosperm phylogeny: palaeobotany. *Ann. Missouri Bot. Gard.* **62,** 801–824.

Institutional Resources

2 | Libraries and Herbaria

J. CULLEN

Royal Botanic Garden, Edinburgh, Scotland

Abstract: Herbaria and libraries are, and have been for the last 200 years, the two basic tools of the taxonomist. In spite of developments in detail, they remain much as they were. This stability at the foundations of the subject leads to a consideration of one of the basic aims of taxonomy – the production of an accurate and precise means of plant identification. This consideration reveals an unsatisfactory situation, particularly on a world scale; some of the factors contributing to the situation are discussed and a plea is made for an attempt to improve it by means of more globally organized research, greater production of user-orientated results and the increasing use of modern methods of data control.

INTRODUCTION

Possibly the most remarkable fact about libraries and herbaria is that, though there have been obvious developments in both, in terms of curation, storage and cataloguing, they remain much as they always have been. If, for example, George Bentham were to come back to Kew today, he would find much that was completely familiar to him. He might be surprised at some things, such as the vast number of specimens now conserved (somewhere of the order of 300 million in the world), how much better (on the whole) the more recent ones are labelled, the vastly improved equipment (such as dissecting microscopes), and the use of the specimens in detailed studies such as phytochemistry, surface structure, palynology, etc. But, in spite of all these developments, I think his major impression would be one of how similar everything was: people working in broadly the same way with broadly the same materials, producing results – revisions and Floras – perfectly comprehensible to him, and, indeed, very similar

Systematics Association Special Volume No. 25, "Current Concepts in Plant Taxonomy", edited by V. H. Heywood and D. M. Moore, 1984. Academic Press, London and Orlando.
ISBN 0 12 347060 9

in most ways to those he produced himself. So, in many ways, the current topics in libraries and herbaria are not particularly new.

This great stability at the foundations of flowering plant taxonomy (for that is what libraries and herbaria are) can be viewed in two ways:

(*i*) that the founding fathers of the subject got it right and that all we have to do is to follow on, developing and adding to the structure which they built – adding new information and new techniques into the firm but flexible framework that they bequeathed to us. This, I think, would be the view of most working taxonomists. Or,

(*ii*) that the discipline has become fossilized, set in its ways, has undergone none of the social, political and scientific stresses of the last 100 years, and is now a pleasant backwater, outside the main stream of biological development. This view is often held by other biologists and scientists (who often dominate the grant-awarding bodies so important in the financing of research) and, to some extent, by educated laymen.

I don't hold completely to either of these views, but I would like to explore some of the contrasts that they expose, using the herbarium and the library to provide examples. These are, as I have said, the basic tools of the taxonomic trade, so, in making this exploration, I am going to deal with very basic problems. The rest of this volume deals with more advanced matters, such as the newer techniques and the newer conceptual frameworks.

The herbarium and library are two of the three main stores of taxonomic information. The third is the garden or living collection, but as this is discussed by Ashton (Chapter 3), I will not consider it further. Stores of *information*; this is a term I would like to stress. Information about plants is the raw material of taxonomy; it is not plants that we classify, but information about them. The material in the herbarium represents that information in its raw state, in the packets (plants) provided by nature. The literature of taxonomy, contained in the library, is that information digested into various formal patterns which can be used for many purposes.

This brings an immediate contrast. The user of the herbarium is generally a trained taxonomist; the user of the literature may be one, but generally he is not. He may be anyone who requires information about plants: other botanists, scientists of various kinds, conservationists, foresters, gardeners, landscape architects, planners, etc.

SERVICE TO THE USER

The first point I would like to explore, then, is how far taxonomy has provided, and is providing, a service to those who need its results, but are not themselves its adepts – a consumer's report on taxonomy, in fact. The major service provided is, very simply, that of identification: what plant is this?, and its corollaries, where does such and such a plant grow?, and what plants can be found in such and such an area? The names of plants are, of course, the keys to any or all of the available information about them; and this service, identification, is the fundamental purpose for which taxonomy exists. Classifications may be employed for all sorts of other purposes, but a classification that does not work in terms of identification is, in my opinion, useless – a fiction with only tenuous connections with reality.

How far, then, does taxonomy today serve the interested, non-taxonomist member of the public in terms of identification? When we look around a large taxonomic library we must be amazed at the volume of literature that exists – revisions and Floras in abundance; and one might well think that there could be no possible problem. Unfortunately, this is not entirely so, and one of the immediate signs of it is the amount of popular literature on identification existing in parallel with the taxonomic literature. Most of this popular literature is produced by non-taxonomists, and the best that can be said for it is that it interprets taxonomy, or is a gloss or commentary upon it. Much of the interpretation is done by pictures, often coloured, and often quite delightful in themselves. Often, also, such works are of value to the taxonomist, who knows what he is looking for. But their value to the person who is not expert, as far as identification is concerned, is limited. Why, then, with the abundance of taxonomic literature, are such productions needed?

Before I attempt to answer this question, which, in fact, links up with another I shall be posing later on, I would like to give you a few examples of identificatory problems. These arise from work at the Royal Botanic Garden, Edinburgh, where identificatory queries arise from two sources: the public and the garden itself. We receive annually about 300 such queries from members of the public, and about 600 from the garden. Kew, I know, deals with greater numbers than these, but our figures are substantial enough to reveal a wide range of problems. The material we have to deal with includes plants originating from anywhere in the world, though often the area of origin is not known. It generally includes merely the flowering stage, though fruits are sometimes available from the garden, and the material from the public is often purely vegetative, as well as fragmentary and in poor condition.

In order to deal with this material, a group of staff, including both taxonomists and horticulturists meets weekly, and all material for identification passes its scrutiny, with the exception of certain groups – Gesneriaceae, Rhododendrons,

Conifers, etc. – which go directly to acknowledged experts. Our experience has been that identification is often extremely difficult and sometimes impossible, even with the considerable expertise and adequate library and herbarium resources that are available.

The reasons for this difficulty are many and various, and I will briefly mention three examples. The first is provided by an *Agave*. The plant in question is unidentified as to species, and its precise origin is not known. It has been growing for a considerable time (the rosette is almost 2 metres across) and has not yet flowered. It may continue in this state for many years to come, as Agaves flower only at long intervals, and several species are strictly monocarpic. *Agave* is not a particularly large genus and it has a relatively restricted distribution – mainly in the southern part of the United States and in Central America, and there is plenty of taxonomic literature available – revisions by Trelease (1910, 1913), Berger (1915), Michotte (1931) and Gentry (1972, 1978), and accounts in various Floras. Unfortunately, all of these classifications are based on details of the inflorescence, flowers and fruits, and so cannot be used for this particular specimen. Yet, having looked at the genus, at least as represented in cultivation, it seems that vegetative characters – size, shape, toothing, thickness and colour of the leaves, presence or absence of offsets, etc. – could be used to distinguish most of the species. But this kind of information is not easily available in the literature at present, as far as I have been able to discover, so, until it is worked out, or until this plant flowers, identification is impossible.

The *Agave* raises a further point. I know it is an *Agave*, and so, I am sure, do you. But members of the public who have asked us to identify *Agave* plants have quite often thought that they were Aloes. Now, it is perfectly simple to distinguish an *Agave* and an *Aloe* in flower; in most taxonomic systems they are placed in different families. But how are they to be distinguished vegetatively? I have not been able to find the distinction explained anywhere in the literature, though I believe that there are several characters which can be used. This problem, of course, arises only with material in cultivation, or in areas like the Mediterranean where species of both genera have escaped to some extent. In the wild the two genera are widely separated geographically.

My second example is provided by a rather undistinguished plant, a small shrub of which seed was collected in Hong Kong in 1976. The plant flowered in 1981, and has unisexual (male), heavily scented flowers. Hong Kong is a small area with a fairly restricted flora, and there is a fair amount of literature (Bentham 1861; Thrower 1971; Walden and Hu 1977), both scientific and popular, so one might expect this plant to be relatively easy to name. But, because the plant is purely male, it is extremely difficult to determine the family to which it belongs. In most of the available family keys the plant can not be run down, or, if it can, it comes out to unlikely families (such as the Olacaceae) which are unknown in Hong Kong. Eventually, by making use of the clue provided by the antipetalous stamens, by the consultation of numerous standard works, and by trial and error in the herbarium, we established that the plant belongs to the Myrsinaceae. From this point, identification was easy, leading to the name *Embelia laeta*. But this identification took 2 taxonomist-days to achieve!

A third example comes from the bulb genus *Ornithogalum*, a plant collected in eastern Turkey in connection with Professor P. H. Davis's project on the "Flora of Turkey". *Ornithogalum* is a notoriously confusing genus, but the plant in question appears to be what has long been known as *O. nanum* Sibthorp & Smith. Some consideration of the various Floras and other publications in which this plant is included, showed that it is part of a complex distributed as shown in Fig. 1. There is a problem here not so much about what this plant is, as what it should be called, as different taxonomists make different judgements about the limits of species within the complex, and the nomenclatural problem becomes acute. Greuter and Rechinger (1967) have shown that the name *O. nanum* Sibthorp & Smith is invalid, and have provided the new name, *O. sibthorpii*, for the Greek end of it. But the species should probably be called *O. sigmoideum* unless taxonomic judgement decides that *O. schmalhausenii* is not a distinct species, when this name will have to be used. In the meantime, the name *O. nanum* is the most useful, as this is the name the plant appears under in 95% of the literature, and there is no chance of confusion with the South African *Eucomis nana* (details on Fig. 1).

I could go on giving similar examples for a long time, but these are enough. In these cases, and in many others, the literature does not match up to the demands made upon it. And this seems to be a most unsatisfactory situation. Of course, not all plants cause difficulty: many, in fact, can be rapidly dealt with. But sufficient plants produce problems for people who are, in general terms at least, taxonomic experts. This is the situation that 200 years' activity in libraries and herbaria has produced. In making this criticism of taxonomy I am not excusing myself – I have been involved in it long enough to have some feelings of conscience about it. Having said that the situation is not good enough, it is clearly incumbent upon me to suggest some ways in which it might be improved.

A recent publication on conservation in the tropics included a paper by Dr G. T. Prance of the New York Botanical Garden entitled "Floristic inventory of the Tropics: where do we stand?" (Prance 1977), and his Chapter 18 in this

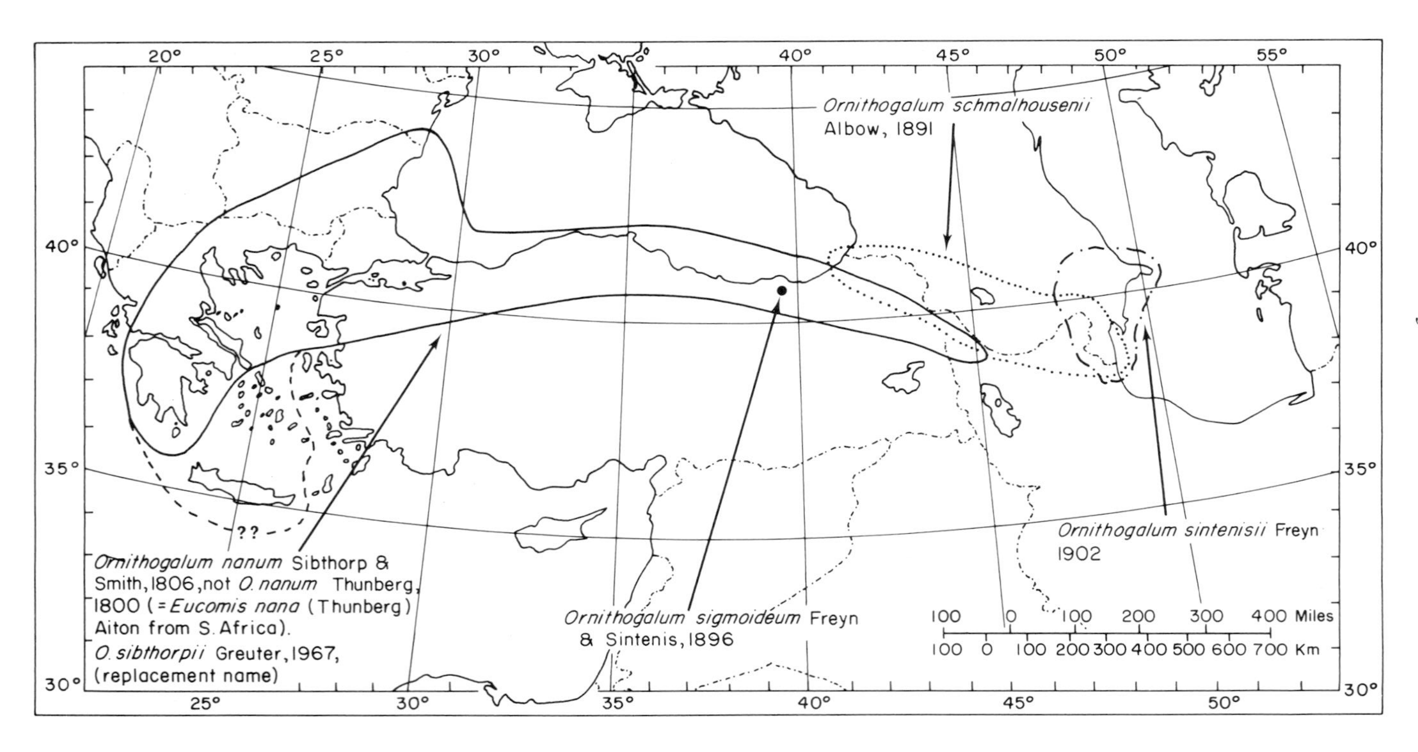

Fig. 1. Distribution of some taxa of *Ornithogalum* (Liliaceae) in the eastern Mediterranean area.

volume is on "Completing the inventory". Yes, indeed: we have a long way to go. We almost need an inventory of what needs to be inventoried. A list of Floras not extant, not complete, or so old as to be inadequate, of comprehensive revisions as yet unwritten. If we are to complete the inventory in a reasonable time, the work must be speeded up and priorities reassessed (bearing in mind some of the problems I have already outlined) and taxonomists and taxonomic institutions all over the world must organize themselves so that the work can be done, speedily and effectively.

ORGANIZATION AND REASSESSMENT OF PRIORITIES

These seem to me to be essential requirements. A forum for organization already exists, in the form of the International Association for Plant Taxonomy, and this should take a more active role than it appears to at present, making use of all the available up-to-date means of communication and information retrieval. Though such organization is difficult, it is less of a problem than the reassessment of priorities, which I would like to concentrate on for the rest of this chapter.

As a starting point we might consider how and why the present situation has arisen. And consideration of this topic leads to the answer to the question I posed earlier about the existence of popular identificatory literature. It is difficult to pin down an answer to the questions stated above; the matter is complex, and has many strands. The conclusion I have come to, and I am not alone in this (cf. Argent 1980), is that about 90% of taxonomic work is done for other taxonomists, with no account taken of any other possible user. My experience of working in botanic gardens and in helping to organize "The European Garden Flora" project (which is a Flora covering all species of flowering plant cultivated for amenity, both out-of-doors and under glass in Europe) has pointed directly to this conclusion. The introversion built up leads inevitably to ritualization, and a mystery cult develops, with its arcana, its sacred texts and rules of procedure, its priesthood and prophets, its orthodoxy, heretics and schisms. Like any cult, its sacred texts are not for the understanding of the uninitiated, so they are written in specialist language under the restriction of evil-eye-averting formalities. Again, like any cult, there are controversies about how many angels can sit on the head of a pin, sectarianism and ancestor-worship.

Now you may think that this is a very extreme picture I am painting; but I think that extremes are necessary now. A Reformation is not only desirable, it is vital. After all, taxonomists have something to contribute to science and to society at large. Only by telling people of its value, and by demonstrating its usefulness, can this be done effectively. And in so doing we do not have to sacrifice scholarship or scientific accuracy; what is necessary is that the results of taxonomic activity should be presented in forms that can be used with understanding and confidence by the large numbers of people who need them.

We taxonomists have to do this, because no-one else is going to do it for us. We have to show that we are aware of our strengths and limitations, aware of what needs to be done and why it needs to be done, and able to organize our manpower effectively so that it is done. If we are to "complete the inventory" world-wide, we have to reassess our priorities in this light, and may have to leave behind some of the habits and rituals that have developed. Taxonomists have laboured long and hard to accumulate and order the information that we have. An important task now is to communicate that information more effectively.

What are these habits and rituals that are slowing down progress? First of all, some matters concerned with herbaria. In dealing with any revision, specimens must usually be borrowed, as visiting all the necessary institutions is prohibitively expensive. But the selecting, packing and despatch of specimens is also time-consuming and becoming ever more costly, and some herbaria are dropping out of the system. It becomes more and more difficult to borrow material, and the delays in obtaining material that is loaned increase and increase. To some extent microfiches or colour films of herbaria can help to overcome the problem, but only a relatively small number has so far been dealt with in this way, and such means are certainly not the complete answer. Indexing of the contents of herbaria will also help, though this in itself is time-consuming, and has been criticized as non-productive and tedious. Still, some work is going on (some of it using untrained staff), and some useful lists have been produced (Clarke 1973; Franks 1973; Crovello 1976). Whether we like it or not, this type of activity is going to become ever more necessary.

Another matter that could improve the situation in herbaria would be the recording and distribution of determinations done on a routine basis. Such lists would be of value in preventing time-wasting duplication of effort, as there is a considerable amount of duplicated material scattered through the herbaria of the world. Some work of this type is done, and the Flora Malesiana organization has produced some very useful lists. As more herbaria are indexed on a continuing basis, this kind of dissemination of information will become both easier and more valuable, and space will be left in publications for information of more general interest and use. Of course, to do any of this requires a considerable body of support staff in a herbarium – and this is a declining asset in most of the herbaria that I know.

In making taxonomic revisions in the herbarium there is often a strong tendency to restrict the project geographically. This is very understandable, as it sets definite limits, and can often be justified in terms of usefulness to the people in the region or country concerned (a justification that sometimes, at least, must be considered spurious). However, this again, can lead to duplication of effort, which I would like to illustrate using the large orchid genus *Dendrobium* as an example.

Dendrobium is a genus of perhaps 1000 species, distributed widely in tropical and subtropical Asia from Nepal eastwards, and extending to New Guinea and Australia. It is a genus of enormous diversity, particularly of floral form, though vegetative features are also variable. The species are distributed in about 50 sections, defined basically for New Guinea by Rudolf Schlechter in 1912, but applicable, with some modification, throughout the range. I have had to deal with the 80 or so species widely grown in Europe as ornamentals; these cover the whole taxonomic and geographical spectra of the genus. There is no overall revision of the genus as a whole, but there are numerous geographical accounts, some excellent, e.g. Seidenfaden and Smitinand's (1960) review of the species in Thailand and Dockrill's (1969) account of the Australian species, others not quite so good.

To attempt to identify a *Dendrobium* of unknown origin, or to try, as I have done, to write an account of the cultivated species, making use of these partial treatments, is a nightmare. The bits and pieces are extremely difficult to fit together. Many specific discriminations have been made in different works with slightly or largely differing results, so that their harmonization is impossible without resort to the specimens. Much work has already been done several times by different taxonomists, and yet I find myself having to do it again.

This duplication of effort is ridiculously inefficient. How much more effective would be complete accounts of the various sections, even though they would not perhaps be of much immediate use to people on the ground, say in Thailand or Australia. It is good to know that work of this type is going on (e.g. Reeve and Woods 1980; Cribb 1981). This is basically the old argument about Floras vs. monographs, which has been so frequently debated. In spite of the pressing need for revisionary work, it is clear that the production of Floras will continue; the trend to Floras of large areas (e.g. "Flora Malesiana", "Flora Europaea" "Flora Neotropica", "Flora Mesoamericana") is at least a step in the right, direction, even though the time scales of such Floras may be very extended (Prance 1981).

When we turn to the taxonomic literature, the library end of my topic, we come up against other problems. Publication in the form we know it is slow, expensive and rigid, and I think the whole matter requires considerable re-thinking. As a profession, we are still tied to the printed book as our main means of communication. But there are now numerous, rapid and relatively cheap means of publication available – at one end of the scale the humble type-writer and duplicating machine, at the other the word-processor, possibly linked to a photo-composer, and the whole area of microfilm technology, to say nothing of the possibility of publication by the storage and exchange of information in and between compatible computer memories. Publication by these means can be cheap enough for small volumes to be revised and reissued as new information appears. No Flora or revision is the last word on any group; revision goes on continually, but its publication is often delayed, sometimes for years, because of the rigid book-format of the original publication; or, if publication of later revision is achieved, it tends to be in easily overlooked supplementary volumes or papers. This matter of cheap and rapid publication is important. It is aided by the production of small volumes or fascicles, rather than full-scale volumes, but, unfortunately, many publishers do not like this method of production. Prance, whom I have quoted frequently already, also touches on this problem (1981). He points out that since the publication of his account of the Chrysobalanaceae for "Flora Neotropica" in 1972, 27 new species (8% of the original total number) have been described. A revised version of the Chrysobalanaceae is therefore much to be desired, but the capital tied up in the first publication presumably makes a revised reissue (which itself may need revising again in 10 years) very difficult if not impossible. The hold that the idea of the printed book has over us is one of the factors that is really slowing down progress and actually preventing effective communication. Many people like books as objects, but there is no doubt that, as far as taxonomy is concerned, the book is an obsolete form.

Another aspect of library work is, or should be, the production of catalogues, bibliographies, indexes, current-awareness lists and the like. Some of these certainly do exist: "Kew Record", "Excerpta Botanica", "Horticultural Abstracts", etc., and we all make good use of them. But many more are necessary. In collaboration with my colleague Dr D. F. Chamberlain and myself, the librarian at the Royal Botanic Garden, Edinburgh, has been compiling an annotated and subject-indexed bibliography to the genus *Rhododendron*. This extends from 1753 to the present, and, even after seven years' work, is still incomplete. Admittedly, *Rhododendron* is a genus with a very extensive horticultural literature, but the man-hours are simply not available now for the completion of the project, and it will have to be published as it is. This is unsatisfactory, but is all that can be done. Such bibliographies can be immensely useful, and can save considerable amounts of time, but, it seems, they have a low priority.

And, of course, the major flowering plant index is the "Index Kewensis", with its 16 supplements. No-one doubts its value, but what an old-fashioned tool it is. I understand that it is to be computer-compiled in the future, but until this is done, we shall have to continue with it as it is. Rouleau's Guide to it (1981) is certainly a great help, but the fact that we need an index to an index shows how bad the situation is.

A further source of delay and confusion in our work is, I am afraid, "The International Code of Botanical Nomenclature" (1978), at least in some of its aspects. Here, I know, I am treading on particularly sacred ground, and I want to make it clear that I am expressing my own opinion and not any official view from the Royal Botanic Garden, Edinburgh. This collection of rules and recommendations attempts to facilitate accurate communication, but it appears to me to delay or prevent it in some cases and, in such matters as the appearing and disappearing letter "i's" of such epithets as *rehderanum/rehderianum*, puts taxonomy in a position of ridicule among the sciences. I have already mentioned the *Ornithogalum* case where procedures in line with the rules help to create confusion. And all such cases seem to me to derive from two factors: the first is the attempt to legislate for everything, exceptions included, and the other, which is much more fundamental, derives from Principle IV of the Code, which I quote from the 1978 version: "Each taxonomic group with a particular circumscription, position and rank can bear only one correct name, the earliest that is in accordance with the Rules, except in specified cases". Why do we ask this of difficult to define biological phenomena when we ask it of practically no other semantic concepts? An organic chemist may talk about alcohol, ethanol, mono-hydroxy ethane, C_2H_5OH, etc., allowing the context to determine the particular choice of name, and the same can be said of other disciplines in which the concepts are not exclusively mathematically symbolized. Even as an ideal, the Principle is flawed, as taxonomists tend to disagree about the limits of taxa and about precisely which higher taxa they belong to. This, nomenclaturists will say, is not a nomenclatural matter; but it is a matter of communication, and it vitiates attempts to achieve the ideal. The name, as such, cannot indicate these complexities, so there will always be a need for synonymy and the ideal will remain unattainable. *Rhododendron polycladum* provides a simple example; it was described in 1886 by Adrian Franchet on the basis of a few specimens collected in western China by Delavay. It was never collected again under that name, and remained a rather "rare" species. In 1916 Balfour and Forrest described *R. scintillans* from the same general area. Plants under this name were widely collected and widely grown in gardens. In 1975 the Philipsons revised *Rhododendron* Subsection *Lapponica*, to which both names belong, and decided that the two were the same species, for which they correctly used the name *R. polycladum*. But not all Rhododendron experts agree that these two are the same species, so the name *R. polycladum* may be used in the literature in two different senses, and for precision it is always necessary to quote the synonym if the wider sense is intended. This situation will persist for years, because most of the information about the species is to be found under the name "scintillans". If we always have to quote the synonym, do we need to make such a fetish of having the puritanically right name? I do not think so. Attempting to achieve this takes up a good deal of time in taxonomic work – for "The Flora of Turkey", on which I once worked, I estimated it at about 15% of the total work time; this is time that seems to me to be largely wasted chasing a will-o'-the-wisp. But the ideal we are unsuccessfully striving for lies at the heart of the Code and permeates all of its rules. Hence the need for the constant juggling with the details of the Code, of which the "now you see them, now you don't" letter "i's" are just a symptom. This is the sort of controversy about how many angels can sit on a pin head that I spoke about earlier – endlessly fascinating, endlessly providing opportunities for barrack-room law, endlessly unanswerable, and not a great deal to do with a world in which species are disappearing faster than we can classify them.

Some reappraisal of the Code seems to me to absolutely necessary, so that time spent on nomenclatural amenities is reduced to the basic necessary minimum. Again, modern methods of data-control must surely have a part to play in this process.

Finally, in this very rapid survey of taxonomic trouble-spots, some thoughts about high-level classification, where taxonomy meets phylogeny, and the resultant structures seem to me to resemble poems far more than they do anything else (one man's view of a part of reality, with more effect on other productions of the same kind than on practical day-to-day concerns). I have nothing to say about the phylogenetic significance of the various classifications that have been proposed, because, fortunately, their impact at the level of taxonomy that I have been discussing is very small. But this impact can be important in terms of the arrangement of herbaria and the arrangement of taxa in floristic works. In both of these cases the phylogenetic content is not significant; what is important is the degree of documentation of the system – is it easy, for example, to find out the family to which any particular genus belongs? Not all systems are equally clear on this point, and it is not helpful for the general user to find that the Flora of one country follows X's system, while that of a contiguous country follows Y's. This is to create difficulty and irritation where none need exist.

CONCLUSIONS

Although I have been critical of many aspects of taxonomic work, this arises from a concern that a very important task is not being done as well or as efficiently as it might be. But I do not want to give the impression that I think the whole of taxonomy is in a mess: I don't. Much valuable work has been done, and is continuing, and I need only instance, as one example among many possible, the completion of "Flora Europaea" in twenty years.

But still my concern remains. Money is short and the competition for it is severe; the number of working taxonomists appears to be slowly declining, as does the number of support staff working in herbaria and libraries. Many people need the information that we can provide, particularly those involved in conservation; but they need it quickly and in a form that they can use. We need to "complete the inventory" as rapidly as we can, before most of what we are inventorying is irretrievably destroyed. I have tried to indicate some of the priorities which I think need reappraisal: we have to decide, individually and corporately what is really important now, and what is really secondary, and to remove as much as possible of the dead weight of formalism that still hangs over the subject.

If we are to do this, we will need, as I have said, to make use, rapidly and on a large scale, of the available data-control and communication equipment. Many of us may not like this, and some of us may feel we are too old to master its complexities, or too put off by some unhappy experiences in the past. But we must overcome these prejudices. The available technology is becoming (relatively) cheaper and of greater and greater capacity, and the associated software of ever-increasing sophistication and flexibility. Surely it is not too much of a pipe-dream to imagine a situation not too far ahead, when one can sit at a terminal in a herbarium or library and obtain a read-out which contains not only facts but opinions – not only where a name was published, by whom, and what the type is, but notification of the facts that Smith says it is a synonym of some other name while Jones considers it a separate species, that Brown puts it into Genus X while Green puts it into Genus Y, etc.? This, after all, is only what we all do in taxonomic libraries, as we flit from volume to volume.

If it is to be done, it requires international agreement, organization and finance. A start has been made at Reading with the European Taxonomic, Floristic and Biosystematic Documentation System. The progress of this project will be watched with the keenest anticipation.

Most readers will, I hope, agree that taxonomy faces a challenge. Not all may agree with my analysis of it or my suggestions for improvement. But if we can all begin to think about what has to be done, *and* for whom it needs to be done, then this diatribe will have been worthwhile.

REFERENCES

Argent, G. C. G. (1980). Review of "The Euphorbiaceae of New Guinea" by H. K. Airy-Shaw. *Notes Roy. Bot. Gard. Edinb.* **38,** 416.

Bentham, G. (1861). "Flora Hongkongensis". Lovell Reeve, London.

Berger, A. (1915). "Die Agaven". G. Fischer Verlag, Jena.

Clarke, G. C. S. (1973). Type Specimens in Manchester Museum Herbarium: Musci. *Manchester Museum Publ.* n.s. **2,** 73.

Cribb, P. J. (1981). A preliminary key to Dendrobium Sect. Latouria. *The Orchadian* **6,** 277–283.

Crovello, T. J. (1976). "The Greene Index: The Edward Lee Greene Herbarium at Notre Dame". University of Notre Dame, USA.

Dockrill, A. W. (1969). "Australian Indigenous Orchids", Vol. 1, pp. 325–507. The Society for Growing Australian Plants, Sydney.

Franks, J. W. (1973). "A Guide to the Contents of the Herbarium of Manchester Museum". *Manchester Mus. Publ.*, n.s. **1,** 73.

Gentry, H. S. (1972). "The Agave Family in Sonora". USDA Agriculture Handbook No. 399.

Gentry, H. S. (1978). "The Agaves of Baja California". Occ. Publ. Calif. Acad. Sci. No. 130.

Greuter, W. and Rechinger, K. H. (1967). Chloris Kythereia. *Boissiera* **13,** 11–206.

International Code of Botanical Nomenclature (1978). Bohn, Scheltema & Holkema, Utrecht.

Michotte, F. (1931). "Agaves et Fourcroyas, Culture et Exploitation" (3rd edn). Paris.

Philipson, M. N. and Philipson, W. R. (1975). A revision of *Rhododendron* Subsection *Lapponicum*. *Notes Roy. Bot. Gard. Edinb.* **34,** 1–72.

Prance, G. T. (1977). Floristic inventory of the Tropics: where do we stand? *Ann. Missouri Bot. Gard.* **64,** 659–684.

Prance, G. T. (1981). Flora Neotropica: where do we stand? *Taxon* **30,** 81–87.

Reeve, T. M. and Woods, P. J. B. (1980). A preliminary key to the species of *Denodrobium* Sect. *Oxyglossum*. *The Orchadian* **6,** 195–208.

Rouleau, E. T. (1981). "Guide to the Generic Names appearing in the Index Kewensis and its Fifteen Supplements". Chatelain, Lac de Brome.

Seidenfaden, G. and Smitinand, T. M. (1960). "The Orchids of Thailand", Vol. 2(2), pp. 185–282. The Siam Society, Bangkok.

Thrower, S. L. (1971). "Plants of Hong Kong". Longman, Hong Kong.

Trelease, W. (1910). Species in Agave. *Proc. Amer. Phil. Soc.* **49,** 232–237.

Trelease, W. (1913). Agave in the West Indies. *Mem. Nat. Acad. Sci.* **11,** 1–55.

Walden, B. M. and Hu, S. Y. (1977). "Wild Flowers of Hong Kong". Sino-American Publishing, Hong Kong.

3 | Botanic Gardens and Experimental Grounds

P. S. ASHTON

Arnold Arboretum, Harvard University, USA

Abstract: The case is made that botanic gardens, while endeavouring to serve the disparate functions of systematic resources and areas of public recreation, have failed to develop their considerable potential in the former. The growing interest in ontogenetic, developmental, chemical and other characters requiring use of live material in plant systematics, and the increasing rate of attrition of the world's Flora, provide important opportunities for botanic gardens. For gardens to come into their own in systematic endeavour, standards of verification and record-keeping need to be further improved and coordinated.

INTRODUCTION

Botanic gardens today are witnessing what I hope will prove to be the end to more than a century of ambivalence, confusion and, increasingly, neglect, uncertain how the balance between their disparate functions should be attained.

The early gardens – Padua, Montpellier, Leiden – were collections of useful herbs and shrubs, laid out in arabesques according to a practical or geographic scheme. Usefulness and aesthetics were both already served, but there was no dualism. In that age of craftsmanship no conflict between the two would have been expected. Indeed. gardens were understood to convey a direct knowledge of God, and their man-induced beauty and plant diversity were joint manifestations of this (Prest 1981). Oxford, Paris, and Uppsala were founded in the following, seventeenth, century on substantially the same model, but aspired to be comprehensive. It was not until the eighteenth century though, with the establishment of the first great tropical colonial gardens and their metropolitan

Systematics Association Special Volume No. 25, "Current Concepts in Plant Taxonomy", edited by V. H. Heywood and D. M. Moore, 1984. Academic Press, London and Orlando.
ISBN 0 12 347060 9

mother-garden at Kew, that the introduction of trees on a big scale into botanical gardens forced fundamental changes to their purpose and design and, ultimately, helped cast the ambivalence that prevails today. These changes coincided with the early successes of the industrial revolution which, by divorcing work from art, was the primary cause of this ambivalence.

The ancient European gardens arose as living pharmacies, under monastic administration. Their colonial successors were testing stations for the natural products sought by the metropolitan powers: not the traditional medicines, though this was still an age of pharmaceutical conservatism, but culinary delights which gave impetus to the first flowering of courtly cuisine in northern Italy, and thence to France. Many of these plants were trees. Trees proved the easiest plants to maintain in cultivation in the humid Tropics. Thus the first tropical garden was established, on Réunion, as a testing station and depot for the developing French spice trade with the East Indies. From there, the most profitable crops were introduced into cultivation in the Antilles.

The major gardens that followed the foundation of Calcutta in Asia, Bogor, Peradeniya, and later Singapore emulated it in being designed according to the new systematic order, but the scale and inherent informality of the predominantly arboreal collections, combined with the prevailing English fashion for landscape gardening, also necessitated a freer style of design.

The ambivalence to which I have referred explicitly arose from this change of scale. The dilemma is more clearly manifest in the parallel, and closely interdependent, history of botanical illustration. Students of historical psychology will not be surprised that Vesalius was illustrating human anatomy from life for physicians in the same decade that Hans Weiditz, artist to physician and herbalist Otto Brunfels, was so illustrating plants; nor that artists such as Jan van Eyck had preceded the scientists in the latter by a full century. As Jacobs (1980) has argued, the later and also simultaneous development of the herbarium as the principal tool of plant taxonomy, the terse Linnaean method of diagnostic description, and the botanical illustration of plants in isolation from both soil and other plant associations, divorced plant systematics from the living world of nature, and also inevitably from the applied biological sciences.

THE DIVORCE BETWEEN SYSTEMATICS AND LIVING PLANTS

Nowhere has the inordinate domination of the herbarium been more unfortunate than in the systematic study of trees. Though the enormous scale of leaf and inflorescence in palms forced botanists to look beyond the herbarium sheet that so patently could not contain them, this lesson was not extended to the study of other trees. Leaves, buds and flowers were minutely described, wood anatomy and fruits less frequently so. But bark, and the dynamics and modes of shoot extension in relation to the mode of construction of whole trees have become the subject of rigorous study and illustration only in the last forty years. The Dutch master Jacob van Ruisdael had painted portraits of temperate tree species with remarkable accuracy in the mid-seventeenth century (Ashton *et al.* 1982), and Frans Post (1612–1680) painted Brazilian trees from life which are recognizable to species; but botanical illustrators never attempted to depict whole trees, transferring instead directly from the herbalistic tradition of depicting branches (sometimes with symbolic roots attached) to semiformalized renderings from twigs mounted in the herbarium. How, then, did the introduction of trees to the botanic garden serve systematic botany?

The separation between systematic botany and plants as living beings in living communities has proven particularly disastrous in the forested regions of the earth, and especially in the tropics. Exacerbated by the dearth of resident systematic botanists, local tropical Floras have too often been reduced to truncated accounts, dwelling on minutiae of variation that frequently have no significance in the living world, based on those features that can be described from a desiccated fragment glued to a card. Thus, the scientists' perception of the botanic garden failed to transcend the confines of the *pulvillus*, or order bed, when the gardens themselves exploded into wooded parks.

Walters (1961) and Raven *et al.* (1971) have pointed out how Linnaeus was influenced in his classification by the northern European folk classification that he inherited. Somewhat analogously, up to the nineteenth century the design and research priorities of tropical botanic gardens were influenced by their role in the introduction and evaluation of useful plant species. This had its advantages, for it sustained the scientific study of whole live plants. During the nineteenth century, though, the thirty major commercial crops of the contemporary world became identified, research was transferred to special stations under the administration of the new state departments of agriculture and forestry, and the botanical gardens were cast adrift from the mainstream of human endeavour to languish in the false security of academia.

Meanwhile, the evolution of the landscape garden in northern Europe during the late eighteenth century had coincided with rapid early advances in herbarium-based, Linnaean systematics. Exotic plants, particularly trees, were being introduced in increasing numbers, and the heyday of botanic gardens as public attractions had arrived. In Europe their purpose was at that time at least nominally to facilitate research and instruction in systematic botany; but also, owing in part to their increasing size and urban location, for ornament and public recreation. Increasingly during the nineteenth century, tropical botanic gardens came to assume these same functions as their utilitarian purposes were shed.

THE ROLE OF GARDENS IN SYSTEMATICS

Today, botanic gardens are touted as invaluable resources for instruction and public enjoyment, but how much are they used, and to what extent will they be needed in future, for research? In those gardens provided with herbaria and experimental grounds, it is there that research has been pursued. Is this inevitable, or is it a manifestation of the constraints imposed by the common perceptions of our own age? Clearly, the answer is of immense importance in view of the great expense of maintaining well-curated living collections.

Plants do not grow in the same manner in cultivation as they do in nature. In an extreme case, freestanding trees adopt a very different habit from those in closed forest. They flower precociously, and ontogeny is telescoped in other ways. Bark manifestations can be substantially, if superficially, different. Phenology can be modified, and synchrony within populations and even individual plants lost. However, the underlying processes remain the same.

There is no doubt that the well-curated and recorded systematic living collections have an enormous, though currently largely unrealized, potential in research into comparative morphology, reproductive biology including embryology and cytology, chemistry, ecophysiology, and especially ontogeny, aspects to which plant systematists have often paid inadequate attention. Recent experience at the Arnold Arboretum, at which individual projects in each of these fields are being carried out, has demonstrated the potential, though few major studies are yet being realized.

An outstanding example of a research opportunity being missed in botanic gardens is in the study of tree architecture. Pioneer advances had been made by Koriba (1958) and by Corner (1952, 1964), exploiting the systematic collections in the Singapore Botanic Garden in novel ways for the time. Hallé and Oldeman, to whom we owe the first comprehensive account and classification of the diversity of tree form (1975), concentrated their studies on the diversity of architecture manifested in saplings within tropical forest. They observed that few of the 24 architectural models they had recognized occur in temperate European trees. Nevertheless, careful study of living collections in botanic gardens would have indicated that some half of these "models" do grow in the temperate zone, and it is instructive that studies initiated in tropical forest in this case have stimulated interest once again in comparative morphological studies in temperate gardens.

The universities must take much of the blame for these missed opportunities. Systematic research above the level of the population has been increasingly neglected by them. There has been a failure to recognize monographic and other fundamental, and especially taxonomic, systematic research as a discipline still capable of advancing biological theory. There is also an unwillingness to take responsibility for training practitioners in plant taxonomy, a critical shortage of which persists in most parts of the world. In some measure, this is the outcome of success. The development and maintenance costs of botanical collections, and in particular living collections, is such that few academic institutions have aspired to comprehensive holdings. These are now the older institutions, and through their efforts the Floras of their regions are well-known. Other priorities, not requiring extensive collections, have arisen. To address them, staff are appointed who, never having had previous opportunities to use research collections, fail to take advantage of them. But it is certain that major new collections will not now be built. The inheritors of the old collections have research opportunities which are uniquely their own – and are saddled with a wider responsibility which they cannot escape. It is not enough for these institutions to merely conserve their collections as repositories for visiting scholars: if they do not foster active research schools in systematic biology, there will in due course be no visiting scholars.

It has so far not been possible, as it should be, to use botanic gardens for systematic research as we use herbaria, that is to consult them as reference libraries with a reasonable expectation that each garden will maintain a reliable catalogue and records, and that these records, and through them the collections at least for the more important accessions, should be published and made generally available to the scientific community. This would enable gardens to pursue a coherent policy for representing, in their respective climates, a major part of the world's flora in culture for future research and use, and thus for complementing and strengthening the value of the growing stockpiles of dead material preserved in the herbaria. The electronic computer has provided an economic, practical means to make good this omission.

The pioneering projects of the Horticultural Society of America's Plant Sciences Data Center, and the Royal Botanic Gardens, Kew, in standardizing living collections records, entering them into computer files, devising multi-access information retrieval systems and, most important, coordinating the records of independent gardens will prove increasingly valuable as systematic botanists are awakened to the new horizons that are thereby being opened to them.

In this context, the potential and currently growing role of botanic gardens in *ex situ* conservation has important implications for plant systematics, for accelerating extinction is permanently depriving us from investigating those attributes of plants which are not preserved in the herbarium specimen, and reducing our scope for applying new techniques.

All the same, though individual plant genotypes can in theory be kept in perpetuity in living collections, their high space requirements and maintenance costs, and the difficulty of sustaining uniform policy and management standards, necessitates that living collections aim at being synoptic rather than comprehensive and emphasizes the need, so far poorly realized, to carefully plan a requisite though modest level of duplication within a carefully coordinated world-wide system. This task would give strengthened purpose and continuity to the International Association of Botanic Gardens. Kew would appear, on account of its resources and traditional associations, to be the obvious lead institution for such a coordination, through eventual extension of its computer records system world-wide, and periodic publication of a records index.

It is, of course, absurd to suggest that the world Flora, even for a few individuals for each species, can be conserved *ex situ* in its entirety in botanic gardens, however well they are organized. Representative taxa from all genera could reasonably be so conserved. Beyond that, advances in comparative and systematic biology are most likely in those taxa which have already received monographic treatment using conventional characters. There is therefore a strong case to permanently maintain the comprehensive living collections of those taxa that have been accumulated by farsighted monographers, such as Burtt's Gesneriaceae at Edinburgh, Foxworthy and Symington's dipterocarp collection at Kepong, or C. S. Sargent's *Crataegus* collection at the Arnold Arboretum, now sadly depleted.

The living collection can best be utilized as a bridge between the herbarium and the wild population, and between systematics and other biological disciplines. It has the advantage over the herbarium that the whole living plant is available for study, the disadvantage that the sample must inevitably be smaller, less representative. It has the advantage over the field population that taxa of distant provenance can be grown side by side in similar conditions for comparison.

It is in these contexts that intrinsically temporary experimental plantings, in independent grounds or plots that are adjuncts to botanic gardens, serve their principal purpose in systematics. From the days of Gaston Bonnier and – and this cannot be an accident – of the Impressionists, experimental plantings of population samples and ecotypes have formed an integral part of the new systematics, in which the systematic botanist at last returned to real living whole plants. Though lacking the potential for the precise standardization and environmental definition and control available in the growth chamber, experimental grounds gain in the greater scale of planting that is possible.

Experimental grounds have provided the means to investigate the systematic and evolutionary significance of those aspects of breeding systems which are characteristic of vascular plants, including polyploidy, apomixis, and self-compatibility. Though current advances are particularly being made through long-term demographic studies of field populations, in which ecophysiological and genetic approaches are combined (e.g. Solbrig 1980), the growing interest in rates of evolution among higher taxonomic categories provides important opportunities for living plant collections. As examples, the single gene differences that appear to exist between some, perhaps many, diploid plant species, and the extraordinary contrast, for instance, between the apparent incompatibility of simultaneously flowering, closely allied rain forest tree taxa on the one hand, and the ability of East Asian and North American species of *Liriodendron*, which have been isolated for perhaps ten million years, to hybridize on the other provide exciting subjects for research in botanic gardens combined with experimental grounds.

CONCLUSIONS

In summary, I see botanic gardens serving in the reinstatement of systematics at the center of botanical research. This they can do by reasserting their integrative function, and by vigorously pursuing higher and more consistent standards of accessions policy, curation, and records keeping. Collaboration between institutions, and particularly the maintenance of free exchange of plants and information between gardens and between nations, is vital.

At the same time, botanic gardens have an opportunity, indeed an obligation which is open to them alone: to bridge between the traditional concerns of systematic biology and the returning needs of agriculture, forestry, and medicine for the exploitation and conservation of biological diversity. The administrative as well as physical separation of plant breeding stations, often specializing on a single crop species, from botanic gardens has cut the former from what should again become the leading repositories of genetic diversity above the species level. It has cut the latter from some of the most active centres of biosystematic research activity. Now, as energy shortages and overpopulation in the poor nations put ever-increasing demands for labor-intensive agriculture, and as marginal lands come under the plough, the need for a broader search for new crop plants and for *ex situ* conservation is ever more pressing. Systematic living collections, and an active community of whole plant systematists are basic resources if these needs are to be fulfilled.

REFERENCES

Ashton, P. S., Davies, A. and Slive, S. (1982). van Ruisdael's trees. *Arnoldia* **42,** 1, 2–31.

Corner, E. J. H. (1952). "Wayside Trees of Malaya" (2nd edn) 2 vols. Government Printer, Singapore.

Corner, E. J. H. (1964). "The Life of Plants". Weidenfeld and Nicholson, London.

Hallé, F. and R. A. A. Oldeman (1975). "Essay on the Architecture and Dynamics of Growth of Tropical Trees" (Trans. B. C. Stone). Penerbit Universiti Malaya, Kuala Lumpur.

Jacobs, M. (1980). Revolutions in plant description. *Landbauhochschool Wageningen Misc. Papers* **19,** 155–181.

Koriba, K. (1958). On the periodicity of tree growth in the tropics. *Gardens Bulletin, Singapore* **17,** 11–82.

Prest, J. (1981). "The Garden of Eden. The Botanic Garden and the Recreation of Paradise". Yale University Press, Conn., USA.

Raven, P. R., Berlin, B. and D. E. Breedlove (1971). The origins of taxonomy. *Science* **174,** 1210–1213.

Solbrig, O. T. (1980). "Demography and Evolution in Plant Populations". Blackwell, Oxford.

Walters, S. M. (1961). The shaping of angiosperm taxonomy. *New Phytol.* **60,** 74–84.

Recent Approaches in Morphology and Anatomy

4 Vegetative Morphology–Some Enigmas in Relation to Plant Systematics

P. B. TOMLINSON

Harvard Forest, Harvard University,
Petersham, USA

Abstract: Vegetative characters in the higher plants are accorded only a limited place in classification, in part for traditional reasons related to a belief in the innate conservatism of reproductive features, and in part due to modular construction. However, since systematics is an intrinsically empirical science, no generalizations of a predictive nature can be made about the systematic value of vegetative features. Vegetative characters can be of primary value in classification but are often ignored or little emphasized. Recent literature shows a growing appreciation for the total organization of the plant in a dynamic context; the seedling shoot may, in this respect, provide good examples of systematically useful characters. This approach puts the subject in a more satisfactory biological context. However, lack of recognition of functional correlations between different parts of organs seems to have been inhibitory to a more extensive use of vegetative features. This further suggests that a more biological approach to plant morphology will ultimately prove beneficial to systematics. Consequent on the modular construction of higher plants, evolution of form in plants can be the result of modification of any stage in the life cycle. Paedomorphosis or neotony seems to be an important process in the evolution of plant form and its more extensive recognition would help systematic interpretation.

INTRODUCTION

A basic enigma in the use of vegetative morphological characters in the systematics of higher plants is that they have been accorded much more limited use than morphological features of the sexual reproductive system. This is reflected

Systematics Association Special Volume No. 25, "Current Concepts in Plant Taxonomy", edited by V. H. Heywood and D. M. Moore, 1984. Academic Press, London and Orlando.
ISBN 0 12 347060 9

initially in the rather few families in which vegetative characters are used as a primary diagnostic feature for the segregation of major groups (Table I). The idea of the innate conservatism of the reproductive organs has been accepted by many systematists who have implicitly followed Jussieu in his "Genera Plantarum" (1789) in ranking characters in order of their usefulness in systematics. This approach is somewhat ironic because Jussieu was the first person to recognize many natural families and emphasized the polythetic basis for this naturalness and reflects, in practice, an *a posteriori* weighting of characters. *A posteriori* weighting of characters is also an essential feature of the Adansonian approach. On the other hand, the idea that there are no consistently conservative (i.e. *a priori*) characters is emphasized by a number of authors (e.g. recently by Stebbins 1974; Davis and Heywood 1963). The latter point out that any character may empirically (i.e. *a posteriori*) be found to be stable and therefore of use in systematics. This point of view goes back at least to the time of Lindley (1830), who strongly emphasized vegetative characters, while also advising a pragmatic approach in their application. Davis and Heywood (1963, 50) are particularly emphatic in their denial of the idea that vegetative characters are less useful in contributing to the development of a natural system of classification:

> The neglect of vegetative characters has been one of the most serious errors in the history of classification and has done much to delay the achievement of a natural system. If we accept the precept that the taxonomic value of a character rests upon its correlation value, then vegetative characters are as valid as floral ones in the delimitation of genera and families and their placing in a natural systèm. Hitherto they have usually been neglected above species and sectional level. And yet the most constant character separating the two great [*sic*] subclasses of the Angiosperms is a vegetative one – the number of cotyledons.

I wish to address this general area with special emphasis to more holistic aspects of vegetative morphology, pointing out the extent to which a more dynamic approach to the study of vegetative characters which has been developed in recent years is proving of increasing value in systematics. I will also suggest ways in which the basic enigma may, at least in part, have evolutionary explanations.

Plant systematics is intrinsically an empirical science so it is relatively easy to argue for any given position on the basis of example – argumentation by advocacy or eclecticism becomes the logical method and evidence for contrasted positions can easily be found. For example, the conservatism of reproductive features which is commonly accepted is easily supported by extreme example. The parasitic vines *Cassytha* and *Cuscuta* show considerable parallelism in relation to vegetative features which give no indication of their respective lauraceous or convolvulaceous affinities, which come entirely from a consideration of floral and fruit characters*. In these examples the reproductive organs are certainly highly conservative relative to the vegetative organs. Another example would be the remarkable contrast in habit between the largest species of the genus *Puya* (especially *P. raimondii*) and the extreme epiphytic species of *Tillandsia*, notably *T. usneoides* (Spanish moss), all members of the natural family Bromeliaceae. This particular example offers evolutionary clues to an explanation of some systematic enigmas, as I will point out later.

However, accepting the axiom of relative conservancy of particular features has had unfortunate consequences in plant systematics, since it may have partly deprived the systematist of a suite of useful diagnostic features. Where this neglect is extreme the systematist himself is the conservative character.

THEORETICAL CONSIDERATIONS

(1) Modular and unitary construction

That there might be innate conservatism of the reproductive organs in plants can be established on theoretical grounds because of the modular construction of the vegetative parts of higher plants, i.e. that the plant body is made up of an indefinite number of repeating units without definite numbers of parts, which can be of various kinds (White 1979). The units may be appendages, like leaves, organ-complexes like shoots or even, at a higher hierarchical level, branch-complexes themselves. The units may collectively be referred to as modules, although the term has a variety of definitions. The reproductive structures, on the other hand, come closer to the unitary construction of the higher animals. A flower, for instance, usually has a fixed number of parts arranged in a constant sequence, and has no capacity for self-repair. Some vegetative modules come close to this unitary constructional principle.

This constructional contrast in turn may be explained in functional terms; the reproductive unit (here the flower) has a set of functions which are invariate and relate to precise objectives: the production of gametes, hence of zygotes and ultimately of embryos. In contrast the vegetative parts have a greater diversity of possible functions such as photosynthesis, mechanical support, protection, storage, anchorage and, also because of the modular construction, they have a great capacity for replication and repair. Furthermore these functions are not so mutually exclusive as are the functions of the reproductive parts.

---

* "The genus *Cassytha*, a parasitical leafless plant, is remarkable for differing from the order Laurinaeae [=Lauraceae] in nothing whatever, except its very peculiar habit" (Lindley 1830).

A given appendage can accommodate a greater diversity of functions, sometimes carrying them out sequentially or even synchronously. Thus a leaf can be a cotyledon, a bud-scale, a photosynthetic appendage, a spine, a tendril, an attracting organ or a bulb-scale. In these different guises the leaf can have differing systematic applications. Often the more specialized the function, the more is the diagnostic value of the structure (cf. Table I).

Unitary construction (cf. that of higher animals) is therefore more frequently represented in the individual flower and it is a consequence of this that the systematist has developed the floral diagram and floral formula as a method for displaying basic information – a representation, in fact, of the floral "*Bauplan*", which may be typological or phylogenetic. Rarely have attempts been made to display information about vegetative construction in the same way, although this is possible in plants with highly organized and regular shoot systems. A good example would be the formulae and diagrams devised by Engler (1877) to express branching patterns in aroids. The concept has had little application elsewhere (for an exception using cotton see Munro and Fairbrother 1969), but might be developed in view of our increased appreciation for the precision of vegetative organization in many plants. I have used it myself in the analysis of shoot construction in *Calyptranthes* (Myrtaceae), which is highly organized (Tomlinson 1980), but shows an interesting range of expression.

(2) Deterministic and opportunistic processes

From these considerations it follows that vegetative structure is rarely totally "deterministic", and in fact "opportunistic" expression in higher plant constructure is one of the adaptive features of their vegetative organization. Elsewhere, I have even suggested that this opportunistic method of construction affords the best comparison with "behaviour" in animals (Tomlinson 1982).

The distinction between deterministic and opportunistic expression of form, although a universal feature of the construction of most higher plants, is most clearly revealed in the Hallé-Oldeman system of description of tree architecture (Hallé and Oldeman 1970; Hallé *et al.* 1978). Here the *architectural model* is the basic abstraction which relates to the developmental blueprint, or genetic ground plan which is ultimately expressed as the visible architecture of the tree itself, while *reiteration*, or repetition of this architecture is the process whereby the organism responds to damage or environmental change at the morphological level. Other aspects of this general appreciation of the difference between deterministic and opportunistic events are found in branching processes; proliferative vs. regenerative branching is an important difference which distinguishes between increase in module number and the simpler process of replacement (as in sympodial substitution). The distinction may seem esoteric to the systematist but has importance in population biology, especially in clonal organisms.

Owing to practical difficulties of observing them in the herbarium it is understandable that the systematist has been little concerned with gross aspects of construction. However, the morphologist has begun to produce analyses which makes possible a discrimination of characters resulting from the two processes of architectural development and reiteration; in this sense a systematically useful advance has been made, and examples can be presented later.

(3) Adaptive radiation of floral mechanisms

This basic developmental contrast accounting for the supposed conservative nature of characters of the reproductive parts remains puzzling, however, if one considers the actual diversity of floral mechanisms in closely related taxa with the same floral "*Bauplan*". In the tribe Rhizophoreae of the Rhizophoraceae (the mangrove species) adaptive radiation of floral mechanisms is considerable but based on a rather restricted structural plan. Nevertheless, there is a range from wind pollination to animal pollination, which itself involves a wide range of pollinating agents, from birds to small moths (Tomlinson *et al.* 1979). In particular a quite complex mechanism of pollen discharge in the genera *Bruguiera* and *Ceriops* accommodates most of this range of flower visitors. The genetic view here might be that the basic floral ground plan involves such a complex developmental control mechanism that the selective process which brings about the functional diversity in mechanisms must also remain undisruptive of the elaborate control process. Do we then have to assume that there is a much less elaborate genetic control for developmental processes in the construction of vegetative parts? I doubt it, but presented in this way the situation does seem enigmatic. The Rhizophoreae affords a particularly interesting example because here the vegetative parts are very conservative in their expression and seem to provide little diagnostic information when examined superficially.

(4) Correlation of characters

A primary explanation for the limited application of vegetative features in plant systematics may be the failure of the morphologist to demonstrate structural and functional correlations between seemingly unrelated characters. This concept has been much more widely applied by systematic zoologists, probably because the unitary construction of most animals facilitates the recognition of this kind of correlation (Mayr 1982). However, I would suggest that the situation in plants is not much less different if only botanists will more widely accept the existence of structural and functional correlations within the plant's organization. It is perhaps only of limited value to exhort the systematist to greater effort by listing morphological characters which might be of systematic diagnostic value, as do Davis and Heywood (1963). A better approach might be to arouse his or her biological curiosity by recognizing and accounting for correlations so that diagnostic features are not seen in isolation. On this basis it might then be possible to establish how variation in one character is influenced by or influences other characters. This is the task of the morphologist and anatomist and specialists in these disciplines have been singularly uninterested in the subject. The net result, of course, would be to reduce the weighting given to correlated characters.

There are such examples. In the basic division of the angiosperms, the unit of monocotyledonous construction compared with that of dicotyledons is expressed by a set of vegetative characters which are structurally and developmentally correlated – alternate leaves, encircling leaf insertion, stem vasculature, absence of an initial vascular cambium. Any one of these can occur as a single character in other groups of plants, but the assemblage ultimately relates to the monocotyledonous condition in which secondary growth is lacking. The way in which these features relate to habit has been explored quite fully, notably by Holttum (1955) so that the growth form of the monocotyledon is itself seen to be an expressed set of functional correlations. Further, our understanding of monocotyledonous vascular systems has increased enormously in the past 20 years, thanks to the application of cinematographic methods for unravelling complex vascular systems (see Zimmermann 1981). Our understanding of monocotyledons is therefore becoming highly integrated and we should certainly be able to develop a dynamic view of monocotyledonous systematics in the future.

Our view of the dicotyledons is less coherent, in part because they are more diverse architecturally, since they lack the growth constraints of monocotyledons, and in part because functional correlations are less easy to appreciate. Perhaps if a more dynamic and integrated approach to the study of bud and shoot morphology is adopted (Tomlinson 1977), some of these functional correlations will become evident. Stipules, for example, are a widely accepted diagnostic feature in plant systematics, but their distribution does not engender obvious "rules". Perhaps if we understood their functions, which must relate to bud dynamics, then our systematic use of them would be sharpened because we might understand their homologies. In this sense correlations between stipular morphology, vernation and shoot dynamics should be sought more actively. Well worked-out examples of this kind might also provide clues as to the relative evolutionary status of plant structures, which still provides a major stumbling block to refined cladistic analysis, as Stevens (1980a) points out.

How a simple set of correlations involving quite distinct characters can be made use of is suggested by the work of Philipson and Philipson (1979) on *Nothofagus*. Here four kinds of leaf vernation are recognized: plicate, plane, revolute and conduplicate. These contrastable characters are individually strongly correlated in various ways with geographical distribution, pollen structure, deciduous condition and cupular morphology and suggest a new subdivision for the genus. Any functional correlation, of course, still has to be made. This is also an example of the way in which leaf vernation has recently been revived as a character offering diagnostic information, and the review by Cullen (1979) provides a valuable summary of appropriate terminologies. The character is essentially a dynamic one and may be difficult to observe in the herbarium. However, it is not a "new" one; systematists have tended to neglect it in recent years, as Cullen points out, since its variation was dealt with as early as 1751 by Linnaeus in his "Philosophia Botanica" and it was used later in several influential textbooks, for example by Lindley (1830) in his "Introduction to the Natural System of Botany" and by Asa Gray (1881) in his "Structural Botany".

ARCHITECTURE IN SYSTEMATICS

(1) Lack of systematic association

Having mentioned tree architecture it seems appropriate to begin a discussion of gross features of vegetative morphology with this topic, which represents plant construction in its most holistic phase. The limited taxonomic usefulness of this "character state" is suggested by Hallé and Oldeman (1970), Hallé *et al.* (1978), when they point out the existence of distinctive tree growth models in quite unrelated families. For example, Nozeran's model, represented by *Theobroma cacao*, is unusual in its sympodial trunk, tiered series of plagiotropic branches and rhythmic growth. It is not a common method of tree construction but occurs in such unrelated families as Sterculiaceae, Euphorbiaceae, Icacinaceae and Olacaceae. On this basis architecture, though distinctive, would seem to have little diagnostic value other than at the specific or generic level. This kind of example can be repeated for other models and other families. This seems rather puzzling since the assembly of developmental characteristics which makes up several tree models is rather complex; we must therefore assume that their repeated occurrence in unrelated groups is an example both of evolutionary convergence and character correlation. Accepting this conclusion would certainly be ironic, since the original concept of architecture had its birth in systematics. A number of authors after whom tree models were named had emphasized the architecture of the tree, or at least its branching pattern, in their systematic studies. We can think of Aubréville's studies on the Sapotaceae, Fagerlind's and especially Petit's studies on the Rubiaceae, Cook's studies on the genus *Theobroma* and Leeuwenberg's studies in the Loganiaceae (for specific citation see Hallé *et al.* 1978). The empirical nature of the approach partly accounts for this disparity, as does the very general nature of the criteria used in discriminating tree architectural models, but the enigma is partly resolved if one establishes at which rank one expects "rules" to apply. Ultimately, however, a satisfactory resolution will come when a more precise appreciation of the adaptive significance of tree form has been gained and character correlation established.

(2) Woody monocotyledons

The group of monocotyledons which have putative affinities with the palms (e.g. the Agavaceae, Araceae, Cyclanthaceae and Pandanaceae), provide a good example of a group in which aspects of vegetative construction are of considerable systematic significance. Each family has a range of architectures expressed by a diversity of architectural models, with some overlap between them (Tomlinson 1982). However, the palms never have terminal substitution of the inflorescence, which is a common feature of most members of the other families. Expressed in architectural terms, Chamberlain's, Leeuwenberg's and Stone's models do not occur in palms. This growth constraint seems somehow related to root development, since the palms do not include root-climbing lianes, present in all other families. The palms also lack "compound" vascular bundles in the stem. These very elementary observations tend to be overlooked when basic homologies of reproductive parts are considered. Once their fundamental nature is appreciated, they become one more set of characters which sets the palms further apart from other monocotyledons. The set of structural correlations can, of course, be expanded. Curiously the Australian genus *Kingia* shares a surprising number of these correlative features with palms. Is it a close relative of the palms?

(3) Shoot morphology

The example of *Viburnum* recently explored by Donoghue (1981, 1982) demonstrates, of course, that architectural analysis does not subvert systematic ends. Dynamic characters of shoot construction are highly diagnostic for several sections of the genus. The characters refer to time of development of renewal shoots in relation to flowering, degree and kind of plagiotropy and the morphological status of the terminal bud. Originally observed in the field and in cultivated specimens, the characters become transposable to the herbarium once they have been recognized because they produce a set of clearly recognizable morphological characters, which experience can translate into a dynamic analysis, even of herbarium material.

(4) Seedling morphology

Renewed interest in seedling morphology augurs well for the application of this character state as a systematic tool, since examples are well-described and terminologies have been elaborated (Ng 1978; DeVogel 1980). In this respect the application by Stevens (1980b) of seedling morphology to the systematics of *Calophyllum* at the species level is noteworthy, especially as he implies that functional differences may be discerned. The use of seedling characters by Léonard (1957) at the generic level in the subdivision of the tribe Cynometreae (Leguminosae) is by now almost a classic and serves as a model for the appreciation of vegetative structures in plant systematics. Quite clearly functional correlations should be examined as part of the further search, since early stages in establishment are likely to be of considerable adaptive significance.

NEGLECTED FEATURES

To venture in a direction which takes the systematist to task in pointing out neglect of systematically useful characters is a dangerous step for a non-systematist, were it not already done by the systematists Davis and Heywood, (1963) in the passage I have quoted earlier. For example, in their survey of the monocotyledons, Dahlgren and Clifford (1982) discuss the distribution of "stipules" in monocotyledons, even though it is debatable if any such structure exists in monocotyledons homologous with the variety of structures called "stipule" in dicotyledons and especially in view of the fact that the literature dealing with ligule, stipule, leaf-sheath and hastula in monocotyledons has lead only to an obfuscation of the structures investigated The vascular and nodal system of monocotyledons is developmentally so different from that of dicotyledons that it is inappropriate to seek homologies with nodal structures in dicotyledons.

On the other hand, the presence of an open or closed leaf-sheath, not considered by these authors, is a character which is, by and large, readily observed and is highly diagnostic for most major groups of monocotyledons. One may contrast in this respect the palms, on the one hand, which without known exception have a closed tubular leaf-sheath with families, on the other hand, which all appear to have open leaf-sheaths (e.g. Cyclanthaceae, Pandanaceae, Agavaceae and perhaps most Araceae). This structural constancy in palms is achieved in spite of a considerable mechanical diversity as the leaf matures, frequently involving the secondary splitting of the tube to accommodate growth processes in the palm crown (Tomlinson 1962). This constancy in basic construction is correlated with the very distinctive and precise mechanism of leaf development in palms; the character complex "leaf construction" is here obviously elaborate and a particular type of leaf-base morphology which is established early in ontogeny may be part of this developmental complex. In contrast where there is diversity in leaf sheath construction, there may be diversity in other aspects of leaf development and morphology, including vernation, as in the genus *Potamogeton*. The contrasted examples of palms and pondweeds certainly underline the empirical nature of systematic vegetative morphology, but it is the central principle of observational methodology which needs to be developed.

Neglect of vegetative features in basic diagnosis may simply be the result of traditional prejudice, or historical precedent. In my work for a Ph.D. thesis I was able to contrast very strongly the Zingiberaceae (*sensu stricto*) with *Costus* and related genera (Costaceae) in features of vegetative morphology (which was also associated with important anatomical differences). Resistance to the recognition of the Costaceae as a natural family is a resistance to the evidence from vegetative morphology. True, *Costus* has a structure called the labellum, like the gingers, although the structure may not be homologous; and both groups have a single stamen enclosing a slender style. In all other respects the two groups are different and it makes no sense to bury the Costaceae simply as a matter of historical precedence (cf. Maas 1972 and Tomlinson 1956).

STRUCTURAL AND FUNCTIONAL CORRELATIONS

I will now further address the problem of the functional significance of structural features and provide examples which suggest that correlations are readily overlooked.

Table I. Vegetative features used in diagnosis of major taxa.

Family	*Subdivision*	*Character*
MONOCOTYLEDONS		
Alliaceae	Tribal	Axis morphology (bulb or not)
Bromeliaceae	Subfamily	Marginal leaf spines
Cyclanthaceae	Subfamily	Leaf morphology
Dioscoreaceae	Subgenus	Direction of stem twining
Liliaceae	Subfamily	Axis, habit, leaf morphology
Marantaceae	Generic	Leaf anisomorphy (homodromous vs. antidromous)
Palmae	Two major groups	Leaf segment plication (induplicate vs. reduplicate)
Zingiberaceae	Subfamily	Plane of leaf distichy (horizontal vs. vertical)
DICOTYLEDONS		
Cactaceae	Tribe	Leaf morphology
Caesalpiniaceae	Tribe	Seedling morphology
Chenopodiaceae	Two major groups	Embryo and cotyledon structure
Cucurbitaceae	Subfamily	Tendril morphology
Epacridaceae	Subfamily	Leaf scar morphology
Ericaceae	Subfamily	Bud-scales and leaf morphology
Euphorbiaceae	Two major groups (unnatural)	Cotyledonary morphology
Gesneriaceae	Subfamily	Cotyledonary symmetry
Lauraceae	Subfamily	Growth habit
Myrtaceae	Subfamily	Phyllotaxis
Umbelliferae	Subfamily	Stipules present or absent

If one examines the few families of flowering plants in which morphological features of the vegetative system have been accorded major significance in the subdivision of the family (Table I), these often seem trivial and without obvious structural or functional benefit. In the palms, for example, the induplicate vs. reduplicate condition simply relates to two alternate methods whereby a multiplicate leaf can become segmented, either splitting along the adaxial or the abaxial fold. The difference seems correlated with some aspects of gross morphology, notably presence or absence of a terminal leaf segment, which is most evident in seedling leaves and appear to be a simple consequence of the distribution of splits in a regularly folded leaf (Tomlinson 1960). Since there seems to be no obvious adaptive significance in either method of splitting, we might cite this as an example of an adaptively neutral character which is correlated with other features which are presumably adaptively significant, possibly as a consequence of pleiotropic gene effects. Plication confers mechanical benefits on broad laminae – this is turn may relate to the closed leaf-sheath. Any correlation which might account for the peculiar distribution of leaf-blade asymmetry in the Marantaceae, which has some constancy in systematic subdivision (Tomlinson 1961), is not at all obvious.

Previously overlooked adaptive significance may alter this neutralistic interpretation dramatically. Thus the contrasted plane of orientation of leaf distichy in different tribes in the Zingiberaceae is hardly likely to be adaptively neutral, since it produces contrasted rhizome systems; those which are deterministically linear (vertical distichy), vs. those which are deterministically branched (horizontal distichy). These seem to reflect two major contrasted patterns of rhizome geometry which are of general occurrence and which seem important in the efficiency with which the substrate is exploited (Bell and Tomlinson 1980). This example suggests that a scrutiny of "function" may have to take in population consequences, and should not be thought of in terms of adaptive benefits bestowed only on individuals or single modules.

AN EVOLUTIONARY PROCESS

The general conclusion, which is hardly novel, must then be that gross features of morphology can be of systematic significance in some groups, but not in others, but that the range of characters is being increased by an interest in dynamic features. The concept of "ground plan" so familiar in unitary organisms does not necessarily exist. The orthodox view that the vegetative features are of limited systematic significance and that natural relationships are indicated primarily by the much more rigid developmental and structural patterns of the reproductive parts should be challenged as a misleading generalization especially if, in developmental terms, it implies that the genetic determinism of the reproductive parts is less variable than that of the vegetative parts. But there still remains the basic enigma that vegetative parts are more plastic and correlate less often with systematics at any rank level. In part this may be because the character which might be used is not so much a single feature, but can be a character complex. However, at the evolutionary level I can suggest that the processes of neoteny and paedomorphosis in modular organisms might be introduced to explain this situation. This application has indeed already been invoked, e.g. by Davis and Heywood (1963), and a useful set of terms (Table II) defined by Gould (1977) has been provided. A single example serves to illustrate how the processes might work in plants, although the terms defined by Gould are actually difficult to apply because they refer to *unitary* and not modular organisms.

Table II. Definition (from Gould, 1977) of evolutionary processes in unitary organisms which may occur in combination in modular organisms.

Paedomorphosis
The retention of ancestral juvenile characters by later ontogenetic stages of descendants
Neoteny
Paedomorphosis produced by retardation of somatic development
Progenesis
Paedomorphosis produced by precocious sexual maturation of an organism still in a morphologically juvenile state
Heterochrony (according to de Beer)
A phyletic change in the onset of timing of development

Note: These definitions do not fully take into account the possibility of interpolation of entirely new appendages and modules, possible only in modular organisms.

(1) Evolutionary morphology of *Tillandsia usneoides*

The vegetative dissimilarity between *Tillandsia usneoides* and other members of the family Bromeliaceae and even all species of its own genus has already been mentioned. One can account for the existence of this growth form in comparative terms, especially as intermediate forms exist and there is also a readily appreciated functional background to the evolutionary interpretation (Tomlinson 1970). In the family Bromeliaceae the reasonable assumption is made that the terrestrial habit is unspecialized (primitive) and the epiphytic habit is specialized (derived). Epiphytes themselves may be specialized in diverse ways to capture and store water in the leaf bases, either as tank or cistern epiphytes, or in the most highly developed types as "dew epiphytes" or "extreme atmospheric epiphytes" which store no free water and are highly drought resistant. They occupy the most exposed sites. In this specialization the root system is progressively reduced and absorption occurs through the leaf surface.

Within the epiphytic genus *Tillandsia* itself there is a range of such types, with the tank epiphyte represented by *Tillandsia utriculata*. Most *Tillandsia* species are of the "dew" type and some are succulent. A progressive reduction series of decreasing size and increasing independence of the geotropic response (which is essential to tank epiphytes) can be seen, as represented by the sequence

T. fasciculata, *T. balbisiana*, *T. circinnata*, *T. recurvata* with the ultimate in both reduction and specialization represented by *T. usneoides* with its pendulous, monopodially proliferated shoots, single-flowered inflorescence and virtual absence of roots. However, this typological sequence of adult forms should *not* in itself be interpreted as an evolutionary one. One can better see this progressive specialization in vegetative morphology truly as a "series of modified ontogenies" but here with increasing elaboration of the juvenile phase and decreasing dependence on free water as the plant matures. This description, of course, is the classic statement about the results of the evolutionary process but it can rarely be seen in such complete terms, because of extinction of intermediates.

An essential feature of this process is that seedlings of species which are tank-epiphytes as adults are effectively dew epiphytes in the juvenile phase; probably a preadaptation which results from the generally xerophytic ancestry. *Tillandsia* species are not easily distinguished as seedlings. The ageotropic response of adult dew epiphytes is simply the persistence of a response found in juvenile forms of congeners (and presumed ancestors). However, the process is not simply the persistence of features from the ancestral state into the adult form of the derived state (paedomorphosis, cf. Table II). There is elaboration of the morphology, notably the elongation of the internodes which becomes a distinctive feature of *T. usneoides*. The pendulous, festooning habit is due to continued monopodial branching, which proliferates the shoot system and facilitates the vegetative dispersal of the species (a structural change). The root system is dispensed with (a functional change). Branching in other *Tillandsia* species is mostly a juvenile feature, as is ageotropy. The floral organs are little modified by these trends and retain the characteristic *Tillandsia* facies; however, sexual reproduction is minimized – the r-strategy of *T. utriculata* (which is monocarpic) is now replaced by an extreme K-strategy since there is a high permanence in vegetative construction, despite the seeming delicacy of *T. usneoides*.

Although this is a distinctive example because transitional forms are so extensive, it does illustrate a process which seems fundamental to the evolution of modular organisms. Developmental processes which influence vegetative features and those which influence reproductive features are essentially independent and can readily be uncoupled in terms of evolutionary modification. The familiar orthodox view of the systematic botanist (if it is true) that sexual reproductive parts are more conservative in their expression than are the vegetative parts is therefore a simple consequence of the modular construction of higher plants and the evolutionary and adaptive features working in contrasted ways.

This permits plasticity of vegetative growth and a greater diversity of responses to environmental conditions. Plasticity of reproductive parts tends to be directed towards variation in numbers (e.g. of flowers, of seeds) which reflects variation in r-orientated or K-orientated strategy, this in its turn must be dependent on a successful vegetative "strategy".

MODULAR PLASTICITY

Modular organization leading to the concept of serial homology in higher plants has consequences which influence the ability of the plant to diversify its vegetative form. Kaplan (1980) has recently demonstrated that the development of each leaf primordium in the ontogenetic sequence of *Acacia* seedlings is governed by processes which must be viewed individually for each appendage. This may be interpreted to mean that the morphological expression of any one appendage at a given node is an evolutionary variable. We are familiar with the variation in appendicular expression in the shoots of plants (e.g. cotyledons, cataphylls, juvenile or adult leaves in seedlings; prophylls, scale-leaves, transitional leaves, foliage leaves in individual lateral axes). These forms are usually developed in a regular sequence which ultimately determines the overall construction of the plant. The developmental plasticity of the appendages are enormous in higher plants. Contrast this with the much more rigid sequence of appendages (bracteoles, sepals, petals, sporophylls) in the flower in which evolutionary change can only be gradual because of the highly specialized processes for which each kind of appendage is developed, and the highly specialized functions of the flower as a whole.

Natural selection thus seizes on the vegetative organs as structures which are highly malleable and ductile in an evolutionary sense. Where the evolutionary process is conducive to permanence in vegetative structures they are relatively fixed and remain as useful diagnostic tools for the systematists, where adaptive advantages can be found in changes in the vegetative morphology, fixity of organization is readily abandoned and newer, more successful modes of development are adopted. The parasitic *Cassytha* provides an extreme example of this evolutionary plasticity.

(1) Homology in modular organisms

An obvious complication arising out of modular plasticity when considered at the phylogenetic level, is that precise homologies are difficult to establish, since the concept of serial homology can hardly apply when entirely new sequences of appendages or modules are interpolated into ontogenies. A strict part-for-part comparison becomes impossible. This message indeed may be extracted from Kaplan's studies of seedling leaf development in *Acacia*, although this is my interpretation and not Kaplan's. Such an extreme view seems completely nihilistic because the concept of serial homology is the foundation for the comparative and evolutionary morphology of higher plants. The important point to emphasize, however, is that with modular organisms we lack the precision which the zoologist is privileged to exploit in his comparison of unitary organisms. The point cannot be elaborated here but a re-examination of the concept of homology in higher plants would benefit systematics considerably, since the neglect of vegetative features in the systematics of higher plants comes from the obvious unreliability of a strict part-for-part comparative analysis.

At this stage I would merely like to suggest that a good basis for the establishment of precise homologies will be the equivalence of primary meristems. Structures are strictly homologous when they originate from the same kind of meristem and meristemoid: for example, the embryonic shoot and root apex, subsequent lateral shoot or root meristems, inflorescence and floral buds, the primordia of appendages, the meristemoid which is associated with the development of a stomatal complex or complex trichome. Of course, such homologies can only be established *a posteriori* and their use in establishing affinities involves circularity of reasoning – but this is exactly the way in which existing homology concepts are applied at the present.

Therefore, the main advance which can be made at this stage is to accept that there is a problem; its resolution would benefit systematics enormously. It is important in this scrutiny to distinguish between the concept of homology, which is accepted as an abstract evolutionary principle, and the actual establishment of the evidence upon which the recognition of homologous structures can be based; this is a point emphasized by Mayr (1982).

CONCLUSION

The conclusion therefore is that the natural system of classification of plants is created by systematists, not by the living organisms themselves. It may be annoying that the full range of morphologies by means of which the present diversity of organization of higher plants has been arrived at in evolutionary time is still too incompletely explored to allow systematists to decipher complex evolutionary pathways, or is simply unavailable by virtue of the normal evolutionary process of extinction. Evolutionary success in higher plants resides in part in their distinctive indeterminate construction. A more general recognition of the consequences of modular construction and a more rigorous examination of the morphology of vegetative parts will certainly facilitate future advances in systematic understanding. The approach must be both holistic and dynamic because however appealing it may be to reduce organisms to single character states whose evolution and systematic usefulness can then be discussed, it is still the organism itself which evolves and it is still the organism we classify.

ACKNOWLEDGEMENTS

Dr Peter Stevens may think his continuing efforts to educate me as a systematist are in vain when he reads this, but I do appreciate those efforts and hope they will continue, as they clearly need to.

REFERENCES

Bell, A. D. and Tomlinson, P. B. (1980). Adaptive architecture in rhizomatous plants. *Bot. J. Linn. Soc.* **80,** 125–160.

Cullen, J. (1979). A preliminary survey of ptyxis (vernation) in the angiosperms. *Notes Roy. Bot. Gard. Edinb.* **37,** 161–214.

Dahlgren, R. T. M. and Clifford, H. T. (1982). "The Monocotyledons: a Comparative Survey". Academic Press, London, New York.

Davis, P. H. and Heywood, V. H. (1963). "Principles of Angiosperm Taxonomy". Van Nostrand, Princeton, N.J., New York.

DeVogel, E. F. (1980). "Seedlings of Dicotyledons". Centre for Agricultural Publication and Documentation, Wageningen.

Donoghue, M. (1981). Growth patterns in woody plants with examples from the genus *Viburnum*. *Arnoldia* **41,** 2–23.

Donoghue, M. (1982). A revision of Central American species of the genus *Viburnum* (Caprifoliaceae). Ph.D. thesis, Department of Biology, Harvard University.

Engler, A. (1877). Vergleichende Untersuchungen über die Morphologischen Verhältnisse der Araceae. *Nova Acta Acad. Leopold. Carol Nat.* **34,** 135–232.

Gould, S. J. (1977). "Ontogeny and Phylogeny". Belknap Press of Harvard University Press, Cambridge, Mass.

Hallé, F. and Oldeman, R. A. A. (1970). "Essai sur l'Architecture et la Dynamique de Croissance des Arbres Tropicaux". Masson, Paris.

Hallé, F., Oldeman, R. A. A. and Tomlinson, P. B. (1978). "Tropical Trees and Forests – an Architectural Analysis". Springer Verlag, Berlin, Heidelberg, New York.

Holttum, R. E. (1955). Growth habits in monocotyledons. Variation on a theme. *Phytomorphology* **5,** 399–413.

Kaplan, D. R. (1980). Heteroblastic leaf development in *Acacia*. Morphological and morphogenetic implications. *La Cellule* **73(2),** 137–203.

Léonard, J. (1957). Genera des Cynometreae et des Amherstieae africaines (Leguminosae – Caesalpinioideae). Essai de blastogénie appliqué à la systématique. *Mem. Acad. Roy. Belg. Coll.* 30 (fasc. 2), 1–305.

Lindley, J. (1830). "An Introduction to the Natural System of Botany". Longman, Rees, Orme, Brown and Green, London.

Maas, P. J. M. (1972). Costoideae-Zingiberaceae in Flora Neotropica. Monograph 8. Hafner Publishing Co., New York.

Mayr, E. (1982). "The Growth of Biological Thought". Belknap Press of Harvard University Press, Cambridge, Mass.

Munro, I. M. and Fairbrother, H. G. (1969). Composite plant diagrams in cotton. *Cotton Growing Rev.* **46,** 261–282.

Ng, F. S. P. (1978). Strategies of establishment in Malayan forest trees. *In* "Tropical Trees as Living Systems" (P. B. Tomlinson and M. H. Zimmermann, eds) chap. 5. Cambridge University Press, Cambridge, New York.

Philipson, W. R. and Philipson, M. N. (1979). Leaf vernation in *Nothofagus. New Zealand J. Bot.* **17,** 417–421.

Stebbins, G. L. (1974). "Phylogeny and Evolution of the Flowering Plants". Belknap Press of Harvard University Press, Cambridge, Mass.

Stevens, P. F. (1980a). Evolutionary polarity of character states. *Ann. Rev. Ecol. Syst.* **11,** 333–358.

Stevens, P. F. (1980b). A revision of the Old World species of *Calophyllum* (Guttiferae). *J. Arnold Arbor.* **61,** 117–699.

Tomlinson, P. B. (1956). Systematic anatomy of the Zingiberaceae. *J. Linn. Soc.* (*Bot.*) **55,** 537–592.

Tomlinson, P. B. (1960). Seedling leaves in palms and their morphological significance. *J. Arnold Arbor.* **41,** 414–428.

Tomlinson, P. B. (1961). Morphological and anatomical characteristics of the Marantaceae. *J. Linn. Soc.* (*Bot.*) **58,** 55–78.

Tomlinson, P. B. (1962). The leaf base in palms, its morphology and mechanical biology. *J. Arnold Arbor.* **43,** 23–46.

Tomlinson, P. B. (1970). Monocotyledons: towards an understanding of their morphology and anatomy. *In* "Advances in Botanical Research" (R. D. Preston, ed.) Vol. 3. Academic Press, London, New York.

Tomlinson, P. B. (1977). Plant morphology and anatomy in the tropics – the need for integrated approaches. *Ann. Missouri Bot. Gard.* **64,** 685–693.

Tomlinson, P. B. (1980). "The Biology of Trees Native to Tropical Florida". Published privately, Allston, Mass.

Tomlinson, P. B. (1982). Chance and design in the construction of plants. *Acta Biotheor.* **31A,** 162–183.

Tomlinson, P. B., Primack, R. B. and Bunt, J. S. (1979). Preliminary observations on floral biology in mangrove Rhizophoraceae. *Biotropica* **11,** 256–277.

White, J. (1979). The plant as a metapopulation. *Ann. Rev. Ecol. Syst.* **10,** 109–145.

Zimmermann, M. H. (1981). The vascular anatomy of the palm *Rhapis excelsa*. I. The mature vegetative axis. II. Vascular development in the crown. Encyclopedia Cinematographica, Göttingen.

5 | The Taxonomic Importance of the Leaf Surface

C. A. STACE

Department of Botany, University of Leicester, England

Abstract: Characters of the leaf are second to only those of flowers and fruits in the extent to which they are used, and in their value, in taxonomic studies. They are also strictly comparable over a wider taxonomic range (all vascular plants) than are floral organs, and moreover they are generally present on the plant for a much greater part of its life-span. For this reason they are valuable not only in making primary taxonomic decisions but also the determination of incomplete plants, e.g. sterile specimens, archaeological remains and fragmentary fossils.

The full value of these features can be realized only if they are studied by all means available, so that their true identity is ascertained and therefore only homologous data are compared. This is equally important for taxonomic studies and for the investigation of phylogenetic relationships. With many leaf-surface characters there is a danger in taking apparent short cuts (e.g. the observation of characters by only low-power microscopy, or the inference of developmental sequences from mature structure), which often lead to wrong conclusions. Nowadays a great many techniques are available to the taxonomist, who is unlikely to be able to avail himself of all of them. An awareness of the full range of possibilities and the avoidance of unjustified conclusions can, however, largely make up for this inevitable short-coming. A widely accepted terminology, sufficiently detailed without being too rigid or complicated, is essential in taxonomic work, and there have been many such systems proposed for leaf-surface characters in recent years – in some instances, e.g. the stomatal complex, perhaps too many.

This chapter discusses the use of leaf-surface characters in plant taxonomy, with special reference to work over the past twenty years and to phenotypic plasticity, developmental studies, evolutionary studies and the choice of techniques. Finally, a brief survey is made of recent trends in the study of the different groups of characters involved.

Systematics Association Special Volume No. 25, "Current Concepts in Plant Taxonomy", edited by V. H. Heywood and D. M. Moore, 1984. Academic Press, London and Orlando.
ISBN 0 12 347060 9

INTRODUCTION

Floral characters were originally considered by deductive reasoning to provide the most valuable indicators of taxonomic affinity; aspects of reproduction were thought to be more "essential" than others and hence to be used preferentially when classifying plants. Nevertheless, this choice has stood the test of time, for they still occupy the greater part of most botanical descriptions and diagnoses. This may perhaps be attributed largely to the fact that floral characters are relatively conservative in variation, so that similarity in floral structure is more likely than that in many other organs to be indicative of actual relationships (be they phylogenetic or phenetic). Hence it is no historical accident that taxonomists even today first look to the flower for taxonomic clues, and (often subconsciously) pin a high degree of trust in the characters it provides.

Characters of the leaf are the most widely used of those of all the non-reproductive organs. Their reputation for a lower degree of reliability in assessing relationships or in identification can be justified by the existence of similar shapes and degrees of subdivision, dissection, thickness and pubescence, etc., in the leaves of plants of widely differing affinities. Leaves should, however, have a number of theoretical advantages over flowers as taxonomic markers. They are strictly comparable over a wider taxonomic range (all vascular plants), and they are generally present on the plant for a much greater part of its life-span. For this reason they are of value not only in making primary taxonomic decisions, but also in the determination of incomplete plants, e.g. sterile specimens, archaeological remains, fragmentary fossils, stomachal or faecal contents, and drugs.

The past twenty years have witnessed a remarkable increase in the attention paid to leaf characters, partly but not primarily due to the advent of the scanning electron microscope (SEM). Much of this work has been aimed at discovering whether or not the microcharacters of the leaf (those not or hardly visible with the naked eye or a simple dissecting microscope) are more conservative than the gross or macrocharacters, and hence more trustworthy as taxonomic indicators. The answer, of course, is that sometimes they are and sometimes they are not. We are not yet in a position to generalize to the same extent as we are able to do with macrocharacters, and a belief that leaf characters have as much to offer as floral characters has certainly not been adopted by the majority of systematists. Hence Takhtajan (1980), in a review of the main trends of evolution in flowering plants, allocated 16 pages to flowers and 2 pages to leaves, and about half of the latter concerned stomata.

This paper is concerned with only the leaf surface of vascular plants, but within that restriction from the macroscopic to the subcellular level. In the past twenty years several valuable general reviews on various aspects of this topic have appeared. The following should be mentioned: Stace (1965) covered characters to be found on isolated cuticles; Napp-Zinn (1966, 1973–4) provided an encyclopaedic review of leaf anatomy, including about 300 pages on leaf-surface features of angiosperms alone; Martin and Juniper (1970) described the structure of the plant cuticle; Heywood (1971) edited a symposium volume on the use of the SEM in taxonomic studies; Preece and Dickinson (1971) edited a symposium volume on leaf-surface micro-organisms, including several chapters relating to the structure of this microhabitat (the phylloplane); Stant (1973) wrote a short review on the use of the SEM in plant anatomy; Dilcher (1974) surveyed characters enabling the identification of angiosperm leaf remains; Dickinson and Preece (1976) provided a sequel to Preece and Dickinson (1971); Barthlott and Ehler (1977) described features of the leaf surface of seed-plants as seen by the SEM; Cutler and Hartmann (1979) summarized the ecological and taxonomic significance of leaf-surface features with special reference to the Liliaceae and Mesembryanthaceae; Wilkinson (1979) reviewed the systematic anatomy of the leaf surface of dicotyledons; Barthlott (1981) summarized and updated the data in Barthlott and Ehler (1977); and Cutler *et al.* (1982) edited a symposium volume on many aspects (including structure) of the plant cuticle.

Obviously, even a superficial review of this subject would fill several hundred pages, and this account is intended only as an assessment of our knowledge of and progress in the application of leaf surface characters to taxonomic and evolutionary studies.

THE USE OF LEAF-SURFACE CHARACTERS

For the sake of clarity and expedience it is convenient to classify leaf surface characters into different categories. Hickey (1979), following his earlier schemes, recognized nine groups of characters concerning "leaf architecture". Most of these relate to gross features of the leaf (subdivision, shape, orientation, etc.), and only four refer to the actual leaf surface: texture, gland position, type of venation, and tooth architecture. The first of these is based upon a number of different microscopic aspects, the second relates equally to the presence of numerous structures other than glands which cannot be differentiated without the use of the microscope, and the fourth can be considered a special aspect of the third. Barthlott (1981) had four categories of characters: the arrangement of cells, the shape of cells (primary sculpture), the fine relief of the outer cell wall (secondary sculpture), and epicuticular secretions (the tertiary sculpture). These two schemes concentrate on opposite ends of the size spectrum (macroscopic and subcellular respectively), and neither one nor both together are applicable to the whole range. An attempt has been made in this paper to provide a list of seven categories which might be of more general use in describing leaf surface characteristics (see below); it is similar to the scheme of Holloway (1971).

Naturally, it is often difficult in practice to place a characteristic in a single category. Of more practical importance, however, is the need to investigate every characteristic fully, i.e. across a range of material, in different stages of development, and by all appropriate techniques, before applying them to taxonomic problems.

Environmental, developmental and technical aspects will be mentioned later under separate headings. At this point it is sufficient merely to underline the point that the same general precautions should be taken in the use of leaf-surface characters as in the use of, for example, floral characters. If ten to a hundred plants need to be sampled in order to gauge the range in petal size or nectary shape, then the same is no less true of stomatal size or wax platelet shape. The implications of this may seem unpalatable in view of the microscopic nature of so many leaf-surface characters, but the catalogue of mistaken conclusions reached by a premature application of unproved characteristics is already far too long. Short cuts are dangerous and sometimes disastrous.

The need to use a microscope so reguarly means, in practice, that leaf-surface features will be used rather sparingly for routine identifications, except when applied to sterile or otherwise incomplete material, where quicker methods might not be available. This, however, in no way reduces the value of the leaf surface in taxonomic decision-making or in evolutionary investigations.

The view is taken here that it is usually not possible to predict the value of a given character, or which of a certain group of characters will prove the most valuable, in any taxon that has not been investigated for such features. Similarly, the level of taxonomic discrimination by a character is unpredictable. Many lists of "the taxonomically most useful characters" in a taxon have been drawn up, e.g. that by Hickey and Wolfe (1975) concerning leaf architecture, but they are valid only as *a posteriori* summaries (hindsights), not as predictions. Reliance on them with respect to future investigations will lead to omissions and errors of judgement. Even when they prove to be useful pointers in general, important exceptions will crop up from time to time, and these will be missed by short cut methods. There are numerous examples of exactly the same character being most useful variously at the familial, generic, specific or infraspecific levels in different taxa. With regard to epidermal characters, Barthlott (1981) considers that their high structural diversity renders them most suitable for classification between the levels of family and species, although "there is also some evidence" for their use above the family level.

One generalization which can, perhaps, be made is that xeromorphic leaves seem to provide more leaf-surface characters of taxonomic importance than do mesomorphic leaves. A range of mesomorphic species is more likely to present a relatively invariable stereotyped pattern than is a range of xeromorphic species, presumably because the latter have responded to the need to combat drought in a variety of ways. The taxonomy of plants with thick, drought-resistant leaves has thus often been greatly aided by the use of leaf-surface characters, e.g. conifers (Alvin *et al.* 1982), mangroves (Stace 1966), Magnoliaceae (Baranova 1972); Ericaceae (Watson 1964), and various succulents (Cutler and Hartmann 1979; Ihlenfeldt and Hartmann 1982).

As with all aspects of taxonomic description, a widely accepted and precise yet flexible terminology for the various structures encountered greatly aids their use in taxonomy and encourages their assimilation into standard taxonomic practice. Whereas twenty years ago few such terminologies existed for leaf-surface features, many are now available for the epidermis (Wilkinson 1979), leaf architecture (Hickey 1979), trichomes (Theobold *et al.* 1979), epicuticular wax (Barthlott and Wollenweber 1981; Barthlott 1981, Baker 1982), and stomata (Cotthem 1970; Fryns-Claessens and Cotthem 1973; Dilcher 1974; Stevens and Martin 1978; Patel 1979; Payne 1979), among others. It is not to be expected that these terminologies are fixed, for little time has elapsed since their proposal. As more taxa are investigated, new situations requiring new descriptors are discovered, and existing schemes with little flexibility have to be rejected. On the other hand, schemes can be made cumbersome by the addition of too many new categories or the subdivision of too many old ones. As stated by Rasmussen (1981), "there are limits to how much information it is feasible for a system of terminology to contain. A gain in the level of information may easily lead to a loss in utility". Such a loss has already occurred in the case of stomata, for which there is now an unhelpfully confusing number of conflicting classifications available.

GENETIC CONTROL AND PHENOTYPIC PLASTICITY

Many early workers were fully aware of the need to determine the genetic basis of leaf-surface characters, in order to allow for the effects of phenotypic plasticity. The knowledge that characters such as trichome and stomatal density and distribution, cuticle thickness, epidermal cell anticlinal wall patterns and cuticular striations had been shown to vary according to environmental conditions (see Stace 1965; Dilcher 1974) had led to a general suspicion of epidermal features as relatively unreliable taxonomic indicators. For example, with regard to papillae, Wilkinson (1979) concluded that their presence and prominence are diagnostically unreliable because they vary with the climate or distribution of the species, despite the fact that several workers have found them to be constant and of taxonomic value. There have been relatively few carefully controlled studies of the effect of the environment on leaf surfaces, except in a small number of important crop plants, etc.

Direct observations of a correlation between a particular anatomy and environmental conditions by no means indicate the existence of phenotypic plasticity. For example, Barthlott (1981) was able to conclude that most herbaceous plants possess a sculptured leaf surface, whereas most woody tropical plants have a smooth surface. Baas (1975) found in the Aquifoliaceae that high mountain species possess thicker cuticles and straighter-sided epidermal cells than species from the humid tropics. But in those cases which were further investigated the characteristics turned out to be genetically controlled adaptations to the environment rather than ecophenetic responses. Although many cases of phenotypic plasticity have been demonstrated, and many epidermal characters do seem genetically to adapt readily to changing conditions, it has often been shown that characters indicating taxonomic relationship equal or outnumber those indicating adaptations to environmental conditions, e.g. in mangroves (Stace 1966) and apophyllous xerophytes (Böcher and Lyshede 1972).

Some of the best evidence for the genetic basis of surface characters has come from studies of their inheritance in hybrids. In intergeneric and interspecific hybrids of *Aloe* and *Gasteria* hybrids are variously intermediate between their parents; the characters of any particular hybrid are unpredictable but very constant (Cutler 1972; Cutler and Brandham 1977; Brandham and Cutler 1978). Moreover, these authors claim to have been able to demonstrate a correlation between particular surface features (lobes overarching stomata, amount of wax, micropapillation of cuticle) and specific characteristics of the karyotype. The hybrid *A. rauhii* × *A. dawei* was closer to *A. dawei*, presumably because the latter is tetraploid and *A. rauhii* diploid, and this tendency was more marked in a plant which was tetrasomic for a particular chromosome (L_1). However, in a plant with a delation of part of L_1 the characters were much closer to those of *A. rauhii*, indicating localization of the relevant genes on L_1. In all these plants variation in environmental factors seems to have extremely little effect on structure.

Baas (1978) found that in interspecific *Ilex* hybrids characters could be dominant, recessive, or either according to the cross, and that sometimes new characteristics appeared in the hybrids. When the observed inheritance was linked to knowledge of the leaf structure of the whole genus information was obtained on the probable role of hybridization in the evolution of *Ilex*.

Much work has taken place on the variation in wax under various conditions. Hunt and Baker (1982) found that in *Pisum* the amount increases in conditions of low soil moisture or air humidity, but the shape and chemical composition did not alter much. In the same species Juniper (1959) had earlier shown a similar reduction in surface wax following soil-feeding with trichloracetic acid. Hallam and Chambers (1970), however, demonstrated a wide range in not only amount but also microstructure and chemical composition in samples of *Eucalyptus camuldulensis* from different areas, but this might well be due to genetical differences, for in general in the genus they found little evidence of phenotypic plasticity within species grown in different geographical areas or in glasshouses. Hall *et al.* (1965) discovered a cline in *Eucalyptus urnigera* which is expressed in abundance, micromorphology and chemical composition of the wax, but again there was no evidence that it is other than genetically based. Species or variants with more glaucous (less wettable) leaves have more wax which is more randomly orientated and in the form of rods or tubes rather than plates.

Jeffree *et al.* (1976) showed that the form in which wax is deposited on the leaf surface is very largely dependent upon its chemical constitution, since it can be dissolved and recrystallized from solution (in isolation from the leaf) in the same form. Mutants with wax of a different chemical composition produce wax of a different crystalline structure, although there was also some evidence for a physical constraint being imposed by the cuticular membrane upon this more obviously genetically-based chemical mechanism.

DEVELOPMENTAL STUDIES

There is now a widespread realization that documentation of leaf-surface characters is not complete without the inclusion of the developmental aspects of variation. Developmental studies often help in understanding the nature and basis of the mature structure and in compensating for observed differences which might be due only to varying ages of the material; in addition extra characters not apparent in the mature structure are often uncovered. In the future it is to be expected that even more attention will be paid to these aspects due to the notion that ontogenetic studies are important in cladistic analysis.

It was, for example, argued by Cowan (1950) that the possession of early stages of development in common by all the very varied trichomes of *Rhododendron* was evidence that they were phylogenetically homologous.

Some features of the leaf surface appear rather late in development, and these are perhaps the ones most affected by environmental factors. Examples are cuticular striations, amounts of epicuticular wax and epidermal cell anticlinal wall undulation, the development of all of which has been studied in some detail.

The development of subsidiary cells around stomata has undoubtedly given rise to more debate in recent years than that of any other leaf-surface feature. It is well known that there is not necessarily any direct correlation between the developmental sequence and the mature appearance of the stomatal complex, but right from the start of modern investigations by Florin fifty years ago these two aspects have been confused. As emphasized by several workers, most recently by Rasmussen (1981), the terminologies relating to mature structure and to developmental sequences should be kept quite separate. The former can be observed in any plant, but the latter has been followed in only a tiny fraction of species. A great many examples could be cited where identical mature structures have arisen by different pathways, and where the same developmental pattern has given rise to different mature structures.

EVOLUTIONARY STUDIES

The renewed interest in angiosperm phylogeny which has become evident in the past ten years or so, and which has been given a new slant and emphasis by cladistic studies, has led to many proposals relating to the evolutionary pathways of leaf-surface features. These have been made with varying degrees of factual basis. For example, Hickey and Wolfe (1975) listed the following six criteria which they used to discern evolutionary trends in leaf architecture:

1. The fossil record.
2. Features possessed by the most primitive living members of a taxon.
3. Features possessed by a number of taxa that are ancestors, direct descendants, or common descendants of the one being analysed.
4. Features possessed by the most primitive members of a number of subdivisions of the taxon under examination.
5. The presence of a feature considered irreversibly lost.
6. A hypothetical combination of features needed to reconcile a number of trends considered divergent from a common ancestor.

Needless to say, the last five of these are simply bold statements attempting to rationalize the subjective reasoning which must always be prevalent in phylogenetic studies, and even the first often comes into this category, since fossils are rarely complete and their identity is usually less than certain. Nevertheless, Hickey and Doyle (1977) have surveyed an extensive series of fossil leaves from the early Cretaceous of North America, and claim that they can establish the following six primitive character-states for angiosperm leaves: leaves simple; venation pinnate; secondary veins camptodromous; separate orders of vein division indistinct; margin entire; stipules present.

Takhtajan (1980), presumably using similar criteria, considers that mesogenous stomata, distinct subsidiary cells and paracytic subsidiary cells are primitive features of angiosperms, although within the monocotyledons he considers perigenous anomocytic stomata to be primitive. Baranova (1972) found that, within the Magnoliales, mesogenous paracytic stomata are prevalent, and she also considered this the primitive type within the angiosperms. However, Payne (1979) concluded that the primitive type of stoma for land plants is mesoperigenous, and that this is the progenitor-type for the angiosperms; it is retained as the primitive type for the monocotyledons but not for the dicotyledons, where the mesogenous paracytic type has become prevalent in the most primitive living taxa. Such equivocal evidence, depending upon the scope of the outgroup considered, could be repeated many times at lower levels of the hierarchy (order, family, genus, etc.).

Netolitzky (1932, *fide* Carlquist 1961) provided principles for deciding the phylogeny of trichome types and claimed the following character-states as primitive: papilla (rather than trichome), unicellular, radially symmetrical, parallel to leaf surface, cell contents the same as those of the epidermal cells, and no differentiated adjacent cells. Barthlott (1981) recognized several evolutionary developments of epidermal characters in relation to adaptation to different environmental situations, but, perhaps commendably, did not identify overall directions for these trends. Tomlinson (1974) claimed that in the monocotyledons there are no major groups of monocotyledonous families characterized by a particular pattern of stomatal development, and that speculations about the phylogenetic significance of these patterns were premature.

There is no doubt that some of the postulated trends mentioned above could be added with advantage to the catalogue of evolutionary dicta listed by authors such as Hutchinson (1926) and Sporne (1977). However, it is clear to anyone who has worked with leaf-surface characters that the taxonomic incidence of most of them is unusually scattered, which clearly indicates a very frequent reversal of the usual evolutionary direction (whether or not the latter can be defined). Moreover, many (?most) structures which appear to have undergone evolutionary reduction do not pass through developmental stages which recall their more complicated ancestors. These factors render the approaches of ontogenetic study and "outgroup comparison", now frequently pursued in cladistic analyses, exceedingly hazardous when applied to many epidermal characters.

TECHNIQUES

There are naturally many different techniques which have been applied to leaf surface characters, but it is not necessary or appropriate to describe them all here. However, a careful consideration of the techniques available is particularly important whenever the absolutely direct method (naked-eye examination) is not applicable. The most important requirement, as noted previously, is that the structure should be *fully* investigated, and this often entails the use of several different techniques, or at least trials with them to ascertain the most suitable one. The link between macroscopic and microscopic observations is important here. Classical taxonomic descriptions of leaves abound with such terms as verrucose, verruculose, papillate and punctate, but it is not possible to use these character-states in comparative studies unless their homologies are established. For example, a "papillate" leaf may owe its appearance to the possession of glands, the bases of broken off trichomes, raised or sunken stomata (or groups of them), cork-warts, raised areas of epidermis overlying crystal-cells in the mesophyll, or other structures. The comparison of non-homologous structures will obviously lead to false conclusions.

For example, if the heading "gland position", as defined by Hickey (1979), were used blindly in the genus *Combretum*, important data would be lost. In this genus glands show two main patterns of distribution: on the veins of the lamina, or in the areolae between the veins (a distinction not actually recognized by Hickey). In general, however, the types of glands showing these two patterns are quite different, stalked clavate glands occurring on the veins and peltate scales in the areolae. But perhaps the single most important taxonomic character relating to gland position is that in two sections (involving only three species of the 200 or more in the genus) there are stalked glands which have the distribution pattern typical for scales (Stace 1973). In this instance a much more precise definition of "gland position" is needed, and for this microscopic study is essential.

An important safeguard is that new techniques should not be used indiscriminantly; innovations are exciting and the temptation to overuse them and over-emphasize their importance can be great. This is particularly true of the scanning electron microscope (SEM). Barthlott (1981) pointed out that "to

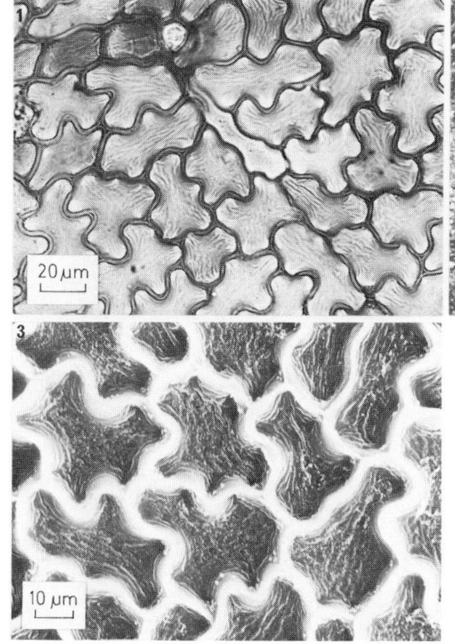

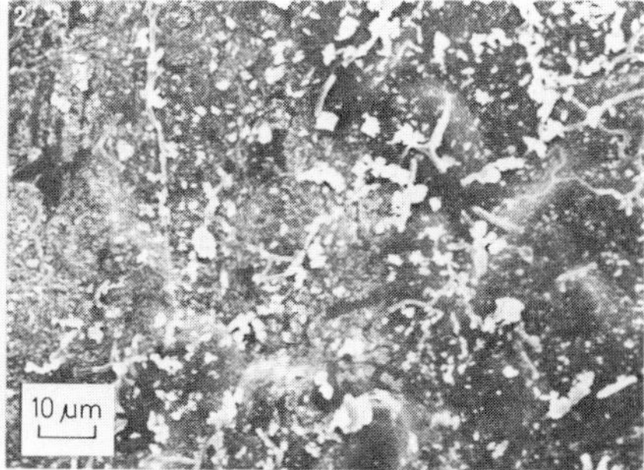

Fig. 1. Cuticular membrane from leaf upper epidermis of *Bucida buceras* L. (Combretaceae), showing cuticular striations. (LM)

Fig. 2. Upper leaf surface of *Bucida buceras*, showing irregular wax deposits and detritus, but no cuticular striations. (SEM)

Fig. 3. Inner surface of cuticular membrane from leaf upper epidermis of *Bucida buceras*, showing cuticular striations. (SEM)

quite an extent SEM micrographs only serve to illustrate characters well-known by light microscopy", and that "the SEM provides no further information to that which could be obtained through transmission electron micrsocopy". Of course, the SEM does have considerable advantages over some aspects of both the light microscope (LM) and the transmission electron microscope (TEM), but the limitations of the method must be weighed against its advantages. In the study of leaf-surface features sometimes one method and sometimes another method will be the most revealing, but usually the use of several methods together will yield the most information (Figs 1–6). This was emphasized by Carr *et al.* (1971), who introduced the term "phytoglyph" to cover the full range of features to be seen on the leaf surface. Even when the same feature is being investigated, the nature of the leaf concerned can often dictate the use of different techniques in different taxa.

With these points in mind, the following list of comments has been compiled:

(*i*) *Intact leaf.* This has been examined by all the traditional methods (dissecting microscopes, etc.), but in addition it is the material now most often viewed with the SEM. In this way every surface contour of the leaf can be analysed satisfactorily. With regard to the epidermal cell outlines, however, in many leaves

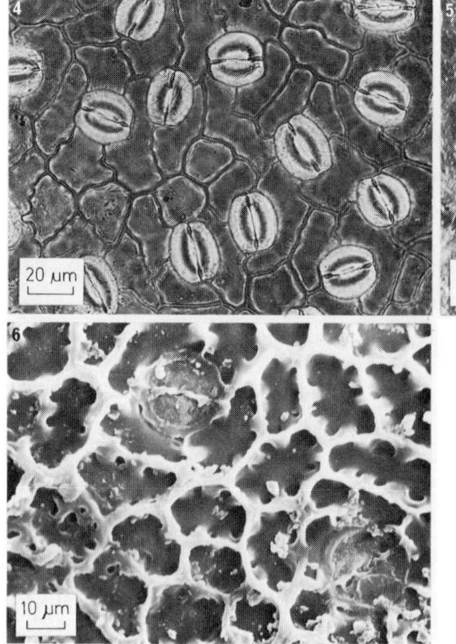

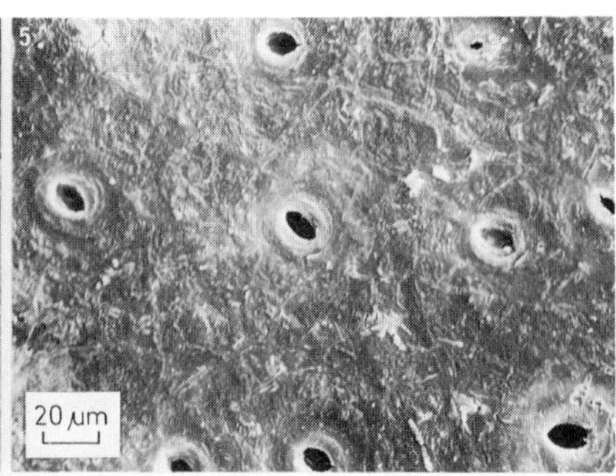

Fig. 4. Cuticular membrane from leaf lower epidermis of *Ramatuella virens* Spruce ex Eichler (Combretaceae), showing cell outlines very clearly. (LM)

Fig. 5. Lower leaf surface of *Ramatuella virens*, showing stomata but few other cellular details. (SEM)

Fig. 6. Inner surface of cuticular membrane from leaf lower epidermis of *Ramatuella virens*, showing cuticular flanges with 'pockets' formed due to the anticlinal wall undulations being confined to the outer part of the wall. (SEM)

with a thick cuticle these outlines are not manifested as surface contours, and in such leaves the appearance under the SEM can be most disappointing (Figs 2, 5 and 16). According to the observations of Barthlott (1981) this is most likely to be true of woody plants from the humid Tropics, in which in addition a smooth cuticle is usual. The SEM is particularly suitable for the examination of epicuticular wax, for which purpose herbarium dried leaves (coated with gold, etc., but otherwise untreated) can be ideal, but fresh or pickled material (carefully dried) can be used as well. Where wax is absent or not being studied, the leaves are often better washed to remove debris and fungi, etc.

(*ii*) *Imprints*. These provide an indirect way of examining the leaf surface. The various methods employed are more time-consuming than direct observation of the leaf surface, but can be valuable either when the leaf material may not be damaged or removed from position or when a transparent object is required. Many materials are available with which imprints can be made; cellulose acetate, resins and nail varnish, etc., are all suitable for light microscopy, and carbon replicas have been used specifically to examine leaf-surface wax (Juniper and Bradley 1958).

(*iii*) *Epidermal strips*. The epidermis can be stripped off many leaves and observed with the light microscope. This method is often valuable for examining delicate trichomes *in situ*, but leaves are not always amenable to this method. However, gentle maceration, e.g. with lactic acid, followed by scraping away the underlying cells can often overcome this problem, as with the usual technique for examining the epidermis of grasses (Figs 7 and 8).

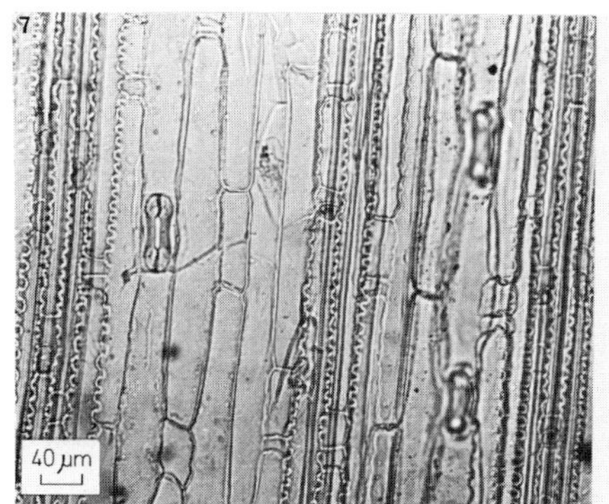

Fig. 7. Lower epidermis of *Cutandia memphitica* (Sprengel) Richter (Poaceae), showing long cells over the veins and in the intercostal zones with undulate and straight walls respectively. (LM)

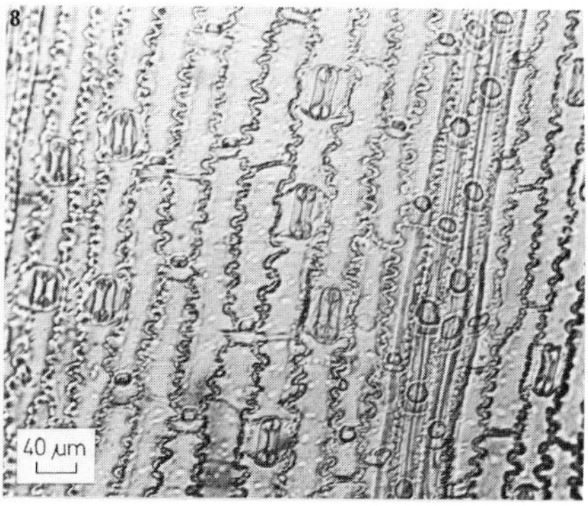

Fig. 8. Lower epidermis of *Wangenheimia demnatensis* (Murbeck) Stace (Poaceae), showing all long cells with undulate walls, and distinctive silica-cells over the veins. (LM)

(*iv*) *Cuticular membrane*. The isolation of the cuticular membrane is effected by maceration techniques, which either dissolve away the cellular layers or separate these from the cuticular membrane. Chemicals (such as strong acids or alkalis) or enzymes (such as pectinases and cellulases) can be employed. Light microscopic study of the isolated cuticular membrane often affords the best means of observing features of the leaf epidermis, since it is transparent, acellular and potentially carries the contours of the leaf surface plus the contours of the epidermal cell-cuticular membrane interface. Even better results are sometimes obtained by the examination of suitably dried and coated cuticular membranes with the SEM, when the inside and outside contours can be examined separately (Figs 4–6, 15–17). Virtually the only limitation of this method is found in leaves with very thin cuticular membranes, where the membrane is too flimsy to allow its isolation as a continuous sheet or to carry the epidermal contours. This might be true of the leaves of the majority of herbaceous plants.

(*v*) *Clearing*. This involves treatment with chemicals which allow the whole leaf to be examined in transparency. The chemicals are actually very mild macerating agents so that the time of clearing has to be controlled carefully (e.g. NaOH, NaOCl); full maceration is not necessary as the tissues become bleached as well. Clearing often enables examination of the epidermis by light microscopy, but its main value is in the examination of the venous system as well as that of relatively opaque structures (Figs 9 and 11). The method is valuable as it enables study of the whole intact leaf down to the cellular level.

(*vi*) *Leaf X-rays*. The exposure to X-rays of whole leaves placed directly on an X-ray plate in order to obtain a photograph of the venous system and gland distribution in species of the Clusiaceae has recently been adopted by Jones (1980). Excellent results comparable at the grosser level with those obtained by clearing have been obtained by Jones and by A. A. Alwan and C. A. Stace (unpublished) (Figs 9–12). The main advantages of this method are that it is much quicker, good X-ray photographs of leaves which are very resistant to clearing can be obtained, and valuable leaves (even when still attached to herbarium sheets) can be used without being damaged. The main disadvantage are the relatively high cost of X-ray plates, the need for access to a suitable machine, and the fact that only contact prints (or enlargements without enhanced resolution) can be obtained.

(*vii*) *Sections*. Frequently it is necessary to cut sections of leaves in order to interpret structures seen in surface views. Many techniques are available for fresh, pickled and dried material, and with or without embedding. The value of leaf sections in the study of leaf-surface features is often under-estimated, and they should be more widely used; the LM and SEM are both suitable for viewing sections.

THE LEAF-SURFACE CHARACTERS

(1) Trichomes

Besides the distribution and anatomy of the actual trichomes, the degree of modification of the adjacent epidermal cells and the scars left by lost trichomes are important aspects of variation. Often developmental studies will provide extra taxonomic evidence. The SEM has provided the means of examining the sculpturing of the trichome surface, which is often quite differently marked from that of the rest of the epidermis.

Fig. 9. Cleared leaf of *Terminalia lucida* Hoffsgg. ex Martius (Combretaceae), showing brochidodromous venation pattern.

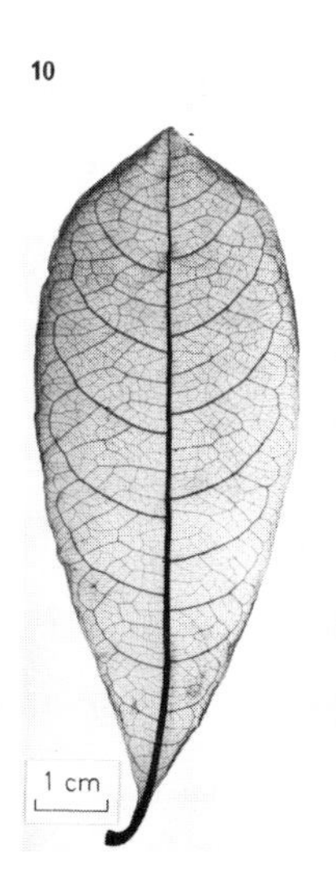

Fig. 10. X-ray photograph of whole leaf of *Terminalia lucida.*

Fig. 11. Cleared leaf of *Terminalia fagifolia* Martius & Zucc. (Combretaceae), showing craspedodromous venation pattern proximally and eucamptodromous venation pattern distally.

Fig. 12. X-ray photograph of whole leaf of *Terminalia fagifolia*, showing small glands where secondary veins meet the leaf margin (not visible in cleared leaves, Fig. 11).

At one end of the scale of variation (papillae) trichomes merge into simple convexities of a single epidermal cell, while at the other end (emergences) they are large outgrowths involving much subepidermal tissue as well. Even ignoring these extremes the range of trichome variation is enormous. Uphof (1962) has provided an excellent survey covering all aspects (except scanning electron microscopy) in 206 pages, backed up by a 42-page survey of the distribution of various trichome types in the angiosperms order by order by Hummel and Staesche (1962). Napp-Zinn (1973–74) devoted 185 pages to papillae, trichomes and emergences, including a systematic survey similar to the last, in his treatise on angiosperm leaf anatomy. Theobald *et al.* (1979) listed the occurrence of different sorts of trichomes in various families of dicotyledons, thus providing a kind of cross-reference to Hummel and Staesche's (1962) list.

The extent to which trichomes can be classified is debatable. Many attempts have been made, for example in recent years by Uphof (1962), Roe (1971), Ramayya (1972) and Theobald *et al.* (1979), but in the writer's view it will generally be necessary for workers to prepare their own individual systems tailored to the special dictates of the taxon under investigation. Any classification attempting to cover all known trichomes is likely to be thwarted by the reticulate (non-hierarchical) pattern of variation encountered. Good examples of special classifications covering particular taxa are those of Seithe (1960, 1962) in *Rhododendron* and *Solanum*, Roe (1971) in *Solanum*, Ahmad (1978) in Acanthaceae, Aleykutty and Inamdar (1980) in Ranales, and Stace (1981) in *Combretum*. Useful glossaries covering trichomes and indumentum were given by Payne (1978).

(2) Venation systems

The study of leaf venation patterns has been given added impetus over the past ten years by the work of L. J. Hickey in Washington (Hickey 1973; Hickey and Wolfe 1975; Doyle and Hickey 1976; Hickey and Doyle 1977; Hickey 1979). His classification and terminology of leaf venation has been challenged by Melville (1976), but it has become adopted virtually universally, and will be used here.

Hickey recognizes six main types of venation system: pinnate, parallelodromous, campylodromous, acrodromous, actinodromous, and palinactinodromous. Several of these categories are subdivided, especially the pinnate pattern, within which about eight further categories can be differentiated. Naturally, many intermediate conditions are found in nature (Fig. 11), but the adoption of a standard classification and terminology does considerably aid the description and comparison of venation patterns. Further aspects of leaf venation which are classified by Hickey include: the numbers and courses of secondary veins, tertiary veins, higher order veins and veinlets (i.e. ultimate branches with free ends); the development and shape of the areolae (ultimate areas of leaf bounded by veins); and the venation system in relation to the leaf margin, especially leaf-teeth. A useful attempt to analyse similarly the venation pattern of ferns has been made by Wagner (1979).

A quite separate topic is the degree to which the venation system is visible by examining the untreated leaf surface. The prominence of the venation pattern is a manifestation of internal leaf anatomy, particularly the extent to which the vascular tissue is separated from the epidermis by non-lignified cells. Where there is a direct connexion between the veins and the epidermis by flanges of mechanical tissue (a very common situation), the epidermal cells are modified (usually elongated parallel to the veins) so that the venation system can be seen, at least in part, by examining isolated epidermides or even cuticular membranes. This feature itself is often a very useful taxonomic character in the dicotyledons, and is a major aspect of the leaf epidermis of the Poaceae (Fig. 7). Hence, in the writer's view, one of the three main subcategories of Hickey's pinnate pattern, the hyphodromous pattern ("all but the primary vein absent, rudimentary or concealed within a coriaceous or fleshy mesophyll"), is totally artificial and should be restricted to situations where the secondary and lower orders of veins are *actually* absent or rudimentary. (If the term is derived from hypha, a web, then surely it also should be changed.)

Hickey's emphasis on the practical difficulties associated with recognizing venation patterns probably stems from his palaeobotanical work. This has also led to his interest in the evolution of venation patterns (see p. 75) and the suggestion that the camptodromous pattern (a sub-category of the pinnate pattern) is the primitive state for the angiosperms. Whether or not this is correct, it seems very likely that camptodromous patterns have arisen from others in some groups of plants (thus perhaps representing a reversion to a primitive state), and it would be unwise to consider that a unidirectional pathway of vein evolution existed within the angiosperms or even dicotyledons.

(3) Epidermal cells

The epidermal cells, other than those modified by their relationship to trichomes, stomata, the venation system or other special structures, provide many characters of taxonomic value. These include their size, shape, orientation, anticlinal wall undulation (Figs 7–8 and 13–14) and periclinal wall curvature, which were

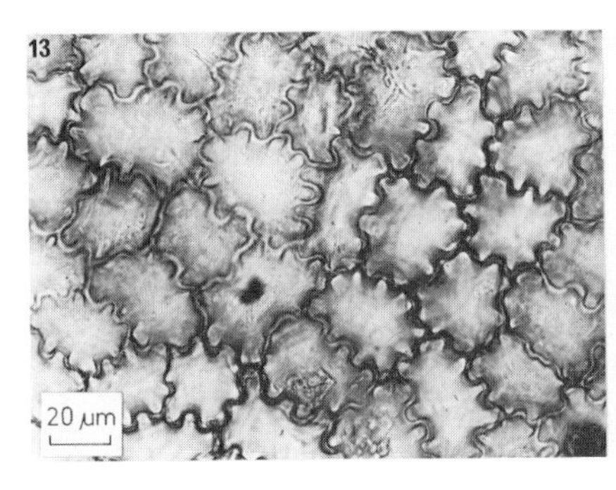

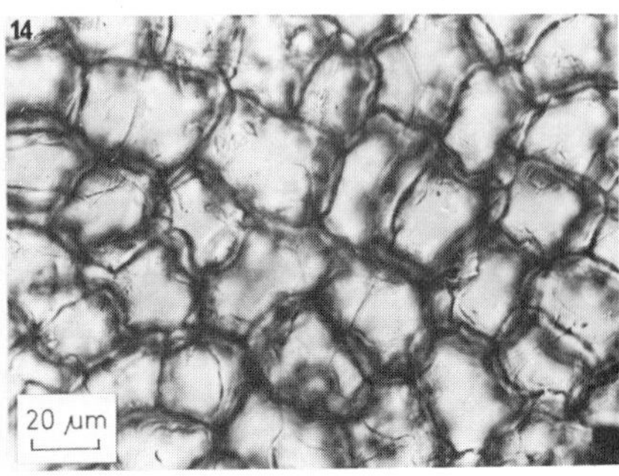

Fig. 13. Cuticular membrane from leaf upper epidermis of *Ramatuella crispialata* Ducke (Combretaceae) at outer level of focus, showing undulate walls. (LM)

Fig. 14. Cuticular membrane from leaf upper epidermis of *Ramatuella crispialata* at inner level of focus, showing straight walls. (LM)

collectively called the primary sculpture of the leaf surface by Barthlott (1981), and the thickness, ornamentation and pitting of the anticlinal and periclinal walls. The ornamentation or fine relief of the outer surface of the outer periclinal wall may or may not be evident on the outer surface of the cuticular membrane (q.v.), but it is generally mirrored at least on the inner surface of the latter, which can afford the best means of observing it (Alvin *et al.* 1982), (Figs 1–6).

Perforations in the anticlinal walls (and in the cuticular flanges between them) are apparently characteristic of most major groups of gymnosperms, but are virtually absent from angiosperms (Barthlott 1981). Pits and other areas of thinness are, however, of very common occurrence, and their taxonomic value has so far been greatly under-exploited. The most complete account available is that given by Napp-Zinn (1973–74), but more recent work with the SEM is providing a great deal of additional information (Figs 15–17).

Modification of epidermal cells occurs not only in relation to trichomes, stomata, the venation system and other special structures (treated separately here), but also on the leaf margin (Stace 1966), and when silicified, calcified, suberized, gelatinized or containing various types of crystals. Silicified and suberized epidermal cells (Fig. 8) are of major taxonomic importance in the Poaceae (Metcalfe 1960).

The value of epidermal cell characteristics to the taxonomist will grow as the genetical basis and extent of ecophenetic variation becomes better understood, and reliance on observed features becomes better based.

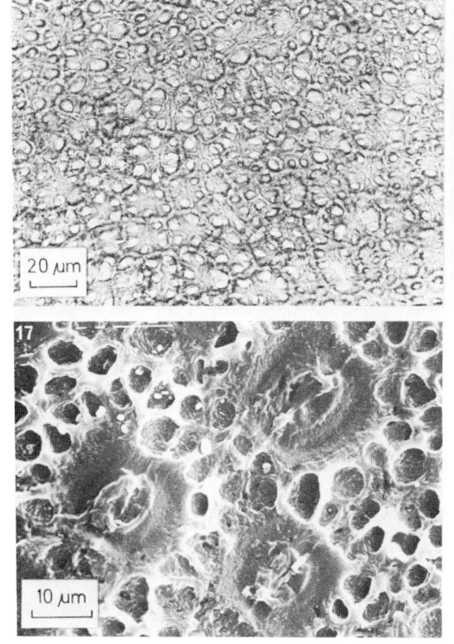

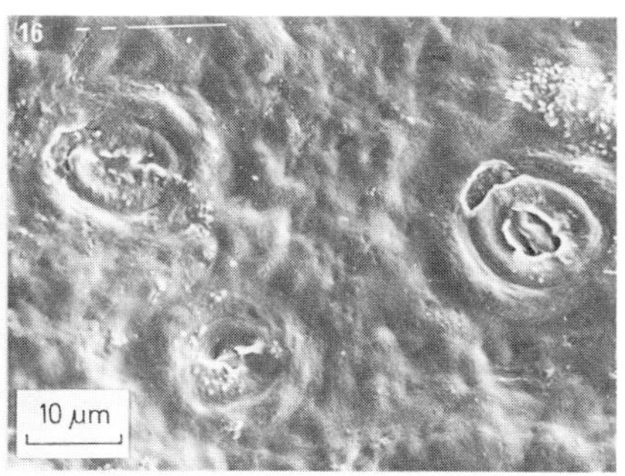

Fig. 15. Cuticular membrane from leaf upper epidermis of *Combretum obanense* (Baker f.) Hutch. & Dalziel (Combretaceae), showing cell outlines and thin areas in outer periclinal walls. (LM)

Fig. 16. Outer surface of cuticular membrane from leaf lower epidermis of *Combretum obanense*, showing stomata but few other cellular details. (SEM)

Fig. 17. Inner surface of cuticular membrane from leaf lower epidermis of *Combretum obanense*, showing stomata and conspicuous thin areas in outer periclinal walls. (SEM)

(4) Stomata

In the past twenty years probably more work has been carried out on stomatal structure and development, and on their use in assessing taxonomic relationships and evolutionary pathways, than on any other leaf character (Napp-Zinn 1973–74; Wilkinson 1979).

The shape, size, distribution and orientation of the stomata, and the various thickenings and ornamentations of the guard-cell walls, are all characters which are frequently of taxonomic value. In addition special stomata may be found in relation to nectaries (nectar-stomata) and hydathodes (water-stomata), etc. By far the most valuable, however, are the epidermal cells surrounding the stomata; if they are unmodified they are known as neighbouring cells, and, if modified, subsidiary cells (Cotthem 1970). The need for simple yet flexible systems of classification, one for structure and one for development, has already been stated.

The existence of so many recent papers on the subject, several of them in the nature of reviews, obviates the necessity of giving here more than a guide to the important developments. The present state of the subject concerning the mature form of the stomatal complex (stoma plus subsidiary cells) can be said to date from Metcalfe and Chalk (1950), who introduced five terms coined for them by H. K. Airy-Shaw. Since that time, according to my calculations, 45 further terms with the suffix "-cytic" have been introduced but, since some of these are group terms (which are subdivided) and others are synonyms, the total of 50 represents only 35 different situations. Actually, a full list of these has never been published, and so for convenience it is given here (Table I). It includes only terms with the suffix "-cytic", and even so a few terms (hypothetical cases, synonyms used once only – sometimes in error, and examples not fitting into the main scheme) are omitted. In my opinion the best scheme is that of Dilcher (1974), who recognized 31 types (the hypocytic type of Cotthem was not mentioned – presumably it was not considered distinct from the diacytic type, and three types were not described at that time). These 31 types were grouped into eight major categories (polycytic, anisocytic, diacytic, paracytic, tetracytic, hexacytic, polocytic and pericytic). A new terminology was introduced by Patel (1979), who recognized 44 categories based upon the suffix "-cyclic", most of which are synonyms or subdivisions of the terms used by Dilcher.

A modern system of developmental categories was founded by Pant (1965), who, following the classic work of Florin, recognized a perigenous pattern, where the stomatal meristemoid gives rise directly to two guard-cells, and a mesogenous pattern, where it first cuts off other cells, which often (but by no means always) become differentiated as subsidiary cells. He also defined a mesoperigenous pattern, where some of the cells adjacent to a stoma are derived from the stomatal meristoid and some are not. Stevens and Martin (1978) proposed a finer definition of Pant's terminology, with a total of seven terms, three of which (agenous, hemiperigenous and euperigenous) are subdivisions of Pant's perigenous category and three (hemimesoperigenous, eumesoperigenous and hemimesogenous) of his mesoperigenous category; the seventh, eumesogenous, corresponds with Pant's mesogenous. This terminology accounts for all the theoretical possibilities, and it was accepted by Rasmussen (1981). Payne (1979), however, used different criteria for classifying ontogenetic patterns, viz. the mode of division of the guard mother cell with respect to the wall which cuts it from the meristemoid. Three main types are recognized (diameristic,

Table I. List of terms ending in "-cytic" used for describing the mature stomatal complex, with synonymy and first known source.

Metcalfe and Chalk (1950): actinocytic[6], anisocytic, anomocytic[6], diacytic, paracytic
Stromberg (1956): encyclocytic[6]
Metcalfe (1961): tetracytic (subdivided later[1])
Stace (1963): cyclocytic (synonym of encyclocytic)
Cotthem (1968): desmocytic, hypocytic, pericytic (subdivided later[2]), polocytic (subdivided later[3]), staurocytic
Cotthem (1970): axillocytic[3], coaxillocytic[3], coparietocytic[3], copericytic[2], duplopericytic[2], eupericytic[2], hemiparacytic, hexacytic (subdivided later[4]), parietocytic[3]
Payne (1970): allelocytic (subdivided[5]), diallelocytic[5], helicocytic, parallelocytic[5]
Dilcher (1974): amphianisocytic (synonym of helicocytic), amphibrachyparacytic, amphibrachyparatetracytic, amphicyclocytic[6], amphidiacytic (synonym of diallelocytic), amphiparacytic (synonym of parallelocytic), amphiparatetracytic[1], amphipericytic (synonym of duplopericytic), anomotetracytic[1], brachyparacytic, brachyparahexacytic – monopolar[4], brachyparahexacytic – dipolar[4], brachyparatetracytic[1], copolocytic (synonym of coparietocytic), parahexacytic – monopolar[4], parahexacytic – dipolar[4], paratetracytic[1], polycytic (group name for four above terms[6])
Patel *et al.* (1975): coperidesmocytic, coperipolocytic (synonym of copericytic), copolodesmocytic, duplopolocytic (synonym of coparietocytic)
Den Hartog and Baas (1978): laterocytic
Brett (1979): pleioparacytic (synonym of laterocytic)

The subdivisions and grouping of terms are indicated (by superscript numbers) only where the group name is *not* exactly repeated by one of its subgroup names. A few terms are deliberately omitted – see text.

parameristic and anomomeristic), the first two of which are each subdivided into mesoperigenous and mesogenous patterns, with lower categories as well in some cases. In comparing the systems of Stevens and Martin and Payne, it should be noted that the second does not recognize the perigenous pattern as a valid category. Which of these two systems will prove to be the more useful must await the test of time.

Diverse and in some cases complicated and conflicting though the above schemes may be, they all have the merit that they make it clear that they are dealing with mature structure *or* developmental sequences only. In some cases it may be desirable to indicate that the stomata of a particular species have a particular type of mature structure and a particular type of development. For these cases Pant (1965) and Stace (1965) suggested that, where known, the developmental type could be added to the mature type as a prefix, e.g. eumeso-paracytic using terms from the Stevens and Martin and Dilcher schemes, but unfortunately this has not been followed. Instead, authors wishing to indicate these binary data have tended to produce elaborate, inflexible schemes which by their very nature are only applicable to species which have been investigated developmentally. Fryns-Claessens and Cottem (1973) used the three developmental categories of Pant, and added prefixes to them denoting the mature structure. In this way they coined 26 terms (6 under perigenous, 9 under mesoperigenous and 11 under mesogenous), which are in a sense the reverse of those suggested by Pant and Stace (the example given above became para-mesogenous in the 1973 scheme), and cannot be readily shortened when only mature structural data are available. Taking account of the unnamed subtypes mentioned by Fryns-Claessens and Cotthem 47 categories are recognized (*fide* Payne 1979), and that number would be greatly increased if all the subsequently discovered situations were incorporated. The complexity of such a system is illustrated by Dilcher's (1974) attempt to reconcile the scheme of Fryns-Claessens and Cotthem with that based on mature structure alone.

Without doubt, adoption of the best systems of classification will greatly aid taxonomic work, as has been admirably demonstrated by Raju and Rao (1977) in the Euphorbiaceae.

(5) Cuticular membrane

The cuticular membrane may be defined as the (usually outer) part of the epidermal cell wall which contains cutin in addition to cellulose and other cell-wall polysaccharides. It is often loosely referred to as the cuticle, but the cuticle proper refers to the outermost part of the cuticular membrane which is more or less free of polysaccharides. The extent of cutinization varies greatly, and may itself be a taxonomic character. At one extreme it occurs in only a small outermost fraction of the thickness of the outer periclinal wall, whereas at the other it extends throughout most of the epidermal cell walls and even to the outermost parts of the subepidermal layer.

Where the cuticular membrane can be isolated from the rest of the leaf it can be used to study a wide range of surface leaf characters, as has been practised by palaeobotanists for well over a century. In particular, in those cases (which are very common) where even rudiments of cuticular flanges (cutinization of the anticlinal epidermal walls) are present, the epidermal cell outlines can be studied in great detail (Stace 1965; Dilcher 1974). In addition, however, there are often many intrinsic characters of the cuticular membrane itself which may prove of no less taxonomic value. Traditionally, these mostly involve the shape and extent of striations, reticulations and micropapillae as observed by the light microscope, but with the aid of the SEM the sculpturing (the secondary sculpturing of Barthlott 1981) on the inside and outside of the cuticular membrane can be easily differentiated (Figs 1–3).

Good reviews of the range of variation of cuticular sculpturing so far as known at present (and the state of our knowledge is still rudimentary) are given by Amelunxen *et al.* (1967); Napp-Zinn (1973–74), Barthlott and Ehler (1977), Wilkinson (1979) and Barthlott (1981 and Chapter 6 in this volume). To judge from the papers of Barthlott and Ehler (1977) and Barthlott (1981), and from unpublished work of A. A. Alwan and K. L. Alvin, the sculpturing of the inside of the cuticular membrane is a rich and so far mainly untapped source of taxonomic data.

The presence of an alveolar layer – an outermost part of the cuticular membrane, apparently composed of cutin and granular or globular in appearance, has been noted by various authors. Often, in cuticular preparations, it obscures the features of the rest of the cuticular membrane underneath it. Bongers (1973) was able to use it very effectively in taxonomic discrimination in the Winteraceae, in which he recognized five patterns of distribution and three types of structure. Koster and Baas (1982) have also found it of value in the Myristicaceae.

(6) Epicuticular secretions

Epicuticular secretions occur in the form of wax in a very wide range of plants. They give a bluish-white (glaucous) appearance to leaves and are involved in the unwettability of leaves and in their protection against pests, etc. Good general review of the chemistry, development and morphology of surface wax are provided by Amelunxen *et al.* (1967), Hallam and Juniper (1971), Jeffree *et al.* (1976), Wilkinson (1979), Barthlott and Wollenweber (1981), Barthlott (1981) and Baker (1982).

The extent of the taxonomic significance of the form of wax on leaf surfaces is only just emerging, despite the fact that Amelunxen *et al.* (1967) provided a working classification of wax forms and Hallam and Chambers (1970) produced an excellent survey of wax in the genus *Eucalyptus*, where they found that its morphology delimited "natural groupings of species", more than ten years ago. Barthlott (1981) described the surface wax as the tertiary sculpturing of the leaf surface. The most important publication on the taxonomic importance of epicuticular wax is that of Barthlott and Wollenweber (1981), who surveyed over 5000 species of vascular plants. They recognized seven categories of wax deposits in two main groups: continuous (including smooth and sculptured forms); and discontinuous (including irregular particles, simple rodlets or threads, simple flakes, compound rodlets and threads and flakes composed of sub-units, and rodlets or threads with a crystalline surface).

According to Barthlott (1981), the presence of secondary (cuticular) and tertiary (wax) sculpturing of leaf surfaces is nearly always mutually exclusive.

(7) Special structures

Apart from all the features mentioned above, many leaves bear special structures of very diverse kinds, which may therefore prove of great diagnostic value.

These include a great variety of secretory (glandular) structures, such as salt-glands, chalk-glands, extra-floral nectaries, hydathodes and leaf-teeth glands. In addition there are hydropoten, domatia, cork-warts and emergences of a great variety. An extensive review of many of these structures is given by Napp-Zinn (1973–74), and a briefer one by Wilkinson (1979).

"Special structures" of these types are put in a separate category here because often little is known about them, so that a more natural classification is not possible. "Warts" or "mounds" on leaf surfaces may be caused by various structures under the epidermis which do not themselves come to the surface (e.g. idioblasts), or their emergence on to the leaf surface might be irregular (e.g. sclereids). Nevertheless, they can often provide the leaf surface with a very characteristic appearance.

ACKNOWLEDGEMENTS

I am indebted to Mr A. A. Alwan and Dr S. W. Jones for allowing me to quote some of their unpublished work.

REFERENCES

Ahmad, K. J. (1978). Epidermal hairs of Acanthaceae. *Blumea* **24,** 101–117.

Aleykutty, K. M. and Inamdar, J. A. (1980). Structure, ontogeny and classification of trichomes in Ranales. *Feddes Repert.* **91,** 95–108.

Alvin, K. L., Dalby, D. H. and Oladele, F. A. (1982). Numerical analysis of cuticular characters in Cupressaceae. *In* "The Plant Cuticle" (D. F. Cutler, K. L. Alvin and C. E. Price, eds) pp. 379–396. Academic Press, London, New York.

Amelunxen, F., Morgenroth, K. and Picksak, T. (1967). Untersuchungen an der Epidermis mit dem Stereoscan-Elektronenmikroskop. *Z. Pflanzenphysiol.* **57,** 79–95.

Baas, P. (1975). Vegetative anatomy and the affinities of Aquifoliaceae, *Sphenostemon, Phelline* and *Oncotheca*. *Blumea* **22,** 311–407.

Bass, P. (1978). Inheritance of foliar and nodal anatomical characters in some *Ilex* hybrids. *Bot. J. Linn. Soc.* **77,** 41–52.
Baker, E. A. (1982). Chemistry and morphology of plant epicuticular waxes. *In* "The Plant Cuticle" (D. F. Cutler, K. L. Alvin and C. E. Price, eds) pp. 139–165. Academic Press, London, New York.
Baranova, M. (1972). Systematic anatomy of the leaf epidermis in the Magnoliaceae and some related families. *Taxon* **21,** 447–469.
Barthlott, W. (1981). Epidermal and seed surface characters of plants: systematic applicability and some evolutionary aspects. *Nordic J. Bot.* **1,** 345–355.
Barthlott, W. and Ehler, N. (1977). Raster-Elektronenmikroskopie der Epidermis-Oberflächen von Spermatophyten. *Trop. subtrop. Pflanzenwelt* **19,** 1–110.
Barthlott, W. and Wollenweber, E. (1981). Zur Feinstruktur, Chemie und taxonomischen Signifikanz epicuticularer Wachse und ähnlicher Sekrete. *Trop. subtrop. Pflanzenwelt* **32,** 1–67.
Böcher, T. W. and Lyshede, O. B. (1972). Anatomical studies in xerophytic apophyllous plants, II. Additional species from South American shrub steppes. *K. Danske Vidensk. Selskab, Biol. Skr.* **18(4),** 1–137.
Bongers, J. M. (1973). Epidermal leaf characters of the Winteraceae. *Blumea* **21,** 381–411.
Brandham, P. E. and Cutler, D. F. (1978). Influence of chromosome variation on the organisation of the leaf epidermis in a hybrid *Aloë* (Liliaceae). *Bot. J. Linn. Soc.* **77,** 1–16.
Brett, D. W. (1979). Ontogeny and classification of the stomatal complex of *Platanus* L. *Ann. Bot.*, n.s. **44,** 249–251.
Carlquist, S. (1961). "Comparative Plant Anatomy". Holt, Rinehart & Winston, New York.
Carr, S. G. M., Milkovits, L. and Carr, D. J. (1971). Eucalypt phytoglyphs: the microanatomical features of the epidermis in relation to taxonomy. *Austral. J. Bot.* **19,** 173–190.
Cotthem, W. R. J. van (1968). "Vergelijkend-morfologische studie van de stomata bij de Filicopsida". Doctoraal Proefschr., Gent.
Cotthem, W. R. J. van (1970). A classification of stomatal types. *Bot. J. Linn. Soc.* **63,** 235–246.
Cowan, J. M. (1950). "The *Rhododendron* Leaf". Oliver & Boyd, Edinburgh.
Cutler, D. F. (1972). Leaf anatomy of certain *Aloe* and *Gasteria* species and their hybrids. *In* "Research Trends in Plant Anatomy – K. A. Chowdhury Commemoration Volume" (A. K. M. Ghouse and M. Yunus, eds). Tata McGraw-Hill, New Dehli.
Cutler, D. F., Alvin, K. L. and Price, C. E. (eds) (1982). "The Plant Cuticle". Academic Press, London, New York.
Cutler, D. F. and Brandham, P. E. (1977). Experimental evidence for the genetic control of leaf surface characters in hybrid Aloïneae (Liliaceae). *Kew Bull.* **32,** 23–32.
Cutler, D. F. and Hartmann, H. E. K. (1979). Scanning electron microscope studies of the leaf epidermis in some succulents. *Trop. subtrop. Pflanzenwelt* **28,** 447–497.
Den Hartog, R. M. and Baas, P. (1978). Epidermal characters of the Celastraceae *sensu lato*. *Acta Bot. Neerl.* **27,** 355–388.
Dickinson, C. H. and Preece, T. F. (eds) (1976). "Microbiology of Aerial Plant Surfaces". Academic Press, London, New York.

Dilcher, D. L. (1974). Approaches to the identification of angiosperm leaf remains. *Bot. Rev.* **40,** 1–157.

Doyle, J. A. and Hickey, L. J. (1976). Pollen and leaves from the Mid-Cretaceous Potomac Group and their bearing on early angiosperm evolution. *In* "Origin and Early Evolution of Angiosperms" (C. B. Beck, ed.) pp. 139–206. Columbia University Press, New York.

Fryns-Claessens, E. and Cotthem, W. R. J. van (1973). A new classification of the ontogenetic types of stomata. *Bot. Rev.* **39,** 71–138.

Hall, D. M., Matus, A. I., Lamberton, J. A. and Barber, H. N. (1965). Infraspecific variation in wax on leaf surfaces. *Austral. J. biol. Sci.* **18,** 323–332.

Hallam, N. D. and Chambers, T. C. (1970). The leaf waxes of the genus *Eucalyptus* L'Héritier. *Austral. J. Bot.* **18,** 335–386.

Hallam, N. D. and Juniper, B. E. (1971). The anatomy of the leaf surface. *In* "Ecology of Leaf Surface Micro-organisms" (T. F. Preece and C. H. Dickinson, eds) pp. 3–37. Academic Press, London, New York.

Heywood, V. H. (ed.) (1971). "Scanning Electron Microscopy. Systematic and Evolutionary Applications". Academic Press, London, New York.

Hickey, L. J. (1973). Classification of the architecture of dicotyledonous leaves. *Amer. J. Bot.* **60,** 17–33.

Hickey, L. J. (1979). A revised classification of the architecture of dicotyledonous leaves. *In* "Anatomy of the Dicotyledons" (C. R. Metcalfe and L. Chalk, eds) (2nd edn) Vol. 1, pp. 25–39. Oxford University Press, Oxford.

Hickey, L. J. and Doyle, J. A. (1977). Early Cretaceous fossil evidence for angiosperm evolution. *Bot. Rev.* **43,** 3–104.

Hickey, L. J. and Wolfe, J. A. (1975). The bases of angiosperm phylogeny: vegetative morphology. *Ann. Missouri Bot. Gard.* **62,** 538–589.

Holloway, P. J. (1971). The chemical and physical characteristics of leaf surfaces. *In* "Ecology of Leaf Surface Micro-organisms" (T. F. Preece and C. H. Dickinson, eds) pp. 39–53. Academic Press, London, New York.

Hummel, K. and Staesche, K. (1962). Die Verbreitung der Haartypen in den natürlichen Verwandtschaftsgruppen. *In* "Handbuch der Pflanzenanatomie. IV (5). Plant Hairs" (J. C. T. Uphof, ed.) (2nd edn) pp. 207–250. Borntraeger, Berlin.

Hunt, G. M. and Baker, E. A. (1982). Developmental and environmental variations in plant epicuticular waxes: some effects on the penetration of naphthylacetic acid. *In* "The Plant Cuticle" (D. F. Cutler, K. L. Alvin and C. E. Price, eds) pp. 279–292. Academic Press, London, New York.

Hutchinson, J. (1926). "The Families of Flowering Plants, I. Dicotyledons". Macmillan, London.

Ihlenfeldt, H.-D. and Hartmann, H. E. K. (1982). Leaf surfaces in Mesembryanthaceae. *In* "The Plant Cuticle" (D. F. Cutler, K. L. Alvin and C. E. Price, eds) pp. 397–423. Academic Press, London, New York.

Jeffree, C. E., Baker, E. A. and Holloway, P. J. (1976). Origins of the fine structure of plant epicuticular waxes. *In* "Microbiology of Aerial Plant Surfaces" (C. H. Dickinson and T. F. Preece, eds) pp. 119–158. Academic Press, London, New York.

Jones, S. W. (1980). "*Garcinia* and the Guttiferae: Morphology and Taxonomy". Ph.D. Thesis, University of Leicester.

Juniper, B. E. (1959). The effect of pre-emergent treatment of peas with trichloracetic acid on the sub-microscopic structure of the leaf surface. *New Phytol.* **58,** 1–4.

Juniper, B. E. and Bradley, D. E. (1958). The carbon replica technique in the study of the ultrastructure of leaf surfaces. *J. Ultrastr. Res.* **2,** 16–27.

Koster, J. and Baas, P. (1982). Alveolar cuticular material in Myristicaceae. *In* "The Plant Cuticle" (D. F. Cutler, K. L. Alvin and C. E. Price, eds) pp. 131–137. Academic Press, London, New York.

Martin, J. T. and Juniper, B. E. (eds) (1970). "The Cuticles of Plants". Edward Arnold, London.

Melville, R. (1976). The terminology of leaf architecture. *Taxon* **25,** 549–561.

Metcalfe, C. R. (1960). "Anatomy of the Monocotyledons, I. Gramineae". Oxford University Press, Oxford.

Metcalfe, C. R. (1961). The anatomical approach to systematics. *In* "Recent Advances in Botany", pp. 146–150. University of Toronto Press, Toronto.

Metcalfe, C. R. and Chalk, L. (1950). "Anatomy of the Dicotyledons". Oxford University Press, Oxford.

Metcalfe, C. R. and Chalk, L. (1979). "Anatomy of the Dictyledons" (2nd edn) Vol. 1. Oxford University Press, Oxford.

Napp-Zinn, K. (1966). "Anatomie des Blattes. I. Blattanatomie der Gymnospermen". *In* "Handbuch der Pflanzenanatomie" (W. Zimmermann and P. G. Ozenda, eds) (2nd edn) Vol. VIII (1). Borntraeger, Berlin.

Napp-Zinn, K. (1973–74). "Anatomie des Blattes. II. Blattanatomie der Angiospermen, 1 and 2". *In* "Handbuch der Pflanzenanatomie" (W. Zimmermann, S. Carlquist, P. G. Ozenda and H. D. Wulff, eds) (2nd edn) Vol. VIII (2A). Borntraeger, Berlin.

Netolitzky, F. (1932). "Die Pflanzenhaare". *In* "Handbuch der Pflanzenanatomie" (K. Linsbauer, ed.), Vol. 4 (29). Borntraeger, Berlin.

Pant, D. D. (1965). On the ontogeny of stomata and other homologous structures. *Pl. Sci. Ser.* **1,** 1–24.

Patel, J. D. (1979). A new morphological classification of stomatal complexes. *Phytomorphology* **29,** 218–229.

Patel, J. D., Raju, E. C., Fotedar, R. L., Kothari, I. L. and Shah, J. J. (1975). Structure and histochemistry of stomata and epidermal cells in five species of Polypodiaceae. *Ann. Bot.*, n.s. **39,** 611–619.

Payne, W. W. (1970). Helicocytic and allelocytic stomata: unrecognised patterns in the dicotyledons. *Amer. J. Bot.* **57,** 140–147.

Payne, W.W. (1978). A glossary of plant hair terminology. *Brittonia* **30,** 239–255.

Payne, W.W. (1979). Stomatal patterns in embryophytes: their evolution, ontogeny and classification. *Taxon* **28,** 117–132.

Preece, T. F. and Dickinson, C. H. (eds) (1971). "Ecology of Leaf Surface Microorganisms". Academic Press, London, New York.

Raju, V. S. and Rao, P. N. (1977). Variation in the structure and development of foliar stomata in the Euphorbiaceae. *Bot. J. Linn. Soc.* **75,** 69–97.

Ramayya, N. (1972). Classification and phylogeny of the trichomes of angiosperms. *In* "Research Trends in Plant Anatomy – K. A. Chowdhury Commemoration Volume" (A. K. M. Ghouse and M. Yunus, eds) pp. 91–102. Tata McGraw-Hill, New Delhi.

Rasmussen, H. (1981). Terminology and classification of stomata and stomatal development – a critical survey. *Bot. J. Linn. Soc.* **83,** 199–212.

Roe, K. E. (1971). Terminology of hairs in the genus *Solanum*. *Taxon* **20,** 501–508.

Seithe, A. (1960). Die Haarformen der Gattung *Rhododendron* L. und die Möglichkeit ihrer taxonomischen Verwertung. *Engler bot. Jahrb.* **79,** 297–393.

Seithe, A. (1962). Die Haartypen der Gattung *Solanum* L. und ihre taxonomische Verwertung. *Engler bot. Jahrb.* **81,** 261–335.

Sporne, K. R. (1977). Some problems associated with character correlations. *Pl. Syst. Evol.*, Suppl. **1,** 33–51.

Stace, C. A. (1963). "Cuticular Patterns as an Aid to Plant Taxonomy". Ph.D. thesis, University of London.

Stace, C. A. (1965). Cuticular studies as an aid to plant taxonomy. *Bull. Brit. Mus. (Nat. Hist.) Bot.* **4,** 1–78.

Stace, C. A. (1966). The use of epidermal characters in phylogenetic considerations. *New Phytol.* **65,** 304–318.

Stace, C. A. (1973). The significance of the leaf epidermis in the taxonomy of the Combretaceae, IV. The genus *Combretum* in Asia. *Bot. J. Linn. Soc.* **66,** 97–115.

Stace, C. A. (1981). The significance of the leaf epidermis in the taxonomy of the Combretaceae, VI. Conclusions. *Bot. J. Linn. Soc.* **81,** 327–339.

Stant, M. Y. (1973). The role of the scanning electron microscope. *Kew Bull.* **28,** 105–115.

Stevens, R. A. and Martin, E. S. (1978). A new ontogenetic classification of stomatal types. *Bot. J. Linn. Soc.* **77,** 53–64.

Stromberg, A. (1956). On the question of classification of stomatal types in leaves of dicotyledonous plants. *Tbilisi sci. Stud. chem.-pharmaceut. Inst.* **8,** 51–66 (in Russian).

Takhtajan, A. L. (1980). Outline of the classification of flowering plants (Magnoliophyta). *Bot. Rev.* **46,** 225–359.

Theobald, W. L., Krahulik, J. L. and Rollins, R. C. (1979). Trichome description and classification. In "Anatomy of the Dicotyledons" (C. R. Metcalfe and L. Chalk, eds) (2nd edn) Vol. 1, pp. 40–53. Oxford University Press, Oxford.

Tomlinson, P. B. (1974). Development of the stomatal complex as a taxonomic character in the monocotyledons. *Taxon* **23,** 109–128.

Uphof, J. C. T. (1962). "Plant Hairs". *In* "Handbuch der Pflanzenanatomie" (W. Zimmermann and P. G. Ozenda, eds) (2nd edn) Vol. IV (5). Borntraeger, Berlin.

Wagner, W. H., Jr (1979). Reticulate veins in the systematics of modern ferns. *Taxon* **28,** 87–95.

Watson, L. (1964). The taxonomic significance of certain anatomical variations among Ericaceae. *J. Linn. Soc., Bot.* **59,** 111–125.

Wilkinson, H. P. (1979). The plant surface (mainly leaf). *In* "Anatomy of the Dicotyledons" (C. R. Metcalfe and L. Chalk, eds) (2nd edn) Vol. 1, pp. 97–165. Oxford University Press, Oxford.

6 | Microstructural Features of Seed Surfaces

W. BARTHLOTT

Institut für Systematische Botanik und Pflanzengeographie der Freien Universität Berlin, Federal Republic of Germany

Abstract: Seeds and small fruits exhibit a complex and high morphological and micromorphological diversity, providing valuable taxonomic information. Their overall shape, colour, size and, in particular, their internal structure (including ultrastructural characters) can be of high systematic significance. In practice, most data have been provided by SEM examination of the epidermal surfaces. These surface features may be grouped into four categories: (1) Cellular arrangement or cellular pattern, (2) Shape of cells (the "primary sculpture" of a surface), (3) Relief of outer cell walls (the "secondary sculpture" superimposed on the primary sculpture), caused mainly by cuticular striations and superficially visible secondary wall thickenings, (4) Epicuticular secretions (the "tertiary sculpture" superimposed on the secondary sculpture), i.e. mainly waxes and related substances.

The systematic application is discussed for each of these structural groups. Seed characters are only slightly influenced by environmental conditions. Their high structural diversity provides most valuable criteria for classification between species and family level. Their application is demonstrated within a dicotyledonous (Cactaceae) and monocotyledonous (Orchidaceae) family.

The application of seed data to taxonomy is rapidly increasing. There is evidence that this source of taxonomic information will play a role as important as pollen morphology has been over the last decades.

INTRODUCTION

Most taxonomists agree that data concerning the structure and microstructure of seeds are of great significance for the classification of angiosperm taxa.

Systematics Association Special Volume No. 25, "Current Concepts in Plant Taxonomy", edited by V. H. Heywood and D. M. Moore, 1984. Academic Press, London and Orlando.
ISBN 0 12 347060 9

However, until recently little research has been done and most of these studies have had little influence upon the shaping of classifications.

In his classical book on seed structure of angiosperms Netolitzky (1926) has admirably summarized our knowledge up to the end of 1923. After a long gap of fifty years Corner published his two volumes on the seeds of the dicotyledons in 1976. These works, complemented by the seed bibliography of Barton (1967) with some 20 000 entries, remain the basis for research in seed structures. Many data on monocotyledonous seeds are provided by Dahlgren and Clifford (1982), two of the few authors who actually integrated seed data in their classification.

During the last decade the use of scanning electron microscopes has greatly increased our knowledge of the microstructure of seeds. Seed-coats are rather thick-walled and stable in a vacuum: this allows quick preparation for SEM examination, without the need for complicated dehydration techniques. The low level of technical expenditure required, in combination with the high structural diversity exhibited and the intuitive ability to understand the "three-dimensional", often aesthetically appealing, microstructures visualized, has turned seed-coat studies into a favourite tool of many taxonomists. A most helpful bibliography of the many papers on the SEM of seeds covering the period from 1967 up to 1976 has been published by Brisson and Peterson (1977); more recent papers have been reviewed by Barthlott and Schill (1982).

The SEM has revealed new structural criteria and provided valuable additional diagnostic and systematic information. General morphological, taxonomic, and ecological aspects of surface sculpturing have been discussed in the following papers, which may be consulted for further references: Stace (1965), Martin and Juniper (1970), Barthlott and Ehler (1977), Wilkinson (1980), and Barthlott (1981).

The terminology used on the following pages has been modified in collaboration with D. F. Cutler (Kew) and is part of a forthcoming joint paper. The brief survey of seed-surface microstructures (including small fruits like, e.g. achenes, which show no basic differences) is based on our SEM studies of the seeds and diaspores of some 3800 species of angiosperms carried out over the last decade within the departments of systematic botany at the universities of Heidelberg and Berlin. One dicotyledon (Cactaceae) and one monocotyledon (Orchidaceae) family, each with some 1000 species examined, have been at the centre of our taxonomic interest. In the following sections we will draw our systematic examples mainly from these two families. For further information and references concerning seed-coat structure and taxonomy of orchids and cacti the following papers may be consulted: Barthlott (1976, 1979), Barthlott and Voit

(1979), Barthlott and Ziegler (1980, 1981), Behnke and Barthlott (1982), Dressler (1981) and Gregory (1981).

Because this work is concerned with microstructural features of surfaces it is necessary to bear in mind that it is only one aspect of the highly complex structure of seeds. Other features are also of high taxonomic significance: seed forms and internal structures (e.g. Corner 1976), seed appendages (Kapil *et al.* 1980), stratification and development of seed coats (e.g. Boesewinkel and Bouman 1980), ultrastructure of protein bodies (Lott 1981).

There is one non-microstructural feature of seeds which should not be omitted in this paper: their colour. This seems to be rather trivial; usually it is neglected in taxonomic considerations. However, in many cases the colours of seeds are of high diagnostic and systematic interest. Huber (1969) and Dahlgren and Clifford (1982) have pointed out the significance of the occurrence of phytomelan in seed-coats for the circumscription of several families and orders. In our studies of the orchids we noticed that the colour of their minute seeds (filled in small glass tubes) is significant whitish, green, yellow, reddish up to brilliant orange, brown or even black. These colours are usually characteristic of particular subfamilies or tribes: dark seeds are encountered in the Vanillinae, bright yellow or orange seeds are found only in certain other groups of the epidendroid and vandoid orchids (e.g. Bulbophyllinae and Dendrobiinae).

ARRANGEMENT OF SEED COAT CELLS

Cellular arrangement or "cellular patterns" and the distribution of idioblastic elements such as trichomes and multicellular appendages can be observed under low magnifications. The cellular arrangement can be of considerable diagnostic and systematic value. Figures 1 and 2 show two particular, easily recognizable, cellular patterns. The seed of *Sceletium compactum* (Aizoaceae; Fig. 1) exhibits a specific arrangement and orientation of smaller and larger cells. Similar arrangements occur within the Caryophyllaceae, Portulacaceae, and other related families: a typical "centrospermoid" pattern, which characterizes many – but not all – taxa of Caryophyllales (compare Fig. 7). The seed coat of *Eschscholzia* (Papaveraceae; Fig. 2) shows a very different arrangement of cells, creating a supercellular, net-like pattern. This striking cellular arrangement was analysed by SEM and described with modifications for all species of *Eschscholzia* by Clark and Jernstedt (1978). However, R. Hooke had illustrated the same pattern some 300 years earlier in the first edition of his "Micrographia" for poppy seeds (*Papaver*): this particular supercellular net-like cell arrangement appears to be a characteristic of several genera in Papaveraceae.

Figs 1–6. Surface sculptures of seeds. Fig. 1. Entire seed of *Sceletium compactum* (Aizoaceae), lateral view showing the arrangement of cells causing the "centrospermoid" appearance, *c.* ×23. Fig. 2. *Eschscholzia californica* (Papaveraceae), seed with a different arrangement of cells, which creates a supercellular net-like pattern, *c.* ×23. Fig. 3. *Aeginetia indica* (Orobanchaceae), seed-coat with isodiametric deeply concave cells and a reticulate secondary sculpture caused by helical secondary wall thickenings, *c.* ×470. Fig. 4. *Matucana weberbaueri* (Cactaceae), single isodiametric tetragonal testa cell with heavy secondary sculpturing caused by the central field type of cuticular ornamentation, *c.* ×940. Fig. 5. *Sceletium compactum* (Aizoaceae), single testa cells with tertiary sculpturing by epicuticular secretions forming long upright rodlets and small rodlets lying on the cell surface, *c.* ×470. Fig. 6. *Jacaranda macrantha* (Bignoniaceae), stellate epicuticular sculpture ("star scale") on the seed-coat, *c.* ×2350.

SHAPE OF CELLS: THE PRIMARY SCULPTURE

The most prominent feature of seed-surface sculpturing is usually the cell shape, particularly the curvature of the outer periclinal wall. Under primary sculpture the following four groups of microcharacters that influence the superficially visible shape are considered:

Outline of cells. Either isodiametric (Figs 3 and 4) or elongated in one direction (Figs 12 and 13). Usually cells are superficially tetragonal to hexagonal (Figs 3, 4 and 14) and ranging between 3- and 30-gonal.

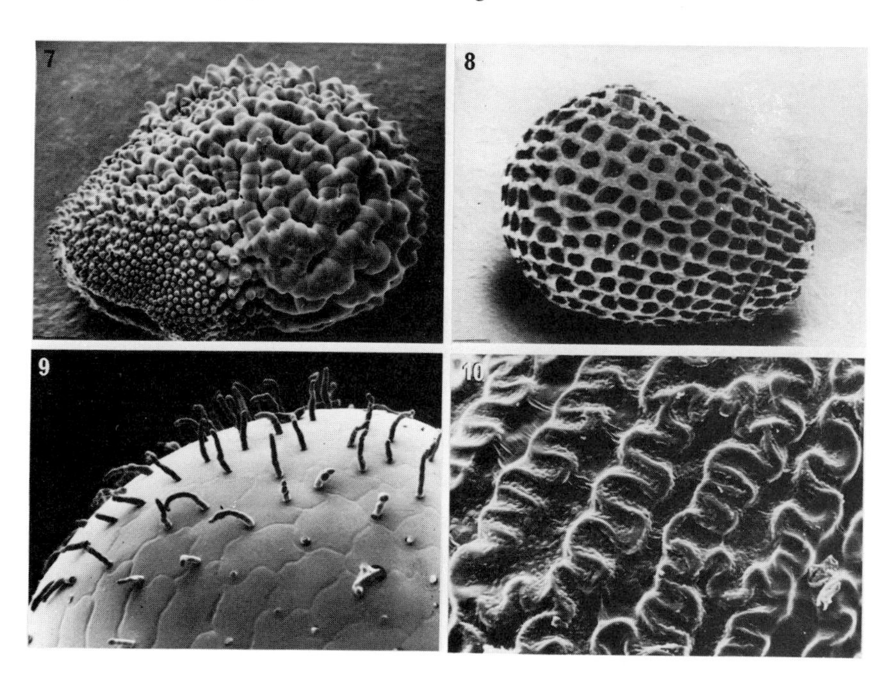

Figs 7–10. Seeds of Cactaceae. Fig. 7. *Browningia candelaris.* Entire seed in lateral view. Convex cells with crater-like sunken anticlinal boundaries and cell junctions, *c.* ×23. Fig. 8. *Mammillaria coahuilensis*, dorsal view of entire seed with deeply concave testa cells, *c.* ×47. Fig. 9. *Blossfeldia liliputana*, each testa cell with a finger-like excentric trichomatous appendix, *c.* ×235. Fig. 10. *Mammillaria johnstonii* with omega-undulated anticlinal boundaries, *c.* ×235.

Anticlinal walls. The superficially visible cell boundaries may be straight (Figs 3, 4), irregularly curved to more or less regularly undulated (sinuated), as in Fig. 10. In descriptive terms, the undulations can be classified into S-, U-, V-, and Omega-types. Usually they are of high taxonomic significance. In the Cactaceae they are restricted to the North and Central American Cactinae and circumscribe this subtribe (Fig. 10). In Orchidaceae regular undulations are restricted to the Australian and South African Diurideae and Diseae and do not occur within the rest of the family.

Relief of cell boundary. The anticlinal boundary may be channelled (Fig. 5) or raised (Fig. 10). The cell boundary may exhibit several additional micro-characteristics (e.g. Barthlott and Ehler 1977; Barthlott and Ziegler 1981). In the seed of *Browningia* (Cactaceae, Fig. 7) some of the anticlinal boundaries and cell junctions are deeply depressed, a characteristic of most ancestral types of cereoid genera, which is completely absent in the derived Cactinae. On the

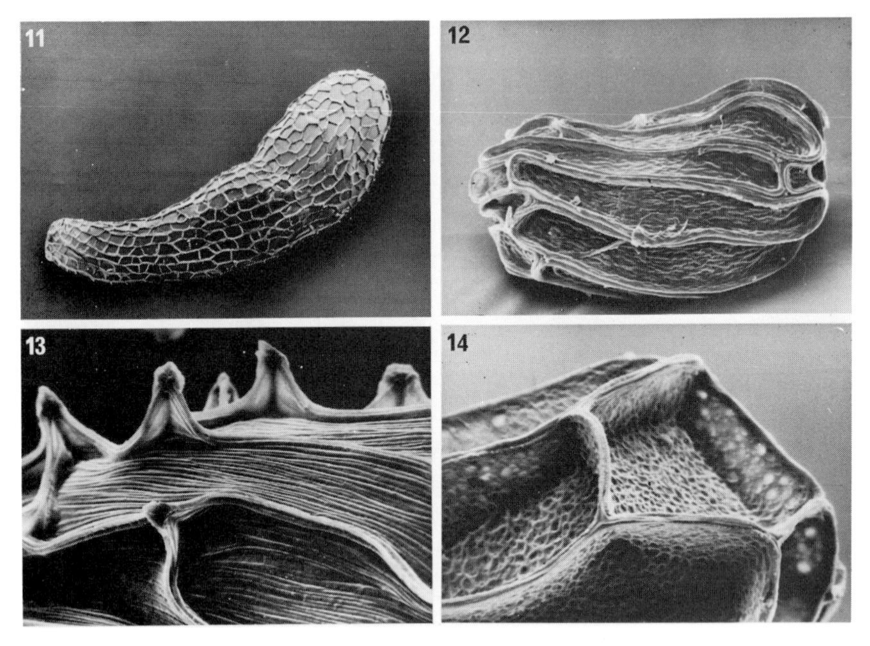

Figs 11–14. Seeds of Orchidaceae. Fig. 11. *Limodorum abortivum*, entire seed. One recognizes the position of the embryo under the thin testa which is uniformly composed of non-structured cells, a characteristic of all neottioid genera, *c.* ×47. Fig. 12. *Dichaea* sp., the entire seed has more or less the length of one cell. Each cell has a heavy marginal thickening and a irregular secondary sculpture caused by wall thickenings, *c.* ×470. Fig. 13. *Catasetum* (*Clowesia*) *russellianum*. Detail of seed-coat with raised cell junctions and longitudinal striations: both characters circumscribe the group of *Catasetum-Cymbidium-Cyrtopodium-Acriops* and related genera, *c.* ×470. Fig. 14. *Chondrorhyncha helleri* with densely reticulate ornamentation, characterizing a group of genera of the Huntleyinae-Zygopetalinae, *c.* ×940.

other hand, raised anticlinal boundaries (Fig. 10) are restricted to the subtribe Cactinae.

Curvature of outer periclinal walls. The curvature is responsible for the often macroscopically visible roughness of the seed-surface: cells may be flat, concave (Figs 3, 8 and 10, 11–14) or convex (Figs 5, 7 and 9). There are many descriptive terms for the frequent convex cell-forms and there exists a continuous transition to unicellular trichomes. These features are of diagnostic value (compare Fig. 7 with Fig. 8). Particular curvatures may be of high taxonomic interest. In the seed coat of *Blossfeldia* (Fig. 9) only one very small excentric portion of the outer wall is curved out into a trichome-like sculpture: this type of cell shape is restricted to the South American Notocactinae within the Cactaceae.

FINE RELIEF OF THE CELL WALL: THE SECONDARY SCULPTURE

The surface of the outer cell wall – if the testa consists of intact cells it means the surface of the cuticle – may be smooth (Figs 9 and 11) or exhibit a micro-ornamentation which, in descriptive terms, could be called striate (Figs 4 and 13), reticulate (Figs 3 and 14) or micropapillate (="verrucose"). Structurally, these striations, reticulations and micropapillations can result from very different portions of the cell wall. For purely descriptive diagnostic purposes it may be sufficient to characterize the microrelief with the descriptive terms mentioned above. For comparative systematic aspects, it is necessary to define these secondary sculptures structurally; their nature can often only be revealed by TEM examinations of thin sections. Major structural categories are:

Cuticular sculptures which usually occur as irregular to very regular striations (Fig. 4). These cuticular "folds" show a high micromorphological diversity (survey in Barthlott and Ehler 1977) and are an angiosperm characteristic. Figure 4 shows a heavily sculptured central field type of cuticular ornamentation in the genus *Matucana* (Cactaceae), which is characteristic for most genera of the Borzicactinae within the family (but again is absent in the Cactinae). Cuticular ornamentation is absent from seed-coats of all Orchidaceae. Generally, cuticular ornamentations may serve as good diagnostic characters, but their systematic significance is, with some exceptions, rather limited.

Secondary wall thickenings occur in helical to reticulate patterns on the inner side of the outer periclinal walls and on the anticlinal walls. In connection with shrinkage-deformation of collapsing desiccated cells of the mature seed coat they become visible on the surface. Usually they occur in the form of reticulations (Figs 3 and 14) or striations (Fig. 13). Secondary wall thickenings are always of high taxonomic significance. In the Orchidaceae, for example, the very dense reticulate pattern illustrated for the genus *Chondrorhyncha* (Fig. 14) occurs in most genera which were formerly put into the subtribe Huntleyinae and which seem to constitute a natural group. The longitudinal striation (caused by underlying secondary thickenings), illustrated for the genus *Catasetum* (Fig. 14), is restricted exclusively to all Catasetinae and *Cymbidium*, *Cyrtopodium*, *Ansellia*, *Acriops*, and closely-related genera. All these form a natural group, which had not been recognized in previous classifications of the family (Barthlott, Dressler and Ziegler, unpublished).

EPICUTICULAR SECRETIONS: THE TERTIARY SCULPTURE

Chemically very different substances may be found on or exude from seed-coats; in particular mucilaginous adhesive substances are very common. Their nature and origin – which can be very different and highly specialized (e.g. in the seed coats of *Cobaea* and *Cuphea*), may be of particular taxonomic and systematic significance. In the following, only the epicuticular waxes and related solid lipophilic substances are considered: they can easily be viewed with the SEM. A survey of the micromorphology and chemistry of these secretions is given in Barthlott and Wollenweber (1981).

In contrast to the surfaces of leaves, stems, and fruits waxy coatings on seeds are an exception. Usually seed-coats exhibit no tertiary sculpturing. Crystalloid waxy secretions occur predominantly on minute water repellent seeds, e.g. of some Droseraceae, Sarraceniaceae, and other families. The micromorphology of waxes and related substances may be of high taxonomic significance and delimit genera, families, and even orders (Barthlott and Frölich 1983). In Orchidaceae, only some epidendroid and vandoid tribes and subtribes possess epicuticular waxes on their seed coats (e.g. common in Bulbophyllinae). Two remarkable cases of non-soluble secretions ("alveolar material" of unknown chemical nature) are illustrated. Figure 6 shows the "star scales" on the seed-coats of the genus *Jacaranda* (Bignoniaceae), which were recently described by Hesse and Morawetz (1980) and which seem to be a characteristic of this genus. Figure 5 shows the insoluable complicated rodlets on the testa cells of *Sceletium*, differentiated in long upright structures and small particles lying on the cell surface. These secretions are characteristic of many genera of Aizoaceae and are restricted to this family (Ehler and Barthlott 1978).

CONCLUSIONS

In earlier literature microstructural features of seed (and epidermal) surfaces have been reported to be of only restricted diagnostic value. However, the above discussion has, again, confirmed the systematic significance of many of these characters. Surface features of seed-coats are surprisingly little affected by the environmental condition in which a plant grows. Differences in seed-coat characters seem always to reflect genetic-phylogenetic differences in the plants concerned. Therefore it is no surprise that, for example, Denford (1980) could indicate a correlation between the flavonol glycoside profiles in leaves and the micromorphology of seed-coats in the genus *Epilobium* (Onagraceae).

Characters like cell-shape and the occurrence and micromorphology of cuticular ornamentation are usually of minor taxonomic importance and are often of little diagnostic significance (for some exceptions see above). Characters connected with the structure of the anticlinal walls (e.g. undulations), secondary wall thickenings, and often epicuticular secretions are usually of high taxonomic/systematic significance. However, like all taxonomic criteria, seed-coat microstructures must be interpreted with great circumspection. For taxonomic studies one should examine seed-coats of all major taxa of the particular plant group concerned and compare the observations with other characters. Many of the papers published within recent years are based on an examination of too few samples.

There is a remarkable contrast between the generally accepted opinion that morphology and micromorphology of seeds are a source of important information for the classification of angiosperm taxa and the way these data are ignored in the existing classifications. Only over the last few years have seed data been increasingly integrated in taxonomic and systematic treatments (e.g. by Dahlgren and Clifford, 1982). As Corner (1976) points out, despite the important work of Netolitzky (1926), great systematists like Engler, Wettstein, Warming, Lotsy, Hallier and others failed to perceive the importance of research into seed structure which French, German and Italian schools had begun to explore in the last century. It would be outside of the scope of this chapter to examine historically why pollen morphology, and, for example, the typology of inflorescences, have influenced classifications whereas seed data have usually been neglected. Compared to the unicellular pollen grains, seeds are highly complex structures exhibiting a vast diversity of taxonomically applicable characters. With today's knowledge there is no evidence that the data from seeds and seed-coats are of less taxonomic significance than for example palynological and other data.

ACKNOWLEDGEMENTS

Grateful acknowledgements are made to Dr David F. Cutler and Miss Mary

Gregory (both of Kew), Dr Robert L. Dressler (Balboa) and our students Gisela Voit and B. Ziegler (both of Heidelberg) for their cooperation. The SEM research was financially supported by the Deutsche Forschungsgemeinschaft.

REFERENCES

Barthlott, W. (1976). Morphologie der Samen von Orchideen im Hinblick auf taxonomische und funktionelle Aspekte. *Proceed. 8th World Orchid Conf. Frankfurt*, 444–445.

Barthlott, W. (1979). "Cacti". Stanley Thornes, Cheltenham.

Barthlott, W. (1981). Epidermal and seed surface characters of plants: systematic applicability and some evolutionary aspects. *Nordic J. Bot.* **1**, 345–355.

Barthlott, W. and Ehler, N. (1977). Raster-Elektronenmikroskopie der Epidermis-Oberflächen von Spermatophyten. *Trop. subtrop. Pflanzenwelt* **19**, 1–110.

Barthlott, W. and Frölich, D. (1983). Mikromorphologie und Orientierungs-Muster epicuticularer Wachs-Kristalloide: Ein neues systematisches Merkmal bei Monokotylen. *Pl. Syst. Evol.* **142,** 171–185.

Barthlott, W. and Schill, R. (1981). Oberflächenskulpturen bei Höheren Pflanzen. *Progress in Botany* **43,** 27–38.

Barthlott, W. and Voit, G. (1979). Mikromorphologie der Samenschalen und Taxonomie der Cactaceae: ein raster-elektronenmikroskopischer Überblick. *Pl. Syst. Evol.* **132,** 205–229.

Barthlott, W. and Wollenweber, E. (1981). Zur Feinstruktur, Chemie und taxonomischen Signifikanz epicuticularer Wachse und ähnlicher Sekrete. *Trop. subtrop. Pflanzenwelt* **32,** 1–67.

Barthlott, W. and Ziegler, B. (1980). Über ausziehbare helicale Zellwandverdickungen als Haft-Apparat der Samenschalen von Chiloschista lunifera (Orchidaceae). *Ber. dt. Bot. Ges.* **93,** 391–403.

Barthlott, W. and Ziegler, B. (1981). Mikromorphologie der Samenschalen als systematisches Merkmal bei Orchideen. *Ber. dt. Bot. Ges.* **94,** 267–273.

Barton, L. V. (1967). "Bibliography of Seeds". Columbia University Press, New York.

Behnke, H.-D. and Barthlott, W. (1983). New evidence from the ultrastructural and micromorphological fields in angiosperm classification. *Nordic J. Bot.* **3,** 43–66.

Boesewinkel, F. D. and Bouman, F. (1980). Development of ovule and seed-coat of Dichapetalum mombuttense Engl. with notes on other species. *Acta Bot. Neerl.* **29,** 103–115.

Brisson, J. D. and Peterson, R. L. (1977). The scanning electron microscope and X-ray microanalysis in the study of seeds: a bibliography covering the period of 1967–1976. *Scanning Electron Microscopy* (IIT Research Inst. Chicago) **2,** 697–712.

Clark, C. and Jernstedt, J. A. (1978). Systematic studies of Eschscholzia (Papaveraceae). II. Seed coat microsculpturing. *Systematic Botany* **3,** 386–402.

Corner, E. J. H. (1976). "The Seeds of Dicotyledons", 2 vols. Cambridge University Press, Cambridge.

Dahlgren, R. M. T. and Clifford, H. T. (1982). "The Monocotyledons". Academic Press, London, New York.

Denford, K. E. (1980). Flavonol glycosides and seed coat structure in certain species of Epilobium – a correlation? *Experientia* **36,** 299–300.
Dressler, R. L. (1981). "The Orchids". Harvard University Press, Cambridge, Mass., London.
Ehler, N. and Barthlott, W. (1978). Die epicuticulare Skulptur der Testa-Zellwände einiger Mesembryanthemaceae. *Bot. Jahrb. Systematik* **99,** 329–340.
Gregory, M. (1981). References to scanning electron microscope photographs of seeds of Cactaceae. *Cactus Succulent J., Gt. Br.* **43,** 114–116.
Hesse, M. and Morawetz, W. (1980). Skulptur und systematischer Wert der Samenoberfläche bei Jacaranda und anderen Bignoniaceen. *Pl. Syst. Evol.* **135,** 1–10.
Huber, H. (1969). Die Samenmerkmale und Verwandtschaftsverhältnisse der Liliiflorae. *Mitteil. Bot. Staatssamml. München* **8,** 219–538.
Kapil, R. N., Bor, J. and Bouman, F. (1980). Seed appendages in Angiosperms. I. Introduction. *Bot. Jahrb. Syst.* **101,** 555–573.
Lott, J. N. A. (1981). Protein bodies in seeds. *Nordic J. Bot.* **1,** 421–432.
Martin, J. T. and Juniper, B. E. (1970). "The Cuticles of Plants". Edward Arnold, London.
Netolitzky, F. (1926). Anatomie der Angiospermen-Samen. *In* "Handbuch der Pflanzenanatomie" (K. Linsbauer, ed.), Vol. X (3). Borntraeger, Berlin.
Stace, C. A. (1965). Cuticular studies as an aid to plant taxonomy. *Bull. Brit. Mus. (Nat. Hist.) Bot.* **4,** 1–78.
Wilkinson, H. P. (1980). The plant surface. *In* "Anatomy of the Dicotyledons" (C. R. Metcalfe and L. Chalk, eds), (2nd edn), Vol. 1, pp. 97–165. Oxford University Press, Oxford.

7 Systematic Anatomy and Embryology–Recent Developments

D. F. CUTLER

Royal Botanic Gardens, Kew, England

Abstract: A very large number of papers on systematic anatomy and embryology has been published since the paper "Current developments in systematic plant anatomy" by C. R. Metcalfe at the Systematics Association conference on Modern Methods in Plant Taxonomy in 1968. It is both a stimulating and difficult task to select for comment particular areas of research for a review paper. The choice depends not only on the relative importance of topics, but also on their interest to the reviewer.

Many of the aspects discussed are the natural development of ideas and investigations that were started long before 1968, but some are new. Some depend on well established methods and others on the application of more recent techniques.

For example, the research on phloem plastids relies heavily on the TEM; surface sculpturing of leaf, seed, petals and so on can be observed readily with the SEM, as can many features of wood structure. But bark, leaf anatomy and stomatal development can be studied satisfactorily with the light microscope. Many modern projects rely on the use of a wide range of techniques, for example the study of transfer cells and contact cells, or stigma types and embryology.

There has been an increased interest in whole vascular systems as indicators of interrelationships, and cine photography has played an important role in these studies. Systems for the nomenclature of leaf venation patterns have been developed.

The computer has not been ignored in anatomical research. There are programmes on a national and international basis for recording and storing characters of wood anatomy. Anatomical characters of grasses have also been recorded for computer analysis.

Ecological or distributional factors may influence the characters used in systematic plant anatomy, as for example the dimensions of vessel elements in secondary xylem. The occurrence of anatomical characters associated with C_4 photosynthesis is also of considerable interest to the systematic anatomist.

Systematics Association Special Volume No. 25, "Current Concepts in Plant Taxonomy", edited by V. H. Heywood and D. M. Moore, 1984. Academic Press, London and Orlando.
ISBN 0 12 347060 9

INTRODUCTION

The starting point for this review is taken as 1968, when C. R. Metcalfe published a paper "Current developments in systematic plant anatomy" in "Modern Methods in Plant Taxonomy" (1968).

A large amount of literature has been published on systematic anatomy and embryology since that time, making selection of topics difficult. There have been several review papers of a general nature, e.g. Philipson (1974, 1975), Babu and Johri (1980) and some of a more specialized type, e.g. Esau (1969) on phloem and Howard (1970, 1974, 1979) on nodal anatomy, which reduce the need for extensive review in their particular areas of study.

In addition, this present volume contains papers on topics that would be relevant to this review (such as those of Tomlinson, Stace, and Barthlott); only brief mention of these subjects is necessary.

Some aspects, e.g. wood anatomy, have been covered more extensively here than others. This is because they have received relatively little attention in recent review papers.

SECONDARY XYLEM

There have been considerable advances in the critical study of wood anatomy during the period under review. The subject needs rather more extensive comment than other aspects of anatomy.

Dickison's (1975) review paid attention to the various characters of wood anatomy, provided lists of cell types useful in taxonomy and looked at taxonomic categories in relation to systematic evidence from wood anatomy. The ground was well covered, and does not need repeating here.

The main, and somewhat controversial area of interest in other recent papers centres on vessel elements (members). In particular, much work has been done on measuring their length and width and relating the results to various external factors. The significance of the different types of perforation plate and lateral wall pitting (and thickening) have also been discussed. General trends as observed particularly by, for example, Bailey (1957) do not seem to be such reliable indicators of phylogenetic advance as was generally accepted. However, taken overall the principles outlined by Bailey still seem to be acceptable.

There have been two main approaches. The one pioneered by Carlquist (1962) is concerned with the apparent juvenile nature of wood found in many island woody plants that come from mainly herbaceous families. The other initiated by Baas (1973) deals with variation in wood structure in relation to altitude and latitude. Both have a bearing on Bailey's (1957) conclusions.

Carlquist has written a series of detailed and interesting papers in which he reported a lot of new information, and developed his ideas. In Goodeniaceae, he found (Carlquist 1969) the xeromorphic species to have narrow, short vessel elements and helical wall thickening. Species from mesic habitats have wider, longer elements. The xeric species lack axial parenchyma but have abundant crystals; ray cells are erect, or the plants are rayless, and the lateral walls of vessel elements have scalariform or elliptical pits. These features of xeric species are regarded as indicators of juvenilism, or paedomorphosis. He did not think that the existence of scalariform plates in low frequency was vestigial, but rather a reversion to an ancestral condition. However, he noted that some juvenile characters, such as the presence of tracheids and fibre tracheids rather than libriform fibres, do seem to persist. The species of this family occurring on remote islands were found to be more woody than those from mainland Australia and New Zealand. *Scaevola* is reported to be the most woody genus. In his study on *Plantago* from islands, Carlquist (1970) found that many of the species are rayless until they are quite old. He noted that raylessness generally occurred where cambial activity is limited or finite and that there seems to be an evolutionary progression towards woodiness where the fusiform initials of the cambium are short and where juvenilism has occurred.

Carlquist's (1975a) paper on the wood anatomy of Onagraceae provided some interesting correlations between wood structure in that family and ecology. Those species with long vessel elements are mesomorphic, whereas xeromorphic species have shorter vessel elements. The xeromorphic species have long-lived libriform fibres, which store starch and may be involved in water storage also. He considered that long, wide vessel elements are of adaptive value when soil moisture is abundant and the relative humidity of the air is more or less constantly high. He thought that vessel element width and length give an accurate indication of the degree of xeromorphy or mesomorphy of a species.

In 1975, Carlquist summarized his views on xylem and ecology in his book "Ecological Strategies of Xylem Evolution" (Carlquist 1975b).

In a 1977 paper, Carlquist further developed the ideas that vessel element diameter and frequency could give an indication of the vulnerability of a species to drought. The "vulnerability index" was expressed as the ratio between vessel diameter: vessels per mm^2. He said that when vessel element length was multiplied by the vulnerability index a measure of mesomorphy could be obtained. (All figures were taken as averages from samples, and related to Australian local Floras (Florulas).) He thought that short vessel elements are less vulnerable to collapse under water stress because of the strength provided by their ends. He found good correlations between rainfall, temperature and other factors and his measures of vulnerability and mesomorphy.

Carlquist (1981) reported results of a study of the wood anatomy of Pittosporaceae. As a result of his observations he thought that species with evergreen leaves have a "buffer" which reduces variation in the wood that could have been expected in response to habitat differences. Once more, narrow vessel elements were seen to relate to xeromorphy, and he developed his views that species from wet habitats are more likely to have scalariform perforation plates than those from dryer sites. He was able to relate the presence of helical thickenings in vessel element walls to xeromorphy.

Mabberley (1974, 1982) in considering Calquist's theory of paedomorphosis found nothing in the arguments to convince him that pachycauls like those worked on by Carlquist are anything but relics in construction. He considered that no theory of juvenilism was needed, and did not agree that there was *de novo* origin of these forms on islands.

Since he observed that the size of vessel elements in a cross-section of a stem was related to the distance from the centre of the stem at which they were sampled, Maberley thought that great care should be exercised when selecting samples for comparison between leptocauls with the comparatively narrow stems and pachycauls. The wider pith of pachycauls causes the growth rings to have a wider diameter for a given age than is so for leptocaul trees, if samples are taken at some distance from the base. He thought this may be what Carlquist had done, and that more suitable pachycaul samples could have been obtained from near their stem bases, where the pith is much narrower.

The discussion and interpretation of the results of Carlquist's studies will undoubtedly continue.

Wood structural variations in response to latitude and altitude have been an area of research interest for Baas and his colleagues. In his study of *Ilex* species, Baas (1973) concluded that changes in vessel element length and bar number in scalariform perforation plates are correlated with altitude and latitude rather than with supposed relationships within the genus.

He found that in temperate zones and in the subtropics *Ilex* species have conspicuous growth rings, numerous narrow vessels with short elements, fewer bars in the perforation plates and conspicuous spirals (helices) on vessel and fibre walls; fibre tracheids with conspicuous bordered pits are present. Species from the tropical lowlands exhibit no growth rings, vessels are wider and less frequent, vessel elements are long and there are many bars on the scalariform perforation plates. Spirals are faint or lacking and fibre tracheids few, with slightly bordered pits. He concluded that a major climatic influence on the wood of this genus was indicated. It appeared to him that the trends from long to short vessel elements and many to few bars in scalariform perforation plates could be reversible.

Van der Graaff and Baas (1974) continued the studies on the effects of latitude and altitude on wood structure. In an examination of 52 species of various genera they found that increasing latitude was correlated with shorter vessel elements, narrower vessels, shorter and sometimes narrower fibres, lower rays, increased vessel frequency and increase in frequency and degree of expression of helical thickenings on the vessel and fibre walls. Increased altitude provided a similar range of correlations, but with weaker expression of each one. Unlike *Ilex*, the number of bars in scalariform perforation plates did not appear to be correlated with either latitude or altitude in the species under study.

Thirty-one *Symplocos* species were examined by van den Oever *et al.* (1981) with respect to the effects of altitude and latitude on their wood structure. The general conclusions reached were similar to those of Van der Graaff and Baas (1974), with the addition that vessel element and fibre wall thickness decrease with latitude, as do fibre length and diameter. Increased altitude did not seem to be correlated with decreased wall thickness.

Van den Oever *et al.* (1981) could see no relationship between ecology and the number of bars in scalariform perforation plates, or the presence or absence of helical wall thickenings. They pointed out that there seemed to be significant variation in certain characters normally used for taxonomy. They believed that because wood characters themselves are normally correlated, functionally adaptive changes brought about during the evolution of one or two wood characters could cause functionally neutral changes to appear in most other characters.

So the studies of both Carlquist and Baas and his colleagues have provided considerable food for thought, and have highlighted the need for more, careful studies on the relationship between habitat and wood structure so that the taxonomic and evolutionary implications can be properly interpreted.

There have also been several extensive monographic studies involving wood anatomy, only a selection of which can be mentioned here.

Gottwald and Parameswaran (1968) reported on 11 species from Ancistrocladaceae and Dioncophyllaceae (Guttiferales). They noted in particular fibriliform vessel elements and "spiral cells" in axial parenchyma (i.e. parenchyma with tertiary helical wall thickening). They reported that all the African *Ancistrocladus* species observed have silica bodies in the ray cells. Dioncophyllaceae all have phloem flecks in the secondary xylem. The close relationship between the two families was indicated, since they were found to share a mixture of primitive and advanced wood characters.

Gottwald (1977) presented the results of a very wide survey of the wood of Magnoliales. He looked at *c*. 700 species from 32 families within the Magnoliales themselves and also representatives of other primitive families. He concluded that there was no compelling evidence for placing Magnoliales as a common base for all modern dicotyledonous families. He recognized two major blocks of families that seemed to be distinct, and that he felt could not be derived one from the other. He termed these the "Theal-Dillenial-Hamamelidal" block and the "Magnolial-Annonal-Myristical" block. He proposed three possible ways of interpreting his data for evolutionary schemes, but did not select one in preference over the others.

Gibson (1973, 1976, 1977) conducted monographic work on the Cactaceae. In his 1973 paper he observed that tall species (in the Cactaceae) have long, wide vessel elements and small species have short, narrow vessel elements. Vessel wall pitting was reported to be scalariform to transitional and vessels and vascular tracheids hav helical thickenings. He stressed the close relationship he had found in Cactoideae between the habit of the plants and wood anatomy. In the Cactoideae there appears to be little correlation between habitat and rainfall, and wood anatomy (in contrast to Carlquist's findings for other families, reported earlier in this present paper).

Features of juvenilism are found in the wood of many of the Cactoideae; they appear to be correlated with succulence.

Gibson (1976) gave a comprehensive account of comparative vascular systems in Pereskioideae and Opuntioideae, and in 1977 described the secondary xylem of 30 *Opuntia* species with cylindrical to globular stems.

Overall, Gibson's work shows that there is little or no help for the systematist in the characters of the secondary xylem of Cactaceae. The complex interactions of habit, succulence and juvenilism tend to obscure any phylogenetic trends that might exist in this family.

The wood anatomy of some woody Rubiaceae (mainly Suriname species) has been documented by Koek-Noorman (1969a, b, 1970, 1972, 1977). Descriptions of the species were provided, with tabulated summaries of the observations. The systematic value of the characters selected is evident from the key based on wood anatomy. Results from the sample have been used in a numerical taxonomic study (Koek-Noorman and Hogeweg 1974).

Mennega (1980) contributed an account of the wood anatomy of genera of the Loganiaceae as part of a multidisciplinary account of the family. She provided descriptions of genera with notes and keys. Mennega, in common with most recent authors on systematic wood anatomy (e.g. Baretta-Kuipers 1981: Leguminosae and Dickison 1980: Cunoniaceae) appreciates the value of combined studies involving several approaches to the taxonomy of a group.

Van Vliet has written a series of papers on the wood anatomy of the Myrtales and Rhizophoraceae (1975–82). His 1975 paper contained an account of the vestured pits in Crypteroniaceae, as revealed by SEM studies. He presented descriptions for 5 genera and considered the systematic value of the characters of wood structure. For example, he thought that vestured types were of diagnostic rather than systematic value, whereas the presence or absence of septate fibres was of use in distinguishing generic groups. The subject of vestured pits was taken up again by van Vliet (1978) in the Combretaceae. He recognized two distinct types which relate to subfamily classification, but he pointed out that in the wider context, particularly because of the incidence of intermediates between vesturing types, the systematic application of these characters was limited. In 1979 van Vliet published the results of a major anatomical survey of the Combretaceae, reporting on the wood anatomy of 90 species from 19 genera. Wood anatomy supports the segregation of two subfamilies.

Within the Combretoideae, the subtribe Combretinae Exell and Stace shows vessels in the rays, a unique character.

Vessel element length was found to be at least in part related to ecology. Anatomical keys are provided.

Ter Welle and Koek-Noorman (1981) examined 47 genera involving 160 species of neotropical Melastomataceae for characters of wood anatomy that might be of systematic interest. They reported no relationships between anatomy and ecology. They found taxonomic value in the shape and size of intervascular pitting, ray features, crystals and parenchyma distribution. They produced a cladistic analysis of their observations on ray and axial parenchyma for some taxa, but noted that since some trends might be reversible and because intermediate forms exist, there could be problems in arriving at a phylogenetic arrangement. One of the interesting conclusions was that in this family, pure axial parenchyma without intermixed fibres is a derived character.

Van Vliet's (1981) paper on 107 species of 36 genera of palaeotropical Melastomataceae complements that of ter Welle and Koek-Noorman (1981). A synthesis of the work is presented by van Vliet *et al.* (1981). Several taxonomic changes were shown to be necessary. The wood anatomy of the three subfamilies was considered. There seems to be a relationship between vessel element length and ecology in some genera of Osbeckieae, with species from drier habitats having shorter vessel elements. The family as a whole could not be used to study latitudinal and altitudinal trends.

Van Vliet (1976) studied the wood anatomy of 65 species representing all 18 genera of the Rhizophoraceae. He was able to divide the genera into four anatomical groups, which corresponded with those proposed by several taxonomists. There did not seem to be any correlation between wood anatomy and the mangrove habitat.

Van Vliet did not consider that scalariform plates gave extra rigidity and hence prevent collapse in the vessel elements as was suggested by Carlquist (1975b).

Van Vliet *et al.* (1981) wrote on wood anatomy and classification of the Myrtales, drawing together data from the papers mentioned above. Two anatomical characters are shared by all families of the order: vestured bordered pits of the xylem and intraxylary phloem. This combination of characters is very rare outside the Myrtales. The contribution of wood anatomy to an understanding of interrelationships within the Myrtales and between the Myrtales and related orders was fully discussed.

Richter (1981) wrote an account of the wood anatomy of Lauraceae.

ULTRASTRUCTURE

There are few characters of systematic value that have been discussed entirely from ultrastructural studies. Those which are available require painstaking preparation of material, and generally a lot of time is used in making them evident. Behnke (1977) commented on the small number of usable ultrastructural characters in his review paper on the TEM in systematics. Apart from structures like the dilated cisternae in the endoplasmic reticulum in Capparales, and pollen wall, he found little other than phloem plastid types to be of taxonomic interest.

In 1971 and 1972 Behnke wrote on the probable taxonomic importance of sieve tube plastids. This was at the start of an interesting series of observations in which the real systematic value of these plastids has been realized. Behnke (1975) reported further observations. The leucoplasts in phloem sieve tube elements can be classified into two major categories, with subgroups; the S-type, or starch accumulators and the P-type, protein accumulators. P-type accumulators may also have some starch associated with the protein crystals, but the S-type do not accumulate protein. Behnke and Dahlgren (1976) related the systematic distribution of plastid types to Dahlgren's classification system. In his 1977 paper, Behnke elaborated the classification of plastid types and gave lists of the families in which the subtypes occur. There was a good correlation between types and recognized groups of families in many instances.

Behnke (1981a, b) gave a detailed analysis of phloem protein and possible evolutionary lines in the monocotyledons. He recognized five subtypes of P-plastids. Subtype P II is the major one confined to the monocotyledons (except for *Asarum* in which it is also found), in which the protein crystals are cuneate. Within the monocotyledons further subdivisions of the subtype P II were recognized, and these occur in a distribution pattern which follows well recognized systematic divisions. Behnke considers that evidence from sieve tube element plastids points to an early origin of the monocotyledons from a generalized angiosperm type.

There seems no doubt that sieve element plastid types are highly significant in systematics (see also Behnke 1976a, b, 1981a).

Possibly of less widespread systematic importance but of considerable anatomical and physiological interest was the investigation by Gunning *et al.* (1968) of the nature of the walls of modified cells in minor leaf veins. They termed these cells "transfer cells", a direct translation of a term used by Fischer (1884) for this type of cell. Fischer did not know about the cell wall modification; he had observed the dense protoplasmic contents of the cells concerned. Transfer cells have elaborate fine projections, either simple or branching and sometimes with fused branches, protruding into the cell lumen, and vastly enlarging the surface area for protoplast-cell wall contact. Gunning *et al.* (1968) thought that the structure of the wall assisted with the active transport of compounds, possibly by accumulation in the transfer cells to a higher concentration than in the surrounding cells, and their symplastic transfer via compound plasmodesmata to other cells.

Gunning and Pate (1969) recorded a wide range of parts of the plant in which transfer cells can be found. The wall ingrowths lack a middle lamella. The cells appeared to favour maximum solute flow, with a minimum of solvent flow, the transfer being mediated through membranes. They showed that by using suitably thin sections (0.5–2.0 μm) the structures could be seen in the light microscope, whereas previously they had been discovered using the TEM.

The list of places in which transfer cells can be seen includes epidermal cells of submerged leaves, nectaries and other glands, vascular parenchyma, embryo sacs, embryos, haustoria, anther tapetum, root nodule pericycle, cells intermediary between xylem and phloem, and bundle sheaths.

A somewhat more taxonomic approach was taken by Gunning *et al.* (1970) in a survey of nodes. They found transfer cells in about half of the species examined. They reported that xylem transfer cells were better developed and more frequent than phloem transfer cells. The latter occurred mostly at the margins of leaf gaps. Transfer cells were seen to be associated with the megaphyllous condition. Vascular transfer cells were shown by Watson *et al.* (1977)

to be a taxonomically valuable character at tribal level in the leaves of Leguminosae-Papilionoideae.

In another paper, Pate *et al.* (1970) reported the presence of transfer cells in the hypocotyls of seedlings.

VASCULAR SYSTEMS

Interest has continued in the study of vascular systems in the monocotyledons. This work has particular significance since there appear to be several distinct types of vascular organization, each with modifications. Some of the main contributions are mentioned here. Zimmermann and Tomlinson (1972) wrote a paper summarizing previous work on palm vasculature. In 1974 Zimmermann *et al.* published on Pandanaceae vasculature and found it to be more complicated than in the palm *Raphis*. French and Tomlinson (1980, 1981a–d) described the vascular systems of Araceae.

NODAL ANATOMY

There have been big advances in the understanding of nodal anatomy during the period under review. In addition to a number of papers on particular taxa, there have been valuable general articles. In an early paper, Howard (1970) laid the foundations for the modern approach to petiole and nodal anatomy; he gave a literature review, and an account of some nodal patterns. He considered the split-lateral and common-gap interpretations of a particular nodal type, and gave examples of this from a number of families, since he thought it to be of taxonomic value. Howard (1974, 1979) wrote on the stem-node-leaf continuum and petiole anatomy of dicotyledons, emphasising the need to section petioles at a number of levels for proper comparative studies. In his 1979 article, Howard gave very comprehensive accounts and an extensive bibliography.

Among other recent works on specific groups, Neubauer (1981) gave an account of the nodal anatomy of Rubiaceae. He found that most species were unilacunar and had one complex trace, but some were trilacunar and with three traces.

LEAF VASCULATURE

Schemes for describing leaf vasculature and architecture have interested botanists for generations, but in recent years new enthusiasm for developing concise, usable systems of terminology has become evident. The incentive for some of this work came from the need to compare fossil leaves with those of present-day species. Systems available in the 1960s were not adequate for the purpose. Hickey (1973) proposed a clear and readily usable system of terminology for leaf architecture in dicotyledons to remedy this situation. In a further paper, Hickey and Wolfe (1975) developed a system and provided a key to subclasses of dicotyledons, and gave the main features of taxa. Among the characters of systematic interest were the configuration of the first three vein orders and characters of the leaf margin.

They noted that a study of vasculature had given evidence that some morphological leaf types had arisen independently from several origins.

Hickey (1979) presented a slightly modified and refined version of his system of terminology in "Anatomy of the Dicotyledons".

Whilst Hickey was gathering information for his system, Melville was actively developing a different set of terms to describe venation patterns. He published the results in 1976. It is unfortunate that two entirely different systems should have appeared almost simultaneously. Melville includes monocotyledonous leaves and those of ferns and pteridosperms. According to Hickey (1979), amongst other things Melville disregards the important and unique characters of angiosperm leaves in comparing vascular patterns, so that his results could be misleading.

It seems that Hickey's system may be used preferentially, as it was taken up by Metcalfe and Chalk (1979). Other authors have found it valuable, e.g. Avita *et al.* (1981) in their paper on leaf architecture in Ranunculaceae.

EPIDERMAL STUDIES

(1) Plant surfaces

Plant surfaces have been the subject of much research. Stace (1965 and Chapter 5, this volume) and Wilkinson (1979) have been involved with developing a modern terminology, as have Barthlott and Ehler (1977). Cutler (1982) has been concerned with establishing the nature of the control of cuticular sculpturing, particularly in Aloes. The paper published in 1982 contains references to earlier work in which it was found that there is strong genetic control over the sculpturing types, with little influence of environment other than very long-term selection pressure. See Stace (this volume, Chapter 5) for further information.

(2) Stomata

Stomatal types have continued to be the subject of numerous papers. Stomata are regarded as useful both taxonomically and diagnostically by many authors. Since Metcalfe and Chalk (1950) broke away from the system of using plant family names to define types, considerable progress has been made. Numerous new types have been named in angiosperms, gymnosperms and ferns.

One of the more important observations has been that whereas mature stomata may be useful for diagnosis, their value in evolutionary studies is strictly limited. In the period under review, the following have made contributions to the subject: Tomlinson (1969, 1974), Paliwal (1969), Payne (1970, 1979), van Cotthem (1970, 1971), Fryns-Claessens and van Cotthem (1973), Patel (1978), Stevens and Martin (1978), Wilkinson (1979) and Rasmussen (1981a, b). It is evident that stomatal development must be studied in each species when statements about interrelationships or phylogeny are to be made. There are several systems that use almost similar terminology amongst those mentioned above. Undoubtedly the simplest possible system that conveys the most information will be given preference by future authors. Probably two parallel systems will emerge: that for use in identification of mature organs, and that for phylogenetic and evolutionary studies.

There have been many papers on stomatal types present in particular taxa; those of Baranova (1969, 1972) will serve as an example. Although they depend on the mature stomatal type, they show the value of stomatal studies in taxonomy, where with the Magnoliaceae, relationships could be indicated.

(3) Trichome types

There is such a wide range of trichome types that classification of them constitutes considerable problems. Trichomes can be of great systematic significance, and often even common types are used for diagnostic purposes in association with other characters. The most recent trichome classification is that devised by Theobald *et al.* (1979). This is practical and straightforward.

THE LEAF

(1) Morphology and development

Some studies, although not directly concerned with systematic problems, have results that are of importance to systematists. Studies involving the understanding of the nature of certain organs are of importance, since they enable the systematist to select homologous structures for comparison and discussion.

Examples of this type of research are provided by Kaplan (1970a, b, 1973,

1975, 1980) and Guédès (1972, 1980). As with modern studies on stomata, Kaplan has shown that it is essential to examine development of leaves in order to understand their structure. This is particularly true of the rachis leaf in Umbelliferae (Kaplan 1970a) and unifacial leaves in monocotyledons (Kaplan 1975).

(2) Photosynthetic pathways

Since the relationship between Kranz anatomy and C_4 photosynthesis was recognized (see, for example, Hatch and Slack (1966, 1970), Laetsch (1969), Tregunna (1970)) a great deal of survey work has been undertaken to identify possible C_4 plants. A useful paper for early references and an account of the systematic position of C_3 and C_4 grasses was written by Carolin and Jacobs (1973) and this has been followed by others, for example Ellis (1974) on *Alloteropsis* and Ellis (1977) on South African Eragrostoideae and Panicoideae. Renvoize (1981) reported on a comprehensive study of the Arundinoideae; he was able to select several stable and taxonomically useful anatomical characters. He discussed the taxonomic grouping of C_3 and C_4 types of species.

Among Dicotyledons, most work has been done on the members of the Chenopodiales. Carolin *et al.* (1975, 1978) contain details of the species with the Kranz or partial Kranz syndrome, and good lists of references. The taxonomic implications of the occurrence of Kranz anatomy in the order Chenopodiales are fully discussed.

Böcher and Olesen (1978) gave a very full anatomical account of the C_4 grass *Sporobolus rigens*, with particular reference to the ecological significance of the anatomical modifications.

(3) Sclereids

Leaf sclereids vary in shape, size and distribution and may have both systematic and diagnostic significance. Recent surveys have been carried out by Rao and Bhattacharya (1978) and Rao and Das (1979). Original observations were included with a summary of observations from other papers.

FLORAL STUDIES

(1) Stigmas

During the course of extensive studies on pollen-stigma interactions, Y. Heslop-Harrison has been able to define different types of stigma, and classify them.

Heslop-Harrison and Shivanna (1977) produced a comprehensive paper in which the two main types of stigma, dry and wet, were defined, together with subdivisions relating to the extent of papillation, if any. This was followed by Heslop-Harrison (1981) in which the results of a survey of *c.* 1000 species from *c.* 900 genera were published, giving some taxonomic implications of the findings.

(2) Nectaries and secretory tissues

Both floral and extrafloral nectaries have been the subject of extensive research in recent years. Interest was reawakened since the TEM has assisted so much in our understanding of these structures. The principal author in the field has been A. Fahn, who in addition to individual papers (e.g. Fahn and Rachmilevitz 1971, on *Lonicera* nectaries) has written a reference book on the subject (Fahn 1979). This latter volume contains a review of relevant literature and reports new observations. The distribution of particular types of secretory structure has taxonomic implications.

(3) Ligules

Baagøe (1977, 1978, 1980) has reported on a new and extensive set of micro-characters of taxonomic significance in the ligules of Compositae; included are cell outlines, cuticle patterns, mesophyll anatomy, location of pigments, presence or absence of stomata, glands and anther rudiments. UV patterns were also discussed.

(4) Petals

Stirton (1981) wrote on petal sculpturing in papilionoid legumes, and Kay *et al.* (1981) on pigment distribution, light reflection and cell structure in petals. These two papers are indicative of a new interest in petal anatomy, and the possible function of the features observed. There is a great potential in this work, as can be seen from the paper by Kay *et al.* (1981), where many applications beyond systematics are discussed.

SEEDS

Barthlott (this volume, Chapter 6) has given an account of seeds and small fruits, so it is inappropriate to develop that aspect here, apart from remarking that the

SEM has opened new vistas in these studies. Barthlott and Ziegler (1981) demonstrate this point themselves in their paper on orchid seeds, in which 270 genera were surveyed, and 1100 species seen, and a classification to subtribe and subfamily level given.

Another particularly interesting paper is that of Kapil *et al.* (1980) in which seed appendages are discussed. This paper shows that developmental studies are necessary for a proper understanding of such structures as arils or raphe and chalaza.

Corner (1976) wrote a monographic treatise on seed morphology and anatomy. This is a source of new information and has an extensive bibliography.

DATA BANKS AND COMPUTER ANALYSIS

Increasing use has taken place of computers for data storage and analysis during the past 14 years. An example of computer analysis can be found in Ambrose (1980) where he reports on a re-evaluation of the Melanthioideae (Liliaceae) using 110 characters. It was possible to suggest new taxonomic groupings, and also to define some characters of diagnostic value. Alvin *et al.* (1982) produced a numerical analysis of cuticular characters in Cupressaceae, involving 83 characters, most with several character states. This type of work, with a comparison between the results of classifications using anatomical or traditional characters, is of increasing importance, and is made more readily practical by the use of computers.

Three other projects should be mentioned. The first of these is the IAWA sponsored wood identification programme: "Standard list of characters suitable for computerized hardwood identification" (1981) compiled by a committee of the IAWA, followed by an explanation of coding procedure (Miller 1981). The objective is to build a large data bank which is available for consultation internationally. The main purpose is to assist in wood identification, but there will be many other uses once the bank is large enough.

The second project is that of Watson and Dallwitz (1981) on grasses. This computerized programme will generate generic descriptions (324 genera) using 150 morphological and 52 anatomical characters; keys can be provided for users. The third project is that of Dolph (1978) who proposed the formation of a data bank of leaf architecture, morphology and cuticle characters.

EMBRYOLOGY

During the period under review there have been some general survey papers, such as those by Palser (1975), Bouman (1971a, b, 1979) and Schulze (1980a, b). Most other papers have concentrated on particular features of the embryo or on smaller groups of species.

Sporne (1969) was concerned with ovule evolution. When data on nucellus type were added to other data to produce an advancement index, he was able to conclude that crassinucellate ovules were more primitive than tenuinucellate ones. He envisaged the ancestral type as being crassinucellate, with 3 envelopes (2 integuments and an aril) each with a vascular supply.

Bouman (1974) gave an excellent historical review of work on seed coats and integuments. He emphasized that developmental studies were essential for correct interpretation of the mature seed structures. This is particularly important if data are to be used in systematics or phylogenetic studies.

Philipson (1974) found a correlation between unitegmic, tenuinucellate ovules and sympetaly. He noted that occasional examples of unitegmic crassinucellate ovules exist.

Guignard (1975) wrote on embryology and embryogenic classification. He stated that embryo form was similar for all systematic groups, and embryogenesis could not afford information of prime importance for the main groups of plant families. He thought that at genus and species level embryogeny was of some systematic value.

In Palser's 1975 survey paper, the author defined lists of embryo and anther characters that distinguished monocotyledons from dicotyledons, but noted that out of all the characters available for comparison, 70% were shared in common. It was concluded that embryo characters are relatively constant at the family level. In families where some variation exists, genera are on the whole constant. The megagametophyte development and its mature form may help in the determination of relationships within families and genera.

(1) Integuments

Bouman continued to report his studies on integuments in a series of papers, either as sole or joint author. Bouman (1975) described the testa organization in 4 members of the Cruciferae. He reemphasized the need for developmental studies, since he found a number of previous descriptions based on mature material to be incorrect.

Bouman (1977) found the ovule ontogeny in *Liriodendron* and *Magnolia* to be very similar. The sclerified outer integument of *Liriodendron* was thought to be a derived character.

Bouman and Calis (1977) discovered a third way in which unitegmy could be derived. They found an unusual possible route from bitegmy to unitegmy in *Eranthus*, *Helleborus* and *Delphinium*. This involved fusion of primordia, shifting of the inner integument and the arrested development of the inner integument.

Bouman and Schlier (1979) found that the integument development in *Gentiana* involved both dermal and subdermal cells.

(2) Ovules

Philipson (1977) arrived at a classification of ovules into 4 types, based on the number of integuments and nucellar thickness.

He stated that there was a complex of orders (mostly Sympetalae) and several polypetalous groups which have a single integument. He equated these in a distinct evolutionary line, called the Unitegminae.

Within the Sympetalae and some Polypetalae unitegmic tenuinucellate ovules were found, so he wondered if this state could have had several origins. Studies of polypetalous families in which unitegmic ovules are constant could in his opinion give better correlations.

(3) Taxonomic applications

Of the more recent works on the taxonomy of specific groups using embryological data, those of two authors have been selected as examples. These are Schulze (1980a, b) and Ly Thi Ba (1981a–d). Schulze (1980b) has studied embryos in the Liliaceae. He defines four major groups. There are two embryo sac types, the "normal" one, which is present in almost all of the genera seen, and one found in *Scilla*. There is not complete correlation between his four main groups and endosperm type (nuclear or helobial). In his paper on Alliaceae (Schulze 1980a) reported that the normal (Liliaceous) type of embryo sac was present in *Agapanthus*, *Muilla* and *Brodiaea* and the *Scilla* type occurred in *Tulbaghia*, *Allium*, *Nothoscordum* and *Leucocoryne*. *Allium* had nuclear endosperm, whereas that in *Nothoscordum* and *Muilla* was helobial. This type of survey will prove to be of considerable interest, when the data are combined with those from vegetative anatomy.

Ly Thi Ba, in a series of papers, has been concerned with the embryology of Ranales and Helobiales, in an attempt to find out if supposed close relationships between these groups was supported or not by embryological data.

In the 1981a paper, arborescent Ranales (Magnoliales) were studied. The embryo was shown to be archaic, with protracted development. A globular tetrad gave rise to a voluminous embryo and bulky suspensor. Trochodendraceae was seen to be isolated, whereas Cercidiphyllaceae showed affinities with Hamamelidaceae, and did not appear to belong with the Magnoliales.

The 1981b paper showed herbaceous Ranales to be homogeneous embryologically. He concluded that the tribe Paeoniae should be given independent family rank. He found Berberidaceae (and to a lesser extent Menispermaceae) to be similar to the Ranunculaceae, particularly the Helleboroideae.

The Nymphaeaceae have a globular tetrad and voluminous embryo. The Cabomboideae, Nymphaeoideae and Nelumbonoideae were thought to be autonomous groups, and Ceratophyllaceae should be included in the Nymphaeaceae. He stated that Nymphaeaceae are definitely Dicotyledons. His conclusions were based on painstaking studies of cell division sequences and wall orientations.

The 1981c paper was on Helobiales. They were shown to have uniform embryonic development with a vesicular basal cell and intermediary albumen type. The pollen was monocolpate. The shoot apex was terminal in Potomogetonales and lateral in Alismatales.

The concluding paper (1981d) contained the main conclusions based on evidence and discussion from the first three. Principally, Ly Thi Ba found no direct parental relationship between Ranales and Helobiales, but thought they were derived from a common ancestral stock. On the basis of his embryological and embryogenetic studies he considered monocotyledons to be derived from dicotyledons.

MAJOR ANATOMICAL SERIES OF PUBLICATIONS

During the period under review, several reference books have been published that cover fields of systematic anatomy. The one series is organized on particular topics, and the other has a taxonomic basis.

Those volumes on topics belong to the "Encyclopedia of Plant Anatomy" series. Von Guttenberg (1968) wrote on primary root structure of Angiosperms and Esau (1969) wrote the volume on phloem. Then followed von Guttenberg's (1971) book on plant movements and organs of perception. Napp-Zinn (1966) produced a volume of leaf anatomy of the Gymnosperms, followed in 1974 by two volumes on leaf anatomy in Angiosperms. Braun (1970) produced a volume on functional histology of secondary xylem. Roth produced two volumes, one on anatomy of fruits (1977) and the other on structural patterns of tropical barks (1981). Embryology is also covered in this series, and Singh (1978) wrote a volume entitled "Embryology of Gymnosperms".

The series of volumes on systematic anatomy with the general title "Anatomy of the Monocotyledons" has been added to over the period. Cutler (1969) wrote on Juncales, Metcalfe (1971) on Cyperaceae and Ayensu (1972) on Dioscoreales. Work is continuing on this series and P. B. Tomlinson's book (1982) has recently been published on Helobiae. The Orchidaceae by E. Ayensu is nearly completed. Various authors are involved in the Liliaceae, Amaryllidaceae and Iridaceae, coordinated by D. Cutler. Araceae is well under way under the authorship of R. Keating.

"Anatomy of the Dicotyledons" (2nd edn) Vol. 1 by Metcalfe and Chalk was published in 1979. The second volume, at proof stage at the time of writing this chapter, deals mainly with wood and ecological anatomy, and contains extensive references.

ACKNOWLEDGEMENTS

I am most grateful to Mary Gregory, who has not only assisted with the selection of suitable references, but has made valuable criticisms of the text.

REFERENCES

Note: Dates in *square* brackets indicate the date of the actual publication

Alvin, K. L., Dalby, D. H. and Oladele, F. A. (1982). Numerical analysis of cuticular characters in Cupressaceae. *In* "The Plant Cuticle" (D. F. Cutler, K. L. Alvin and C. E. Price, eds) pp. 379–396. *Linn. Soc. Symp.* Series 10. Academic Press, London, New York.

Ambrose, J. D. (1980). A re-evaluation of the Melanthioideae (Liliaceae) using numerical analyses. *In* "Petaloid monocotyledons" (C. D. Brickell, D. F. Cutler and M. Gregory, eds) pp. 65–81. Academic Press, London, New York.

Avita, S., Rao, N. V. and Inamdar, J. A. (1981). Studies on the leaf architecture of the Ranunculaceae. *Flora* **171,** 280–298.

Ayensu, E. S. (1972). "Anatomy of the monocotyledons. VI. Dioscoreales", 182 pp. Clarendon Press, Oxford.

Baagøe, J. (1977). Taxonomical application of ligule microcharacters in Compositae. I. Anthemideae, Heliantheae, and Tageteae. *Bot. Tidsskr.* **71,** 193–223.

Baagøe, J. (1978). Taxonomical application of ligule microcharacters in Compositae. II. Arctotideae, Astereae, Calenduleae, Eremothamneae, Inuleae, Liabeae, Mutisieae, and Senecioneae. *Bot. Tidsskr.* **72,** 125–147.

Baagøe, J. (1980). SEM-studies in ligules of Lactuceae (Compositae). *Bot. Tidsskr.* **75,** 199–217.

Baas, P. (1973). The wood anatomical range in *Ilex* (Aquifoliaceae) and its ecological and phylogenetic significance. *Blumea* **21,** 193–258.

Babu, C. R. and Johri, B. M. (1980). Plant systematics – then and now. *In* "Glimpses in

Plant Research. V. Modern Trends in Plant Taxonomy" (P. K. K. Nair, ed.) pp. 315–343. Vikas, India.

Bailey, I. W. (1957). The potentialities and limitations of wood anatomy in the study of the phylogeny and classification of angiosperms. *J. Arnold Arbor.* **38,** 243–254.

Baranova, M. A. (1969). A comparative stomatographic investigation of the genus *Manglietia* Bl. *Bot. Zh. SSSR* **54,** 1952–64. (Russ., Eng. summ.).

Baranova, M. (1972). Systematic anatomy of the leaf epidermis in the Magnoliaceae and some related families. *Taxon* **21,** 447–469.

Baretta-Kuipers, T. (1981). Wood anatomy of Leguminosae: its relevance to taxonomy. *In* "Advances in Legume Systematics" (R. M. Polhill and P. H. Raven, eds) Vol. 2, pp. 677–705. Royal Botanic Gardens, Kew.

Barthlott, W. and Ehler, N. (1977). Raster-Elektronenmikroskopie der Epidermis-Oberflächen von Spermatophyten. *Trop. subtrop. Pflwelt* **19,** 367–467.

Barthlott, W. and Ziegler, B. (1981). Mikromorphologie der Samenschalen als systematisches Merkmal bei Orchideen. *Ber. dt. Bot. Ges.* **94,** 267–273.

Behnke, H.-D. (1971). Sieve-tube plastids of Magnoliidae and Ranunculidae in relation to systematics. *Taxon* **20,** 723–730.

Behnke, H.-D. (1972). Sieve-tube plastids in relation to angiosperm systematics – an attempt towards a classification by ultrastructural analysis. *Bot. Rev.* **38,** 155–197.

Behnke, H.-D. (1975 [1976]). The bases of angiosperm phylogeny: ultrastructure. *Ann. Missouri Bot. Gard.* **62,** 647–663.

Behnke, H.-D. (1976a). Die Siebelement-Plastiden der Caryophyllaceae, eine weitere spezifische Form der P-typ Plastiden bei Centrospermen. *Bot. Jb.* **95,** 327–333.

Behnke, H.-D. (1976b). Ultrastructure of sieve-element plastids in Caryophyllales (Centrospermae), evidence for the delimitation and classification of the order. *Pl. Syst. Evol.* **126,** 31–54.

Behnke, H.-D. (1977). Transmission electron microscopy and systematics of flowering plants. *Pl. Syst. Evol.*, Suppl. 1, 155–178.

Behnke, H.-D. (1981a). Sieve-element characters. *Nordic J. Bot.* **1,** 381–400.

Behnke, H.-D. (1981b). Siebelement-Plastiden, Phloem-Protein und Evolution der Blütenpflanzen: II. Monokotyledonen. *Ber. dt. Bot. Ges.* **94,** 647–662.

Behnke, H.-D. and Dahlgren, R. (1976). The distribution of characters within an angiosperm system. 2. Sieve-element plastids. *Bot. Notiser* **129,** 287–295.

Böcher, T. W. and Olesen, P. (1978). Structural and ecophysiological pattern in the xero-halophytic C4 grass, *Sporobolus rigens* (Tr.) Dsev. *Biol. Skr. Dan. Vid. Selsk.* **22(3),** 1–48.

Bouman, F. (1971a). Integumentary studies in the Polycarpicae. I. Lactoridaceae. *Acta Bot. Neerl.* **20,** 565–569.

Bouman, F. (1971b). The application of tegumentary studies to taxonomic and phylogenetic problems. *Ber. dt. Bot. Ges.* **84,** 169–177.

Bouman, F. (1974). Developmental studies of the ovule, integuments and seed in some angiosperms. Thesis, University of Amsterdam, 180 pp. Los, Naarden.

Bouman, F. (1975). Integument initiation and testa development in some Cruciferae. *Bot. J. Linn. Soc.* **70,** 213–229.

Bouman, F. (1977). Integumentary studies in the Polycarpicae. IV. *Liriodendron tulipifera* L. *Acta Bot. Neerl.* **26,** 213–223.

Bouman, F. (1979). Ovule initiation, ovule development and seed coat structure in angiosperms. *Vistas in Plant Sciences* (T. M. Varghese, ed.) **5,** 1–73.

Bouman, F. and Calis, J. I. M. (1977). Integumentary shifting – a third way to unitegmy. *Ber. dt. Bot. Ges.* **90,** 15–28.

Bouman, F. and Schier, S. (1979). Ovule ontogeny and seed coat development in *Gentiana*, with a discussion on the evolutionary origin of the single integument. *Acta Bot. Neerl.* **28,** 467–478.

Braun, H. J. (1970). "Funktionelle Histologie der sekundären Sprossachse. I. Das Holz", 190 pp. *In* "Handbuch der Pflanzenanatomie", Vol. IX (1). Borntraeger, Berlin, Stuttgart.

Carlquist, S. (1962). A theory of paedomorphosis in dicotyledonous woods. *Phytomorphology* **12,** 30–45.

Carlquist, S. (1969). Wood anatomy of Goodeniaceae and the problem of insular woodiness. *Ann. Missouri Bot. Gard.* **56,** 358–390.

Carlquist, S. (1970 [1971]). Wood anatomy of insular species of *Plantago* and the problem of raylessness. *Bull. Torrey bot. Cl.* **97,** 353–361.

Carlquist, S. (1975a). Wood anatomy of Onagraceae, with notes on alternative modes of photosynthate movement in dicotyledon woods. *Ann. Missouri Bot. Gard.* **62,** 386–424.

Carlquist, S. (1975b). "Ecological Strategies of Xylem Evolution", 259 pp. University of California Press, Berkeley, Los Angeles, London.

Carlquist, S. (1977). Ecological factors in wood evolution: a floristic approach. *Amer. J. Bot.* **64,** 887–896.

Carlquist, S. (1981). Wood anatomy of Pittosporaceae. *Allertonia* **2,** 355–392.

Carolin, R. C. and Jacobs, S. W. L. (1973). The structure of the cells of the mesophyll and parenchymatous bundle sheath of Gramineae. *Bot. J. Linn. Soc.* **66,** 259–275.

Carolin, R. C., Jacobs, S. W. L. and Vesk, M. (1975). Leaf structure in Chenopodiaceae. *Bot. Jb.* **95,** 226–255.

Carolin, R. C., Jacobs, S. W. L. and Vesk, M. (1978). Kranz cells and mesophyll in the Chenopodiales. *Austral. J. Bot.* **26,** 683–698.

Corner, E. J. H. (1976). "The Seeds of Dicotyledons", 2 Vols, 311 pp. and 552 pp. Cambridge University Press, Cambridge.

Cotthem, W. R. J. van (1970). A classification of stomatal types. *Bot. J. Linn. Soc.* **63,** 235–246.

Cotthem, W. R. J. van (1971). Vergleichende morphologische Studien über Stomata und eine neue Klassifikation ihrer Typen. *Ber. dt. Bot. Ges.* **84,** 141–168.

Cutler, D. F. (1969). "Anatomy of the Monocotyledons. IV. Juncales", 357 pp. Clarendon Press, Oxford.

Cutler, D. F. (1982). Cuticular sculpturing and habitat in certain *Aloë* species (Liliaceae) from southern Africa. *In* "The Plant Cuticle" (D. F. Cutler, K. L. Alvin and C. E. Price, eds) pp. 425–444. Academic Press, London, New York.

Dickison, W. C. (1975 [1976]). The bases of angiosperm phylogeny: vegetative anatomy. *Ann. Missouri Bot. Gard.* **62,** 590–620.

Dickison, W. C. (1980). Comparative wood anatomy and evolution of the Cunoniaceae. *Allertonia* **2,** 281–321.

Dolph, G. E. (1978). A proposal for data banking leaf information. *Courier Forschungsinst. Senckenberg* **30,** 159–164.

Ellis, R. P. (1974). The significance of the occurrence of both Kranz and non-Kranz leaf anatomy in the grass species *Alloteropsis semialata*. *S. Afr. J. Sci.* **70,** 169–173.

Ellis, R. P. (1977). Distribution of the Kranz syndrome in the southern African Eragrostoideae and Panicoideae according to bundle sheath anatomy and cytology. *Agroplantae* **9,** 73–110.

Esau, K. (1969). "The Phloem", 505pp. *In* "Handbuch der Pflanzenanatomie", Vol. V(2). Borntraeger, Berlin, Stuttgart.

Fahn, A. (1979). "Secretory Tissues in Plants", 302pp. Academic Press, London, New York.

Fahn, A. and Rachmilevitz, T. (1970). Ultrastructure and nectar secretion in *Lonicera japonica*. *In* "New Research in Plant Anatomy" (N. K. B. Robson, D. F. Cutler and M. Gregory, eds). Academic Press, London, New York; *Bot. J. Linn. Soc.* **63,** Suppl. 1, 51–56.

Fischer, A. (1884). "Untersuchungen über das Siebrohren-System der Cucurbitaceen". Borntraeger, Berlin, Stuttgart.

French, J. C. and Tomlinson, P. B. (1980). Preliminary observations on the vascular system in stems of certain Araceae. *In* "Petaloid Monocotyledons" (C. D. Brickell, D. F. Cutler and M. Gregory, eds) pp. 105–116. Academic Press, London, New York.

French, J. C. and Tomlinson, P. B. (1981a). Vascular patterns in stems of Araceae: subfamily Pothoideae. *Amer. J. Bot.* **68,** 713–729.

French, J. C. and Tomlinson, P. B. (1981b). Vascular patterns in stems of Araceae: subfamilies Calloideae and Lasioideae. *Bot. Gaz.* **142,** 366–381.

French, J. C. and Tomlinson, P. B. (1981c). Vascular patterns in stems of Araceae: subfamily Monsteroideae. *Amer. J. Bot.* **68,** 1115–1129.

French, J. C. and Tomlinson, P. B. (1981d). Vascular patterns in stems of Araceae: subfamily Philodendroideae. *Bot. Gaz.* **142,** 550–563.

Fryns-Claessens, E. and Cotthem, W. van (1973). A new classification of the ontogenetic types of stomata. *Bot. Rev.* **39,** 71–138.

Gibson, A. C. (1973). Comparative anatomy of secondary xylem in Cactoideae (Cactaceae). *Biotropica* **5,** 29–65.

Gibson, A. C. (1976). Vascular organization in shoots of Cactaceae. I. Development and morphology of primary vasculature in Pereskioideae and Opuntioideae. *Amer. J. Bot.* **63,** 414–426.

Gibson, A. C. (1977). Wood anatomy of Opuntias with cylindrical to globular stems. *Bot. Gaz.* **138,** 334–351.

Gottwald, H. (1977). The anatomy of secondary xylem and the classification of ancient dicotyledons. *Pl. Syst. Evol.*, Suppl. 1, 111–121.

Gottwald, H. and Parameswaran, N. (1968). Das sekundäre Xylem und die systematische Stellung der Ancistrocladaceae und Dioncophyllaceae. *Bot. Jb.* **88,** 49–69.

Graaff, N. A. van der and Baas, P. (1974). Wood anatomical variation in relation to latitude and altitude. *Blumea* **22,** 110–121.

Guédès, M. (1972). Contribution à la morphologie du phyllome. *Mém. Mus. natn. Hist. nat., Paris (N.S.)* sér. B, **21,** 1–179.

Guédès, M. (1980). Architecture de la feuille équitante d'Iris. *C. r. Acad. Sci., Paris, D.* **290,** 131–134.

Guignard, J.-L. (1975). Embryogénie et classification embryogénique. *Bull. Soc. bot. Fr.* **122,** 281–294.
Gunning, B. E. S. and Pate, J. S. (1969). "Transfer cells". Plant cells with wall ingrowths, specialized in relation to short distance transport of solutes – their occurrence, structure and development. *Protoplasma* **68,** 107–133.
Gunning, B. E. S., Pate, J. S. and Briarty, L. G. (1968). Specialized "transfer cells" in minor veins of leaves and their possible significance in phloem translocation. *J. Cell Biol.* **37(3),** C7-C12.
Gunning, B. E. S., Pate, J. S. and Green, L. W. (1970). Transfer cells in the vascular system of stems: taxonomy, association with nodes, and structure. *Protoplasma* **71,** 147–171.
Guttenberg, H. von (1968). "Der primäre Bau der Angiospermenwurzel" (2nd edn) 472pp. *In* "Handbuch der Pflanzenanatomie", Vol. VIII (5). Borntraeger, Berlin, Stuttgart.
Guttenberg, H. von (1971). "Bewegungsgewebe und Perzeptionsorgane", 332pp. *In* "Handbuch der Pflanzenanatomie", Vol. V (5). Borntraeger, Berlin, Stuttgart.
Hatch, M. D. and Slack, C. R. (1966). Photosynthesis by sugar-cane leaves. A new carboxylation reaction and the pathway of sugar formation. *Biochem. J.* **101,** 103–111.
Hatch, M. D. and Slack, C. R. (1970). The C4-dicarboxylic acid pathway of photosynthesis. *Progress in Phytochemistry* **2,** 35–106.
Heslop-Harrison, Y. (1981). Stigma characteristics and angiosperm taxonomy. *Nordic J. Bot.* **1,** 401–420.
Heslop-Harrison, Y. and Shivanna, K. R. (1977). The receptive surface of the angiosperm stigma. *Ann. Bot.* **41,** 1233–1258.
Hickey, L. J. (1973). Classification of the architecture of dicotyledonous leaves. *Amer. J. Bot.* **60,** 17–33.
Hickey, L. J. (1979 [1980]). A revised classification of the architecture of dicotyledonous leaves. *In* "Anatomy of the Dicotyledons" (C. R. Metcalfe and L. Chalk, eds) (2nd edn), Vol. I, pp. 25–39. Clarendon Press, Oxford.
Hickey, L. J. and Wolfe, J. A. (1975 [1976]). The bases of angiosperm phylogeny; vegetative morphology. *Ann. Missouri Bot. Gard.* **62,** 538–589.
Howard, R. A. (1970). Some observations on the nodes of woody plants with special reference to the problem of the "split-lateral" versus the "common gap". *In* "New Research in Plant Anatomy" (N. K. B. Robson, D. F. Cutler and M. Gregory, eds) pp. 195–214. Academic Press, London, New York.
Howard, R. A. (1974). The stem-node-leaf continuum of the Dicotyledoneae. *J. Arnold Arbor.* **55,** 125–181.
Howard, R. A. (1979 [1980]). The stem-node-leaf continuum of the Dicotyledoneae. *In* "Anatomy of the Dicotyledons" (C. R. Metcalfe and L. Chalk, eds) (2nd edn), Vol. I, pp. 76–87. Clarendon Press, Oxford.
Howard, R. A. (1979 [1980]). The petiole. *In* "Anatomy of the Dicotyledons" (C. R. Metcalfe and L. Chalk, eds) (2nd edn), Vol. I, pp. 86–96. Clarendon Press, Oxford.
IAWA Committee (1981). Standard list of characters suitable for computerized hardwood identification. *Intn. Ass. Wood Anat. Bull.* (NS) **2,** 99–145.
Kapil, R. N., Bor, J. and Bouman, F. (1980). Seed appendages in angiosperms. I. Introduction. *Bot. Jb.* **101,** 555–573.

Kaplan, D. R. (1970a). Comparative development and morphological interpretation of "rachis-leaves" in Umbelliferae. *In* "New Research in Plant Anatomy" (N. K. B. Robson, D. F. Cutler and M. Gregory, eds) pp. 101–125. Academic Press, London, New York.

Kaplan, D. R. (1970b). Comparative foliar histogenesis in *Acorus calamus* and its bearing on the phyllode theory of monocotyledonous leaves. *Amer. J. Bot.* **57,** 331–361.

Kaplan, D. R. (1973). The monocotyledons: their evolution and comparative biology: VII. The problem of leaf morphology and evolution in the monocotyledons. *Quart. Rev. Biol.* **48,** 437–457.

Kaplan, D. R. (1975). Comparative developmental evaluation of the morphology of unifacial leaves in the monocotyledons. *Bot. Jb.* **95,** 1–105.

Kaplan, D. R. (1980). Heteroblastic leaf development in *Acacia*. Morphological and morphogenetic implications. *Cellule* **73,** 135–203.

Kay, Q. O. N., Daoud, H. S. and Stirton, C. H. (1981). Pigment distribution, light reflection and cell structure in petals. *Bot. J. Linn. Soc.* **83,** 57–84.

Koek-Noorman, J. (1969a). A contribution to the wood anatomy of South American (chiefly Suriname) Rubiaceae. I. *Acta Bot. Neerl.* **18,** 108–123.

Koek-Noorman, J. (1969b). A contribution to the wood anatomy of South American (chiefly Suriname) Rubiaceae. II. *Acta Bot. Neerl.* **18,** 377–395.

Koek-Noorman, J. (1970). A contribution to the wood anatomy of the Cinchoneae, Coptosapelteae and Naucleeae (Rubiaceae). *Acta Bot. Neerl.* **19,** 154–164.

Koek-Noorman, J. (1972). The wood anatomy of Gardenieae, Ixoreae and Mussaendeae (Rubiaceae). *Acta Bot. Neerl.* **21,** 301–320.

Koek-Noorman, J. (1977). Systematische Holzanatomie einiger Rubiaceen. *Ber. dt. Bot. Ges.* **90,** 183–190.

Koek-Noorman, J. and Hogeweg, P. (1974). The wood anatomy of Vanguerieae, Cinchoneae, Condamineae and Rondeletieae (Rubiaceae). *Acta Bot. Neerl.* **23,** 627–653.

Laetsch, W. M. (1969). Relationship between chloroplast structure and photosynthetic carbon-fixation pathways. *Sci. Progr.* **57,** 323–351.

Ly Thi Ba (1981a). Embryogénie comparée et phylogénie des Ranales et des Hélobiales. I. Les Ranales arborescentes (Magnoliales). *Rev. gén. Bot.* **88,** 43–82.

Ly Thi Ba (1981b). *Idem.* II. Les Ranales herbacées (Ranunculales). *Rev. gén. Bot.* **88,** 105–197.

Ly Thi Ba (1981c). *Idem.* III. Les Hélobiales. *Rev. gén. Bot.* **88,** 201–252.

Ly Thi Ba (1981d). *Idem.* IV. Les conclusions. *Rev. gén. Bot.* **88,** 347–373.

Mabberley, D. J. (1974). Pachycauly, vessel-elements, islands and the evolution of arborescence in 'herbaceous' families. *New Phytol.* **73,** 977–984.

Mabberley, D. J. (1982). On Dr Carlquist's defence of paedomorphosis. *New Phytol.* **90,** 751–755.

Melville, R. (1976). The terminology of leaf architecture. *Taxon* **25,** 549–561.

Mennega, A. M. W. (1980). Anatomy of the secondary xylem. *In* "Die natürlichen Pflanzenfamilien" (A. Engler and K. Prantl, eds) (2nd edn) **28b,** 1, pp. 112–161, Angiospermae: Ordnung Gentianales. Fam. Loganiaceae. Duncker & Humbolt, Berlin.

Metcalfe, C. R. (1968). Current developments in systematic plant anatomy. *In* "Modern

Methods in Plant Taxonomy" (V. H. Heywood, ed.) pp. 45–57. Academic Press, London, New York.
Metcalfe, C. R. (1971). "Anatomy of the Monocotyledons. V. Cyperaceae", 597pp. Clarendon Press, Oxford.
Metcalfe, C. R. and Chalk, L. (1950). "Anatomy of the Dicotyledons", 2 vols. Clarendon Press, Oxford.
Metcalfe, C. R. and Chalk, L. (1979 [1980]). "Anatomy of the Dicotyledons. I. Systematic Anatomy of Leaf and Stem, with a Brief History of the Subject" (2nd edn), 276pp. Clarendon Press, Oxford.
Miller, R. B. (1981). Explanation of coding procedure. *IAWA Bull.* **2,** 111–145.
Napp-Zinn, K. (1966). "Anatomie des Blattes. I. Blattanatomie der Gymnospermen", 369pp. *In* "Handbuch der Pflanzenanatomie", Vol. VIII (1). Borntraeger, Berlin, Stuttgart.
Napp-Zinn, K. (1973, 1974). "Anatomie des Blattes. II. Blattanatomie der Angiospermen. A. Entwicklungsgeschichtliche und topographische Anatomie des Angiospermenblattes", Vols 1 and 2, 1424pp. Borntraeger, Berlin, Stuttgart.
Neubauer, H. F. (1981). Der Knotenbau einiger Rubiaceae. *Pl. Syst. Evol.* **139,** 103–111.
Oever, L. van den, Baas, P. and Zandee, M. (1981). Comparative wood anatomy of *Symplocos* and latitude and altitude of provenance. *Int. Ass. Wood Anat. Bull.* (NS) **2,** 3–24.
Paliwal, G. S. (1969). Stomatal ontogeny and phylogeny. I. Monocotyledons. *Acta Bot. Neerl.* **18,** 654–668.
Palser, B. F. (1975 [1976]). The bases of angiosperm phylogeny: embryology. *Ann. Missouri Bot. Gard.* **62,** 621–646.
Pate, J. S., Gunning, B. E. S. and Milliken, F. F. (1970). Function of transfer cells in the nodal regions of stems, particularly in relation to the nutrition of young seedlings. *Protoplasma* **71,** 313–334.
Patel, J. D. (1978). How should we interpret and distinguish subsidiary cells? *Bot. J. Linn. Soc.* **77,** 65–72.
Payne, W. W. (1970). Helicocytic and allelocytic stomata: unrecognized patterns in the Dicotyledonae. *Amer. J. Bot.* **57,** 140–147.
Payne, W. W. (1979). Stomatal patterns in embryophytes: their evolution, ontogeny and interpretation. *Taxon* **28,** 117–132.
Philipson, W. R. (1974). Ovular morphology and the major classification of the dicotyledons. *Bot. J. Linn. Soc.* **68,** 89–108.
Philipson, W. R. (1975). Evolutionary lines within the dicotyledons. *New Zealand J. Bot.* **13,** 73–91.
Philipson, W. R. (1977). Ovular morphology and the classification of dicotyledons. *Pl. Syst. Evol.*, Suppl. 1, 123–140.
Rao, T. A. and Bhattacharya, J. (1978 [1979]). A review of foliar sclereids in angiosperms. *Bull. bot. Surv. India* **20,** 91–99.
Rao, T. A. and Das, S. (1979). Leaf sclereids – occurrence and distribution in the angiosperms. *Bot. Notiser* **132,** 319–324.
Rasmussen, H. (1981a). The diversity of stomatal development in Orchidaceae subfamily Orchidoideae. *Bot. J. Linn. Soc.* **82,** 381–393.

Rasmussen, H. (1981b). Terminology and classification of stomata and stomatal development – a critical survey. *Bot. J. Linn. Soc.* **83,** 199–212.

Renvoize, S. A. (1981). The subfamily Arundinoideae and its position in relation to a general classification of the Gramineae. *Kew Bull.* **36,** 85–102.

Richter, H. G. (1981). "Die Anatomie des sekundären Xylems und der Rinde der Lauraceae". Paul Parey, Berlin.

Roth, I. (1977). "Fruits of Angiosperms", 675pp. *In* "Handbuch der Pflanzenanatomie", Vol. IX (1). Borntraeger, Berlin, Stuttgart.

Roth, I. (1981). "Structural Patterns of Tropical Barks", 609pp. *In* "Handbuch der Pflanzenanatomie", Vol. IX (3). Borntraeger, Berlin, Stuttgart.

Schulze, W. (1980a). Beiträge zur Taxonomie der Liliifloren. V. Alliaceae. *Wiss. Z. Friedrich-Schiller-Univ., math.-naturw. Reihe* **29,** 595–606.

Schulze, W. (1980b). Beiträge zur Taxonomie der Liliifloren. VI. Der Umfang der Liliaceae. *Wiss. Z. Friedrich-Schiller-Univ., math.-naturw. Reihe* **29,** 607–636.

Singh, H. (1978). "Embryology of Gymnosperms", 302pp. *In* "Handbuch der Pflanzenanatomie", Vol. X (2). Borntraeger, Berlin, Stuttgart.

Sporne, K. R. (1969). The ovule as an indicator of evolutionary status in angiosperms *New Phytol.* **68,** 555–566.

Stace, C. A. (1965). Cuticular studies as an aid to plant taxonomy. *Bull. Brit. Mus. (Nat. Hist.) Bot.* **4(1),** 3–78.

Stevens, R. A. and Martin, E. S. (1978). A new ontogenetic classification of stomatal types. *Bot. J. Linn. Soc.* **77,** 53–64.

Stirton, C. H. (1981). Petal sculpturing in papilionoid legumes. *In* "Advances in Legume Systematics" (R. M. Polhill and P. H. Raven, eds) pp. 771–788. Royal Botanic Gardens, Kew.

Theobald, W. L., Krahulik, J. L. and Rollins, R. C. (1979 [1980]). Trichome description and classification. *In* "Anatomy of the Dicotyledons" (C. R. Metcalfe and L. Chalk, eds) (2nd edn), Vol. I, pp. 40–53. Clarendon Press, Oxford.

Tomlinson, P. B. (1969). "Anatomy of the Monocotyledons. III. Commelinales – Zingiberales", 446pp. Clarendon Press, Oxford.

Tomlinson, P. B. (1974). Development of the stomatal complex as a taxonomic character in the monocotyledons. *Taxon* **23,** 109–128.

Tomlinson, P. B. (1982). "Anatomy of the Monocotyledons. VII. Helobiae." Oxford University Press, New York, London.

Tregunna, E. B., Smith, B. N., Berry, J. A. and Downton, W. J. S. (1970). Some methods for studying the photosynthetic taxonomy of the angiosperms. *Canad. J. Bot.* **48,** 1209–1214.

Vliet, G. J. C. M. van (1975). Wood anatomy of Crypteroniaceae *sensu lato*. *J. Microsc.* **104,** 65–82.

Vliet, G. J. C. M. van (1976). Wood anatomy of the Rhizophoraceae. *In* "Wood Structure in Biological and Technological Research" (P. Baas, A. J. Bolton and D. M. Catling, eds) pp. 20–75. *Leiden Botanical Series*, No. 3. Leiden University Press.

Vliet, G. J. C. M. van (1978). Vestured pits of Combretaceae and allied families. *Acta Bot. Neerl.* **27,** 273–285.

Vliet, G. J. C. M. van (1979). Wood anatomy of the Combretaceae. *Blumea* **25,** 141–223.

Vliet, G. J. C. M. van (1981). Wood anatomy of the palaeotropical Melastomataceae. *Blumea* **27,** 395–462.

Vliet, G. J. C. M. van and Baas, P. (1982). Wood anatomy and classification of the Myrtales. *Ann. Missouri Bot. Gard.* **69,** 1–22.

Vliet, G. J. C. M. van, Koek-Noorman, J. and Welle, B. J. H. ter (1981). Wood anatomy, classification and phylogeny of the Melastomataceae. *Blumea* **27,** 463–473.

Watson, L., Pate, J. S. and Gunning, B. E. S. (1977). Vascular transfer cells in leaves of Leguminosae-Papilionoideae. *Bot. J. Linn. Soc.* **74,** 123–130.

Watson, L. and Dallwitz, M. J. (1981). An automated data bank for grass genera. *Taxon* **30,** 424–429 (and microfiche).

Welle, B. J. H. ter and Koek-Noorman, J. (1981). Wood anatomy of the neotropical Melastomataceae. *Blumea* **27,** 335–394.

Wilkinson, H. P. (1979 [1980]). The plant surface (mainly leaf). *In* "Anatomy of the Dicotyledons" (C. R. Metcalfe and L. Chalk, eds) (2nd edn), Vol. I, pp. 97–165. Clarendon Press, Oxford.

Zimmermann, M. H. and Tomlinson, P. B. (1972). The vascular system of monocotyledonous stems. *Bot. Gaz.* **133,** 141–155.

Zimmermann, M. H., Tomlinson, P. B. and Leclaire, J. (1974). Vascular construction and development in the stems of certain Pandanaceae. *Bot. J. Linn. Soc.* **68,** 21–41.

8 | Pollen Features and Plant Systematics

STEPHEN BLACKMORE

Department of Botany, British Museum (Natural History), London, England

Abstract: Current palynological methods and approaches relevant to plant systematics are reviewed with special reference to examples from tribe Lactuceae (Compositae). The role of scanning electron microscopy (SEM) in palynology is discussed together with associated techniques such as sectioning and ion beam etching which can reveal details of exine morphology and ultrastructure usually studied by transmission electron microscopy (TEM). A broader-based approach becoming prevalent in systematic palynological studies may contribute to the rationalization of the cumbersome terminology of the subject and to an improved understanding of pollen characters. In this approach the implications of investigations into ontogeny, function, morphology of untreated pollen and fundamental exine structures are considered in the interpretation of pollen characters.

INTRODUCTION

The history of palynology is almost as long as that of microscopy. The taxonomic applications of the subject predate pollen analysis but, despite such early demonstrations of the value of pollen characters as the work of Brown (1811) on Proteaceae, they did not become widely adopted until some years after Erdtman (1934) devised the acetolysis technique. This method of preparation reveals exine morphology in greater detail and produces more permanent preparations with relative ease. However, it also removes both the cellular contents and the intine layer of the pollen wall and may produce unnatural shaped pollen grains, especially when delicate apertural regions rupture. In the period before the acetolysis technique was developed, which ended with

Systematics Association Special Volume No. 25, "Current Concepts in Plant Taxonomy", edited by V. H. Heywood and D. M. Moore, 1984. Academic Press, London and Orlando.
ISBN 0 12 347060 9

the classic text of Wodehouse (1935), pollen grains were prepared in a variety of ways, including mounting in water or glycerine jelly or drying in air, but were usually examined intact. In studying whole pollen grains the early palynologists were frequently inclined to draw conclusions about the functions of the structures they observed.

In some ways systematic palynology has now come full circle so that, although acetolysis remains the most important technique, the advantages of examining whole pollen grains are again being appreciated. This has come about largely because of the stimulating research into pollen physiology and ontogeny by Heslop-Harrison (1968), Rowley (1976), Dickinson (1976) and others. The result is that current research in systematic palynology is much broader-based, using a wider variety of techniques and making full use of electron microscopy. In this chapter I will discuss some recently developed techniques and outline some of these broader aspects of palynology which have an important bearing on the taxonomic use of pollen characters.

Not the least of the benefits of this broader approach is that it may provide an opportunity for the unnecessarily complicated terminology of palynology to be rationalized since it is increasingly important to be able to compare findings from morphological, ontogenetic, physiological and palaeobotanical studies. Furthermore, now that electron microscopy has been employed by palynologists for some years the proliferation of new terms and the complication of existing ones, which took place with the advent of the electron microscope, has now almost ceased. Discussions at an international level through an International Commission for Palynology working group have achieved some promising steps in the right direction (Nilsson and Muller 1978). Hopefully it will eventually prove possible to achieve a state where descriptive palynological works can be understood much more readily by the non-specialist, thus making pollen characters more accessible to the plant taxonomist.

SOME RECENT TECHNIQUES IN PALYNOLOGY

So many techniques are now used to study pollen grains that it is impossible to consider them all here and I intend instead to discuss mainly those relating to scanning electron microscopy.

Electron microscopy had an enormous impact on palynology, enabling features which were barely visible with light microscopy to be interpreted accurately. The meticulous study of Compositae exines by Stix (1960), for example, pushed light microscopy to its limits but was overshadowed only a few years later by the transmission electron microscopic work of Skvarla and

Larsen (1965) and Skvarla and Turner (1966). However, as Heywood (1968) predicted, the scanning electron microscope has proved even more useful an instrument to palynologists primarily concerned with taxonomy. Indeed with the variety of preparation techniques now available and the much improved resolution of the latest models it can now furnish virtually all the morphological data that transmission electron microscopes can, often with less effort and with results which are easier to interpret. This has been made possible by a variety of techniques for fracturing or sectioning pollen grains, most of which have been discussed by Hideux (1972) or Hideux and Marceau (1972). Some of the methods used to fracture pollen grains are extremely simple but very effective. Van Campo and Sivak (1972), for instance, used fine glass needles to tease off the air sacks of gymnosperm pollen so that their internal surfaces could be examined. Sectioning techniques generally give better results because of the cleanly cut surfaces they yield. Freezing microtomes provide one of the best means of sectioning pollen for SEM and, if the pollen is embedded in a medium such as gelatin (Blackmore and Dickinson 1981), can reveal at least as much information as TEM. This method has been used in a palynological survey of tribe Lactuceae (Blackmore 1981, 1982) and enables fine details such as the minute spaces, termed internal foraminae, within the ektexine elements of *Catananche caerulea* pollen (Figs 1–2) to be observed. This means that the "Anthemoid", "Helianthoid" and "Senecioid" patterns of exine described by

(*Figs 1–10 overleaf*)

Figs 1–4. Scanning and transmission electron micrographs of *Catananche caerulea* L. pollen. Fig. 1, transmission electron micrograph of acetolysed exine, post-fixed with osmium tetroxide, showing internal foraminae in ektexine. Fig. 2, scanning electron micrograph showing internal foraminae in ektexine. Fig. 3, detail of section similar to Fig. 2, showing boundary of foot layer and endexine (arrow), pitted endosculpture (e), and the apparently flexible columellae (c) traversing the cavus (ca). Fig. 4, detail of section similar to Fig. 2 showing internal foraminae and the rounded, free ends of columellae (arrowed) below the internal tectum (it). Scale lines all 5 μm.

Figs 5–10. Scanning electron micrographs of acetolysed *Scorzonera graminifolia* L. pollen. Fig. 5, equatorial view with an aperture at the right. Fig. 6, section of exine to show arrangement of columellae, note the pitted endosculpture (e). Fig. 7, detail of sculpture showing anastomosing ridges and lacunar floors with microperforate tectum. Fig. 8, area comparable to that in Fig. 7 after ion beam etching for 30 min. Fig. 9, detail of Fig. 8 showing feature revealed: large channels below spines (s), columellae, internal tectum (it) and patches of exposed foot layer (f). Fig. 10, area comparable to Fig. 8 showing four layers in the lacuna at the left: tectum, columellae, internal tectum and foot layer. Scale lines all 5 μm.

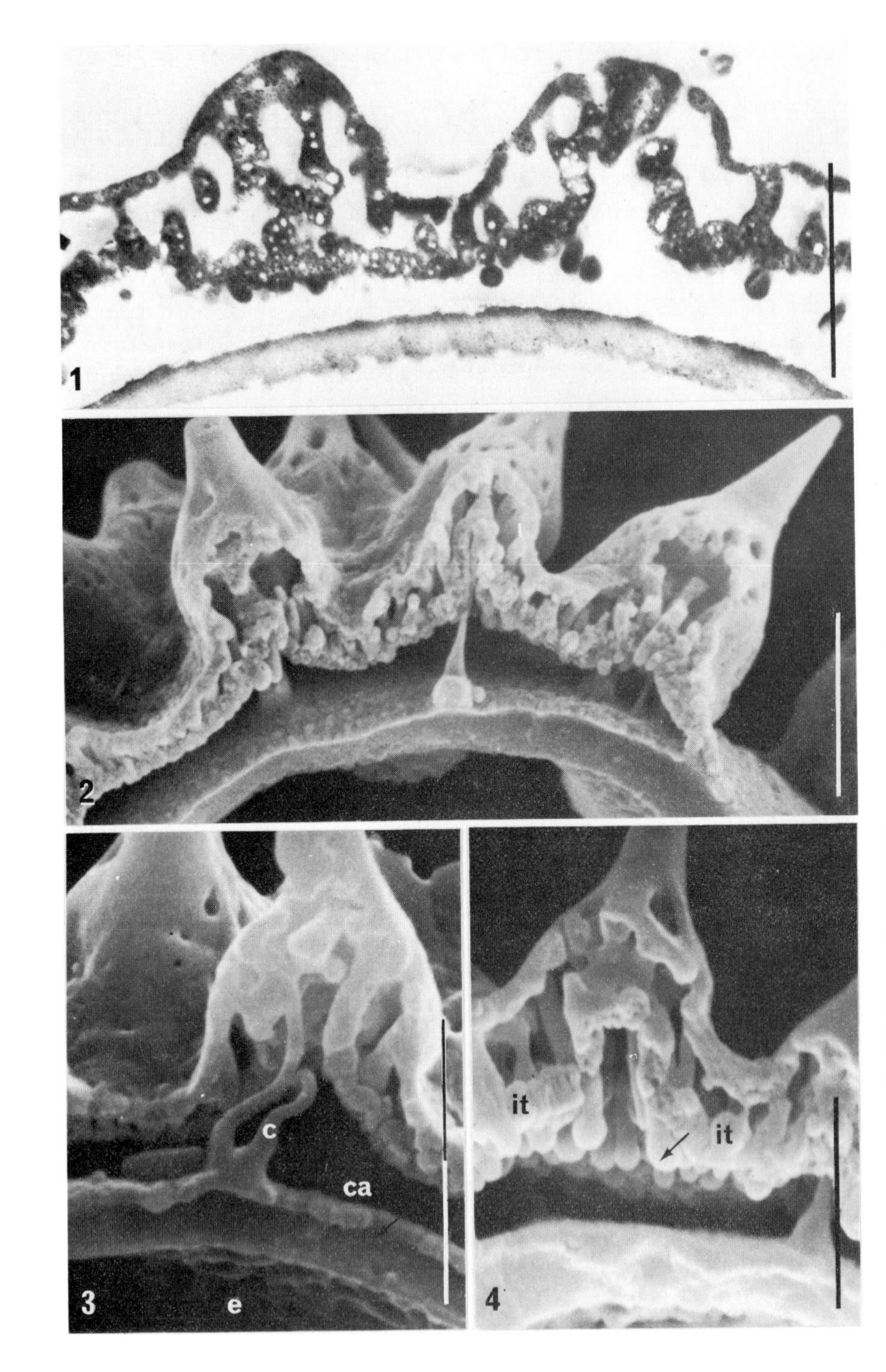
1
2
3
c
ca
e
4
it
it

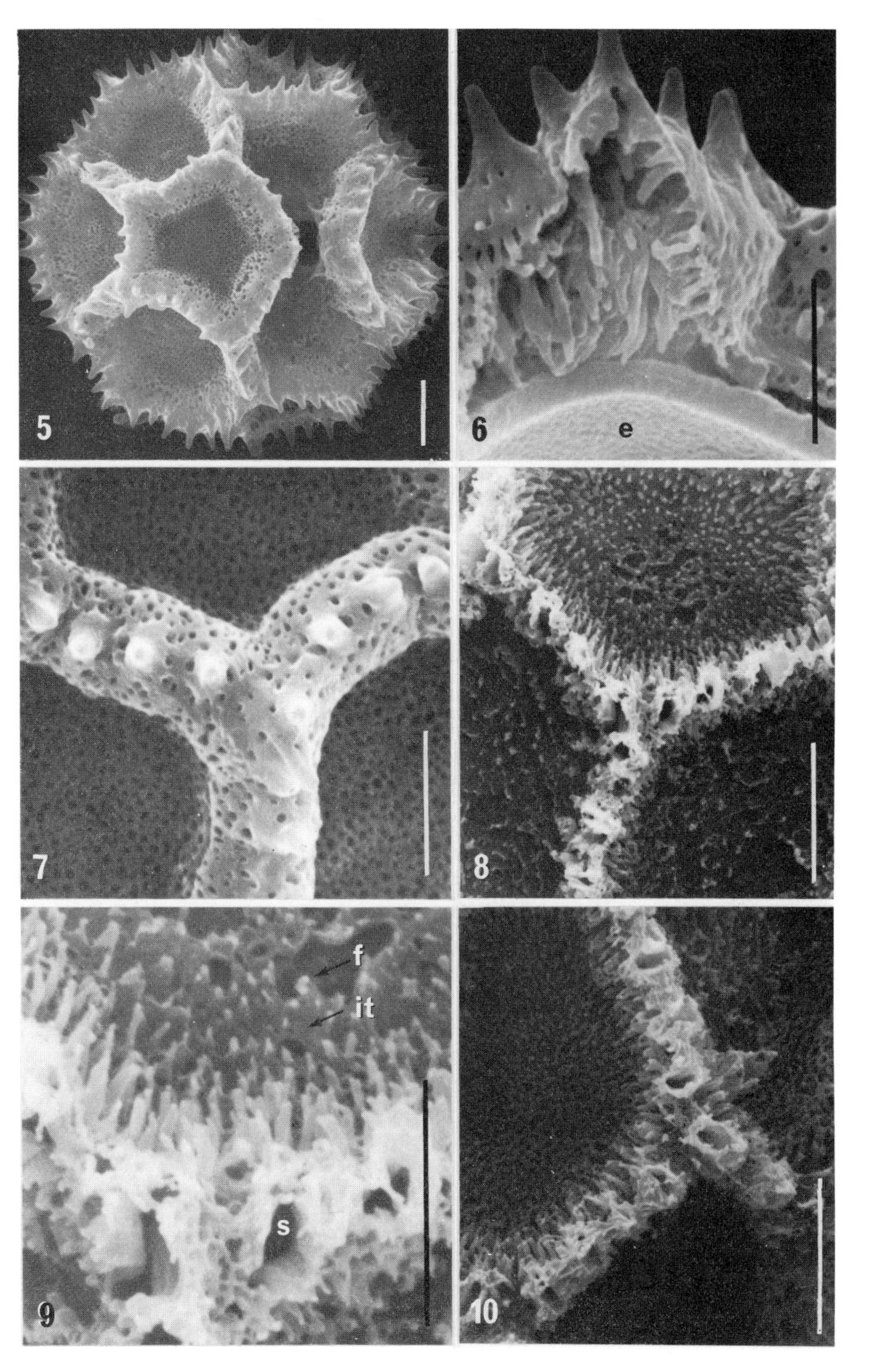
5
6
e
7
8
f
it
s
9
10

Skvarla *et al.* (1977) on the basis of TEM can be distinguished by SEM. The boundary between the endexine and ektexine is frequently seen (Fig. 3) in such sections, perhaps because of a difference in hardness between the two layers. It is thus possible to distinguish between the layers without using the staining techniques by which they were originally recognized. Furthermore in interpreting the complex exines of Compositae pollen SEM investigations show the spatial arrangements of the exine components much more clearly because of the great depth of field available with the instrument. Details of the inner surface of the endexine, or endosculpture (Van Campo 1978), can provide useful taxonomic characters when observed by scanning rather than TEM. The endosculpturing of many Lactuceae pollen grains consists of a fine pitting (Fig. 6) which is more pronounced towards the apertures and is particularly distinct in echinolophate (i.e. having spines borne on ridges) pollen.

Another technique which is particularly useful for studying complex exines consists of eroding the pollen surface by means of an ion beam etching device (Barthlott *et al.* 1976). This can be likened in effect to a very fine sand blaster capable of eroding the specimen surface by a few micrometres at a time. Trials of this method (Figs 5–10) show its potential as a means of dissecting or exposing exine structures which are otherwise only apparent in occasional tangential sections where they are difficult to interpret. One weakness inherent in this method is that, as in sand blasting, softer regions are eroded faster than hard and a misleading surface relief could be formed. In practice, however, it seems that exines, which are composed mainly of sporopollenin, are fairly uniform in hardness. It is thus possible, in Lactuceae pollen, to remove the outer layer, or tectum, exposing the rod like columellae below (Fig. 9). The morphology of the columellae, which in Lactuceae pollen may be a simple or complex anastomozing structure, is probably better studied by sectioning (Fig. 6) but ion beam etching permits their density and relationship to the spine channels (Fig. 9) to be observed. This technique is probably most suited to pollen grains which have an almost continuous or unperforated tectum since the tendency to produce artifacts would be much greater in pollen with an open reticulum.

Another method for observing very cleanly cut sections by SEM, which is well suited to the study of small quantities of material, has been described by Ferguson (1978). The acetolysed pollen grains are embedded in epoxy resin and sectioned by ultramicrotomy for TEM. After this the surface of the resin block containing the pollen is transferred to a SEM stub and the resin is partially dissolved away by rinsing in sodium methoxide. The exposed sectioned exine surface can then be coated and examined by SEM permitting a direct comparison between the results obtained with the two different instruments.

Freeze drying of either fresh or acetolysed pollen grains prior to SEM can be of advantage in certain cases. It can be used, for example, to prevent thin-walled pollen grains from collapsing in the high vacuums used for coating and microscopy. Similarly, critical point drying can be used to much the same effect and is routine in SEM of biological specimens which do not have such hard outer surfaces as pollen grains. Pacqué (1980) studied Passifloraceae pollen critical point dried from various states of hydration to demonstrate the operation of the remarkable range of opercular structures in the family. One problem with this method of preparation is that a large proportion of the sample may be lost during preparation. Nagarajan and Bates (1981) have described a simple method for bonding pollen grains to a coated glass cover slip before critical point drying to overcome this problem.

CURRENT APPROACHES IN PALYNOLOGY

(1) Pollen Floras

Although Erdtman (1970) launched an ambitious series of palynological monographs entitled "The World Pollen Flora", which has been continued as the "World Pollen and Spore Flora" (Henrickson 1973), this did not constitute a Flora in the usual sense, since keys were not always included and not all species were studied in the families covered. More recently (Janssen *et al.* 1976) work has started on a "Northwest European Pollen Flora" and accounts of 28 families have now been published. This project undertakes, for the first time, to provide a comprehensive account of the palynology of all the species within the geographical range of the Flora. Keys and ample illustrations are provided to enable isolated pollen grains belonging to plants of the families covered to be identified and each account provides useful data for taxonomic purposes. The geographical area of the Flora is, palaeopalynologically, the most intensively studied region of the world and thus there is considerable demand for such a detailed work. It is to be hoped that comparable pollen Floras will follow for other regions of the world not least because of the data they provide for taxonomists. It is interesting that some Floras include palynological data, as for example in the account of Gentianaceae by Maguire (1981). Although the inclusion of such data would be impractical in many Floras and would make publishing costs prohibitive it would certainly be worthwhile where the information is available and space permits.

(2) Ontogenetic studies

There have now been numerous studies of pollen ontogeny, ranging from

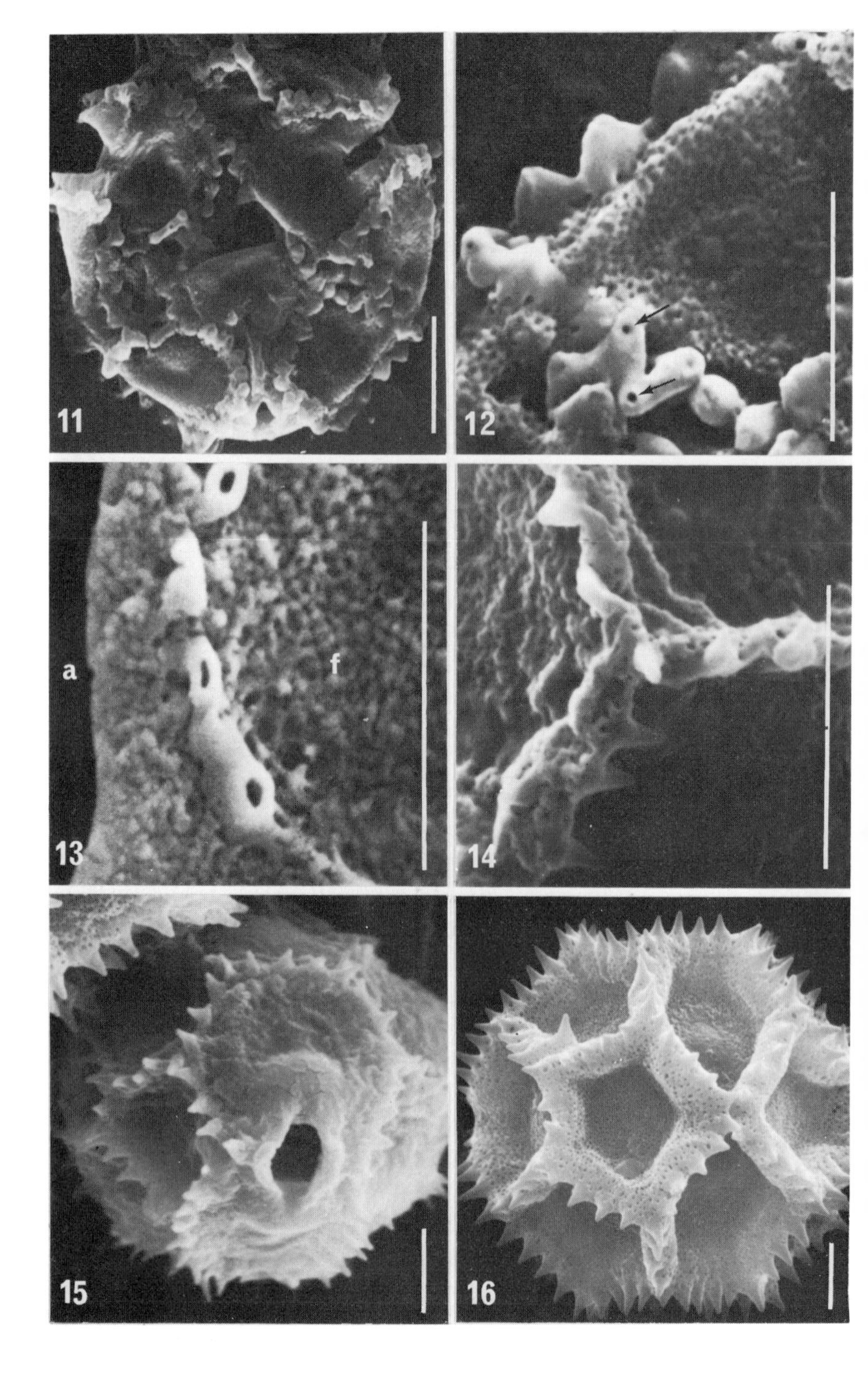
11
12
a
f
13
14
15
16

highly detailed investigations such as that of Rowley and Dahl (1977) on *Artemisia* to short studies designed to reveal the major developmental processes in a species and determine whether it fits into any of the documented sequences of pollen ontogeny. The importance of such studies to systematic palynology is that they give a valuable insight into the homology of features used as pollen characters and explain how the variations found in some characters arise. The main tool in such research is the transmission electron microscope but once again scanning electron microscopy has an increasing part to play. In ontogenetic studies of Liliaceae pollen, for example, Dickinson (1976) used SEM to demonstrate the development of the nexine of *Lilium* and Takahashi (1980) the development of the reticulate ornamentation. Observations of weakly acetolysed, immature *Scorzonera hispanica* pollen grains (Figs 11–16) reveal details that are not visible in light microscopy of the same preparations. The final pattern of ridges, which enables this pollen to be recognized as that of a *Scorzonera* species, is clearly visible while the developing pollen grains are still in tetrads. The first sporopollenous, or at least acetolysis resistant, parts of the exine to appear are the foot layer and part of the structure of the spine channels (Figs 12–13) although the condition of the endexine, which would be visible by sectioning, has not been observed at this stage. The columellae and tectum are only apparent in more mature pollen (Figs 14–16) with those of the ridges seeming to develop before those of the lacunae.

A detailed study of the processes involved in development in this species and *S. humilis*, which has strikingly different pollen morphology, is underway and will enable a more complete understanding. The brief SEM survey demonstrates that the pattern of ridges, which is an important source of characters in the Lactuceae (Blackmore 1982), is established at a very early stage and remains unchanged despite considerable growth of the pollen grains during maturation. Whilst the tectum and columellae of the ridges develop at quite an early stage those of the lacunae appear much later and this may account for one of the

Figs 11–16. Scanning electron micrographs of acetolysed *Scorzonera hispanica* L. pollen. Fig. 11, polar view of collapsed grain immediately after separation from tetrad, the final pattern of ridges is already apparent. Fig. 12, detail from Fig. 11 showing the developing ektexine elements of the ridges. The openings (arrows) appear to correspond to the channels below the spines of mature pollen. Fig. 13, surface of slightly more mature pollen with part of an aperture (a), a ridge and a granular region (f) which is probably the developing foot layer. Fig. 14, ridges of more mature pollen with spines. Fig. 15, apertural view of pollen grain more mature than Fig. 14 but with perforate tectum not yet apparent. Fig. 16, mature pollen grain. Scale lines all 5 μm.

characters which distinguishes *Scorzonera* from the closely related genus *Tragopogon*. *Tragopogon* pollen, in all species so far studied, lacks these last-formed layers in the lacunae and one explanation of this difference is that the development of the pollen may be arrested at an earlier stage in *Tragopogon*. Thus, here it may be possible to detect differences in ontogeny which result in different morphologies and, in this case, a character for distinguishing genera. There is evidence (Blackmore unpublished) that the resulting differences are closely correlated to the means by which the pollen grains adapt to changes in volume caused by changes in hydration. One further point demonstrated by this study is that it is necessary to take very small buds indeed if immature pollen grains are to be found. There is therefore little need for concern that the pollen samples used in routine preparations for taxonomic studies are likely to be immature.

Dickinson (1976) and Rowley (1981) have shown that even differing pollen ontogenies may have a number of common features. Whilst it would be an oversimplification to suggest, for example, that all reticulate exines develop in the same way it does seem that certain processes may be common to all and that some of the known varieties of reticulate sculpturing might be traceable to slight variations in the formative processes. Dickinson (1970, 1976) demonstrated how the columellae and muri of the reticulate exine of *Lilium longiflorum* form by polymerization of sporopollenin on a primexine. If such a process were to be continued for a longer period of time the additional sporopollenin would result in a different morphology, perhaps the beaded muri that occur in a variety of families or simply thicker muri. The occurrence of beaded muri has been reported in a variety of families including Liliaceae, Cistaceae, Plumbaginaceae and others. Similarly the muri of some pollen grains from a wide variety of families have very small spinules along their surfaces which may all result from similar developmental processes with the size of the final feature being dependent on how long sporopollenin deposition continues.

A similar explanation may account for the difficulty in interpreting the internal foramina occurring in some Compositae exines. As mentioned earlier the presence or absence of such foramina is used as an important character in recognizing exine types by Skvarla *et al.* (1977) yet these authors have reservations about the value of such a character since they suggest internal foramina may be present, to some extent, in all Compositae. Amongst the Lactuceae only *Catananche* has distinct internal foramina but, as Tomb (1975) notes, they are also present in *Tragopogon* pollen and, in fact, occur in a weakly developed form throughout subtribe Scorzonerinae (Blackmore 1982). It could well be that this character simply expresses the completeness of sporopollenin deposition. It is interesting however that in the two species of Compositae which have been studied in detail, *Cosmos bipinnatus* (Dickinson and Potter 1976) and *Artemisia vulgaris* (Rowley and Dahl 1977), internal foramina are not apparent in the developing pollen. In *Cosmos*, where the mature pollen is of the Helianthoid type with distinct internal foramina, they appear only to be recognizable in virtually mature pollen.

An important, but largely unanswered, question is to what extent other factors external to the normal pattern of pollen development can influence the final form of the developing pollen. Various workers, including Van Campo (1976), Melville (1981) and Rowley (1981), have discussed the importance of the anther locule environment to pollen morphology and Cousin (1980) has investigated the production of pollen by diseased plants. There is considerable scope for experimental work in this field which would produce information of value to pollen morphologists.

(3) Pollen form and function

The function of pollen grain features is a subject which is currently receiving great attention. At present our understanding of pollen form and function is at an early stage but it is becoming increasingly possible to recognize characters which can be linked to a particular function in much the same way as, for example, the well known xeromorphic features of vegetative parts of plants. Just as recognizing that xeromorphic features are present in many families as a result of convergence can prevent classifications being based on superficial resemblances so an understanding of pollen function should enable pollen characters to be used more discerningly.

As dispersed units of only a few cells pollen grains are exposed to conditions very much drier than the anther locule after dehiscence. The greatest problem that they have to overcome is therefore desiccation and collapse. As Rowley (1978) has pointed out the highly waterproof and elastic sporopollenin wall of pollen grains is pre-adapted to cope with such conditions. Muller (1979) emphasized that the ratio of surface area to volume in pollen grains has an important influence on their shape and ability to accommodate volume changes. One outcome of this is that colpi are the most successful apertures for prolate-shaped pollen (with an equatorial axis much shorter than the polar axis) and pores for oblate (with an equatorial axis much greater than the polar axis) pollen, at least as far as the important function of accomodating volume changes goes. This is very relevant to two of the most frequently encountered patterns of evolution in angiosperm pollen, the "successiform" and "breviaxe" series of Van Campo

(1966, 1976). Successiformy results in the evolution of an increasing number of apertures and a spherical shape and occurs as a consequence of convergence in such diverse families as Convolvulaceae, Cactaceae, Caryophyllaceae and Geraniaceae. In the breviaxy pattern pollen grains become increasingly porate rather than colpate and oblate in shape. This pattern of evolution has also been recognized in a variety of families including Rubiaceae, Apocynaceae, Betulaceae and Myrtaceae. Van Campo suggested that the two patterns are repeated so often throughout the angiosperms because they reflect the results of physical factors such as the relative position of the pollen grains in tetrads and nomothetical factors which produce regular patterns of morphology. Such factors are, as Heslop-Harrison (1972) pointed out, not under direct genetical control but have a direct influence on the final form of the pollen grain. Kuprianova (1979) and Melville (1981) have discussed the importance of the form of the tetrad, which can be alterred by changes in pressure, in influencing the number and position of apertures providing an excellent example of how such physical forces operate. Clarke (1975) described how, in certain *Hypericum* species, individual plants produce a range of different pollen types equivalent to many of the stages in the successiform series. He concluded that in these species the factors which normally determine aperture configurations did not appear to operate and suggested that these factors were physical, resulting from irregularities in meiosis and tetrad formation, rather than genetic. There is thus considerable evidence that the successiform, and perhaps also the breviaxe, pattern of evolution began with dramatic changes in pollen morphology resulting from changes in tetrad formation. It is not yet clear, and will be difficult to establish, to what extent functional constraints dictate which out of a range of theoretically possible forms appear. It can be seen that throughout these two evolutionary patterns there is a close correlation between aperture configurations and shape which could be predicted in the light of the pollen function studies of Payne (1972, 1981) and Muller (1979).

It is interesting to contrast the regularly patterned pollen grains of the Lactuceae with those of the successiform series. Their almost crystalline appearance led Wodehouse (1935) to conclude that the geometry of the tetrads controlled the positioning of the ridges and lacunae, in other words that these were not genetically controlled. Detailed ontogenetic investigations are needed to try to establish exactly what controls the morphology of these complex pollen grains. I strongly suspect that in the case of the Lactuceae and other Compositae with echinolophate pollen functional considerations are of prime importance and that the patterns of ridges and lacunae are the result of selection for functional efficiency although the tetrad configuration may also exert some influence. One line of evidence for this is simply that there are examples of every intermediate stage between echinate, subechinolophate and echinolophate pollen in the Lactuceae, although there is little variation in the pollen of particular species. Furthermore there are a number of distinct arrangements of lacunae, relative to the position of the apertures, within the tribe. Neither of these conditions suggests a change from echinate to echinolophate pollen brought about by reorganization of the terads in the manner described by Melville (1981) for other pollen grains.

There is considerable evidence (Blackmore unpublished) that echinolophate pollen combines a high degree of mechanical strength with an efficient means of accommodating volume changes. Bolick (1978, 1981) in a study of Vernonieae pollen has shown how the ridges of their echinolophate pollen provide a structure which is very resistant to collapse but uses wall materials economically. This work has drawn attention to several closely correlated characters. The very thick columellae found in the ridges of such diverse genera as *Vernonia*, *Scolymus*, *Pyrrhopappus* and *Rothmaleria* are interpreted as being the end point in a trend towards increased reinforcement of the ridges. The pollen of all these genera also have very narrow ridges surrounding large lacunae, virtually no cavus* and frequently have reduced or absent tecta and columellae in their lacunae. Studies of Lactuceae pollen suggest that these are all character states associated with volume change accommodation, or "harmomegathy" (Wodehouse 1935), by means of the lacunar floors bulging inwards or outwards. Obviously such a method of harmomegathy is most efficient where the lacunae cover a large proportion of the surface area and the ridges as small a portion as possible. For narrow ridges to resist collapse they must have thicker columellae and in such cases the cavus ceases to play a part in harmomegathy. In echinate pollen such as that of *Catananche* the cavus is more extensive and does contribute to accommodation of volume changes since the columellae present are narrow and, apparently, flexible (Fig. 3). The most important method of harmomegathy in such cases is the folding inwards of the pollen grain along its colpi which results in a dramatic change in shape (Figs 23–24). The colpi of *Catananche*, and certain other Lactuceae, pollen grains are simple (Fig. 17) with no intrusion of ridges over their surface, to limit this folding action. In subtribe Scorzonerinae each colpus is characteristically divided into two halves, or lacunae (Fig. 18), by short ridge like projections which restrict the extent to which the colpus can fold

* *Editor's note:* The author's use of the term *cavus* and the incorrect plural *cavea*, now commonly used in palynology, is respected. The correct Latin for a hole or cavity is *cavum* (plural *cava*) but even if cavus (which is the adjective from *cavum*, meaning hollow) is accepted as a substantive, the plural would be *cavi* and in no circumstances *cavea*.

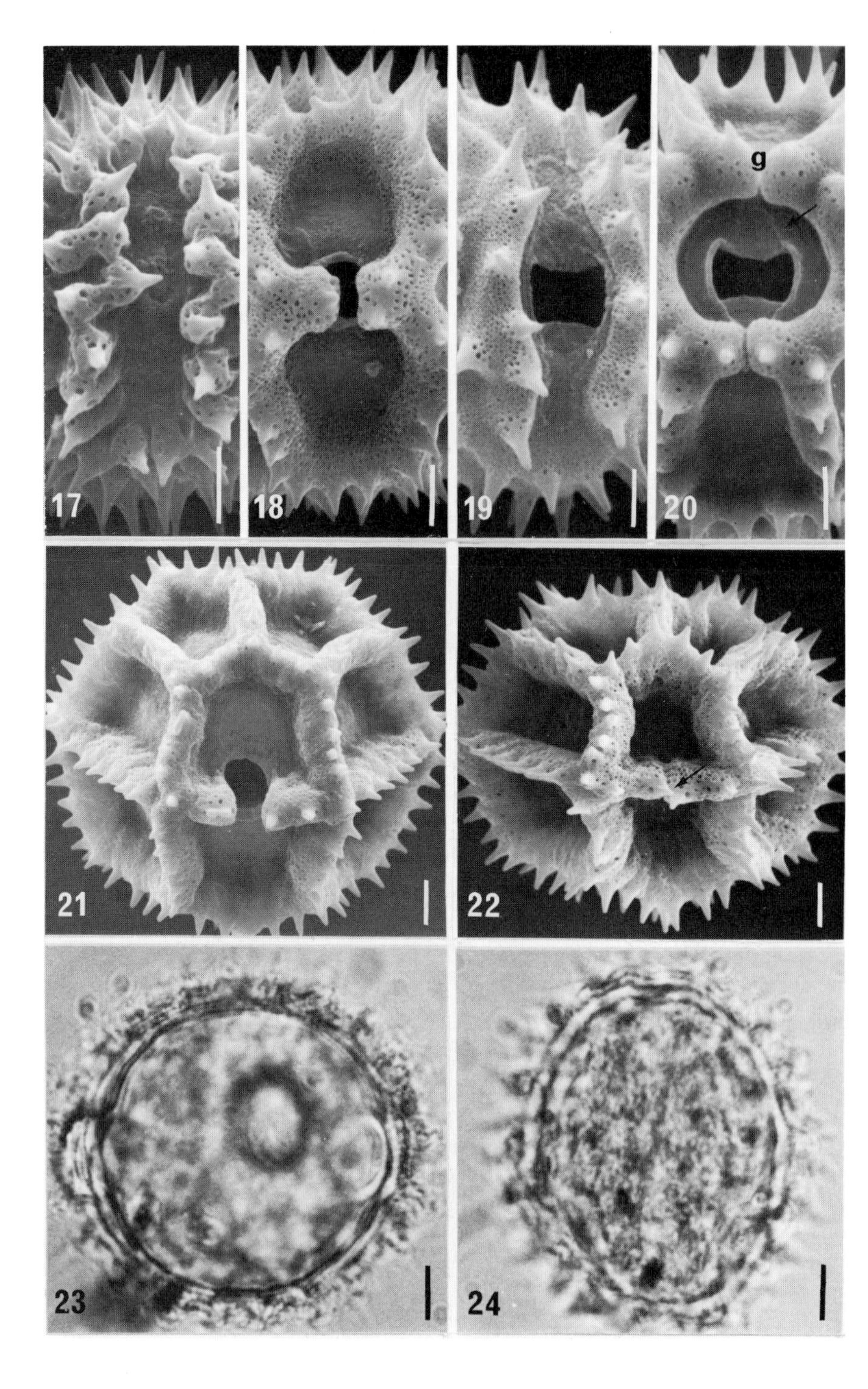
g
17
18
19
20
21
22
23
24

inwards. In the genus *Epilasia* (Figs 21–22) these ridges are particularly prominent and very little folding is possible along the colpi. Here the bulging movements of the floors of the extensive lacunae take over the harmomegathic function. In all subtribes other than Scorzonerinae the colpi are either slightly (Fig. 19) or distinctly (Fig. 20) divided into three lacunae by ridges which act in a similar way to those of *Epilasia* pollen. In all cases where these ridges are well developed the lacunae are large and the ridges narrow. Pollen of the genus *Scolymus* and certain *Vernonia* species take the trend of immobilizing the colpi to its limit, with the ridges fusing together. In *Scolymus* this character is correlated with the presence of thick columellae, large lacunae, narrow ridges and reduced tectum and columellae in the lacunae; all features which point to volume changes accommodated by movements of the lacunar floors.

Whilst the cavea play little part in the harmomegathy of Lactuceae pollen Payne (1972) has shown that they are of great importance in this respect in highly caveate pollen such as that of *Ambrosia*. After dehiscence *Ambrosia* pollen changes little in external size and shape but the volume of the cavea increases as air is drawn into them by the contraction of the pollen cytoplasm. This function of the cavus, together with its role as a repository for tapetally derived substances furnishes an answer to the interesting question raised by Tomb (1975), "why have a cavus?"

There can be little doubt that studies of pollen function will continue to provide fascinating results for some time to come.

(4) Studies of unacetolysed pollen

Whilst acetolysis provides the basic technique used by virtually all pollen morphologists there are certain additional characters which cannot be observed

Figs 17–24. Light and scanning electron micrographs of Lactcueae pollen, all except Figs 23–24 are of acetolysed pollen. Figs 17–20 aperture morphology. Fig. 17 *Catananche caeulea* L. in which the ectoaperture is a simple colpus. Fig. 18, *Tragopogon ruthenicus* Bess. ex Claus. in which the colpus is divided into three lacunae. Fig. 19, *Soroseris gillii* (Moore) Stebbins has colpi which are weakly divided into three lacunae. Fig. 20, *Cichorium intybus* has three distinct lacunae per colpus, separated by inter lacunar gaps (g). Fig. 21, *Epilasia mirabilis* Lipsch., whole pollen grain in apertural view. Fig. 22, as Fig. 21 but fortuitously showing the dehydrated appearance of the pollen, with the ridges closing together (arrow) over the mesoaperture. Fig. 23, fresh pollen of *Catananche caerulea* L. prior to dehiscence. Fig. 24, fresh pollen of *C. caerulea* after dehiscence, showing the distinct change in shape which accompanies dehiscence. Scale lines all 5 μm.

after this treatment. The most important of these are features of the apertures which are usually modified by acetolysis. Operculae covering the apertures of fresh pollen are generally removed during the process and the delicate aperture membranes often rupture. Furthermore, as Thanikaimoni (1978) has stressed, the morphology of the intine layer of the pollen wall, which is not acetolysis-resistant, can be vital to the correct interpretation of pollen apertures. The intine is generally thickened in apertural regions and such thickenings can be used to detect the presence of apertures in some apparently inaperturate pollen grains. Thanikaimoni demonstrated, for example, that *Tiliacora* in the Menispermaceae has three apertures rather than none. Since the apertures are the most important source of taxonomic characters such correct interpretation is plainly important.

The number of nuclei present in mature pollen, usually 2 or 3 in angiosperms, can be used as a character. Brewbaker (1967) and other workers have surveyed the nuclear condition of many families and the character has been used, for example, by Dahlgren and Clifford (1982) in their recent study of the monocotyledons.

Acetolysis is not necessary for taxonomic studies based, as many investigations by non-specialists are, on scanning electron microscopy alone. Lynch and Webster (1975) described a preparation technique which gives excellent results without acetolysis but many workers use much simpler methods such as rinsing pollen in a solvent to remove the oily pollenkitt.

(5) Fundamental structure of the exine

Recently Rowley and his collaborators (Rowley *et al.* 1981a, b) have postulated that exines are composed of units called "tufts" made up of coiled helical subunits wound around central helical subunits. The implications of the existence of such exine units would be far reaching. Rowley (1981) has suggested that such diverse features of pollen grains as the columellae of *Artemisia* and members of many other families, the ornamentation elements of *Quercus* pollen and microechinae of some *Saxifraga* pollen grains are all composed of single exine units. If exines are indeed made up of such units their structure and size would impose certain constraints on the corresponding size, and possibly shape, of pollen features. However, whilst these studies demonstrate convincingly that exines are not simply composed of sporopollenin and that there may be regular units present at certain stages of development it is far from certain that the model proposed is appropriate. To demonstrate the existence of tufts, exines have been subjected to various drastic techniques designed to degrade the mature exine (Rowley *et al.* 1981b) and also to reveal the remnant of the glycocalyx around which sporopollenin deposition takes place in developing pollen grains. Rowley (1981) went so far as to suggest that determining the constituents of the glycocalyx might eventually supplant the use of pollen wall morphological characters. Whilst such a development seems unlikely to diminish the utility of pollen wall characters it may provide valuable additional characters in the future.

CONCLUSIONS

A number of general conclusions can be drawn from this survey of some current aspects of palynology.

Developmental studies provide information useful to taxonomists yet so far only a small proportion of families have been studied and there have been virtually no studies of closely related taxa with different morphologies. The data from such studies is needed to clarify the homology of pollen features in different plant taxa and to provide a basis for understanding variations in pollen morphology.

The influence of the environment on developing pollen needs to be investigated to improve our knowledge of the applicability of pollen characters.

Pollen features are beginning to be understood in functional terms and this will enable strongly correlated characters to be recognized and help in the detection of convergence and parallelism.

In many investigations there are good reasons for examining unacetolysed pollen grains as well as acetolysed samples.

Current research into exine structure may reveal that the exine is constructed of units with a similar structure throughout the higher plants.

In considering the application of pollen characters in plant taxonomy it is now possible, by considering pollen ontogeny, function and other aspects of pollen grains, to use pollen characters more discerningly.

ACKNOWLEDGEMENTS

I am grateful to Graham Heath for assistance with pollen preparations, Theresa Smith for transmission electron microscopy and to Don Claugher for collaborating with ion beam etching of pollen.

REFERENCES

Barthlott, W., Ehler, N. and Schill, R. (1976). Abtragung biologischer Oberflächen durch hochfrequenzaktivierten Sauerstoff für die Raster-Elektronmikroskopie. *Mikroskopie* **32,** 35–44.

Blackmore, S. (1981). Palynology and intergeneric relationships in subtribe Hyoseridinae (Compositae: Lactuceae). *Bot. J. Linn. Soc.* **82,** 1–13.

Blackmore, S. (1982). Palynology of subtribe Scorzonerinae (Compositae: Lactuceae) and its taxonomic implications. *Grana* **21,** 149–160.

Blackmore, S. and Dickinson, H. G. (1981). A simple method for sectioning pollen grains. *Pollen Spores* **23,** 281–285.

Bolick, M. R. (1978). Taxonomic, evolutionary and functional considerations of Compositae pollen ultrastructure and sculpture. *Pl. Syst. Evol.* **130,** 209–218.

Bolick, M. R. (1981). Mechanics as an aid to interpreting pollen structure and function. *Rev. Palaeobot. Palynol.* **34,** 61–79.

Brewbaker, J. L. (1967). The distribution and phylogenetic significance of binucleate and trinucleate pollen grains in the angiosperms. *Amer. J. Bot.* **54,** 1069–1083.

Brown, R. (1811). On the Proteaceae of Jussieu. *Trans. Linn. Soc. Lond.* **10,** 15–226.

Clarke, G. C. S. (1975). Irregular pollen grains in some *Hypericum* species. *Grana* **15,** 117–125.

Cousin, M.-Th. (1980). Changes induced by mycoplasma-like organisms (M.L.O.), etiologic agents of the Stolbur disease, in the different tissues of the anther of *Vinca rosea* L. (Apocynaceae). *Grana* **19,** 99–125.

Dahlgren, R. M. T. and Clifford, H. T. (1982). "The Monocotyledons: A Comparative Study". Academic Press, London, New York.

Dickinson, H. G. (1970). Ultrastructural aspects of primexine formation in the microspore tetrad of *Lilium longifluroum*. *Cytobiologie* **1,** 437–439.

Dickinson, H. G. (1976). Common factors in exine deposition. *In* "The Evolutionary Significance of the Exine" (I. K. Ferguson and J. Muller, eds) pp. 67–89. Academic Press, London, New York.

Dickinson, H. G. and Potter, U. (1976). The development of patterning in the alveolar sexine of *Cosmos bipinnatus*. *New Phytol.* **76,** 543–550.

Erdtman, G. (1934). Über die Verwendung von Essigsäureanhydrid bei Pollenuntersuchungen. *Svensk bot. Tidskr.* **28,** 354–358.

Erdtman, G. (1970). Introductory notes on the World Pollen Flora. *In* "World Pollen Flora, 1" (G. Erdtman, ed.) pp. 5–14. Munksgaard, Copenhagen.

Ferguson, I. K. (1978). Technique utilisant le méthylate de sodium comme solvent de la résine epoxy des blocs d'inclusion "type MET" pour observations de l'exine des grains du pollen. *Ann. Min. Belg.* **2,** 153–157.

Henrickson, J. (1973). Fouquieriaceae. *In* "World Pollen and Spore Flora, 1" (S. Nilsson, ed.) pp. 1–12. Almqvist and Wiksell, Stockholm.

Heywood, V. H. (1968). Plant taxonomy today. *In* "Modern Methods in Plant Taxonomy" (V. H. Heywood, ed.) pp. 3–12. Academic Press, London, New York.

Heslop-Harrison, J. (1968). Pollen wall development. *Science* **161,** 230–237.

Heslop-Harrison, J. (1972). Pattern in plant cells. *Proc. R. Inst. Gt. Br.* **45,** 335–351.

Hideux, M. (1972). Techniques d'étude du pollen au MEB: effets comparés des différents traitements physico-chemiques. *Micron* **3,** 1–31.

Hideux, M. and Marceau, L. (1972). Techniques d'étude du pollen au MEB: méthode simple de coupes. *Adansonia* **12,** 609–618.

Janssen, C. R., Punt, W. and Reitsma, T. (1976). Introduction. *In* "The Northwest European Pollen Flora, I" (W. Punt, ed.) pp. 1–4. Elsevier, Amsterdam.

Kuprianova, L. A. (1979). On the possibility of the development of tricolpate pollen from monosulcate. *Grana* **15,** 117–125.

Lynch, S. P. and Webster, G. L. (1975). A new technique of preparing pollen for scanning electron microscopy. *Grana* **15,** 127–136.

Maguire, B. (1981). The botany of the Guayana highland. XI. Gentianaceae. *Mem. N.Y. Bot. Gard.* **32,** 330–387.

Melville, R. (1981). Surface tension, diffusion and the evolution of pollen aperture patterns. *Pollen Spores* **23,** 179–203.

Muller, J. (1979). Form and function in Angiosperm pollen. *Ann. Missouri Bot. Gard.* **66,** 593–632.

Nagarajan, P. and Bates, S. (1981). A rapid poly-l-lysine schedule for SEM studies of acetolysed pollen grains. *Pollen Spores* **23,** 273–280.

Nilsson, S. and Muller, J. (1978). Recommended terms and definitions. *Grana* **17,** 55–58.

Pacqué, M. (1980). The structure of apertures in some Passifloraceae and their possible functioning. *Vth Int. Palynol. Conf. Abstracts,* 298. International Commission for Palynology.

Payne, W. W. (1972). Observations of harmomegathy in pollen of Anthophyta. *Grana* **12,** 93–98.

Payne, W. W. (1981). Structure and function in Angiosperm pollen wall evolution. *Rev. Palaeobot. Palynol.* **35,** 39–59.

Rowley, J. R. (1976). Dynamic changes in pollen wall morphology. *In* "The Evolutionary Significance of the Exine" (I. K. Ferguson and J. Muller, eds) pp. 39–65. Academic Press, London, New York.

Rowley, J. R. (1978). The origin, ontogeny and evolution of the exine. *Proc. IVth Int. Palynol. Conf., Lucknow* **1,** 126–136.

Rowley, J. R. (1981). Pollen wall characters with emphasis on applicability. *Nordic J. Bot.* **1,** 357–380.

Rowley, J. R. and Dahl, A. O. (1977). Pollen development in *Artemisia vulgaris* with special reference to glycocalyx material. *Pollen Spores* **19,** 169–284.

Rowley, J. R., Dahl, A. O. and Rowley, J. S. (1981a). Substructure in exines of *Artemisia vulgaris* (Asteraceae). *Rev. Palaeobot. Palynol.* **35,** 1–38.

Rowley, J. R., Dahl, A. O., Sengupta, S. and Rowley, J. S. (1981b). A model of exine substructure based on dissection of pollen and spore exines. *Palynology* **5,** 107–152.

Skvarla, J. J. and Larsen, D. A. (1965). An electron microscopic study in the Compositae with special reference to Ambrosiinae. *Grana* **6,** 210–269.

Skvarla, J. J. and Turner, B. L. (1966). Systematic implications from electron microscopic studies of Compositae pollen – a review. *Ann. Missouri Bot. Gard.* **53,** 220–256.

Skvarla, J. J., Turner, B. L., Patel, V. C. and Tomb, A. S. (1977). Pollen morphology in the Compositae and in morphologically related families. *In* "The Biology and Chemistry of the Compositae" (V. H. Heywood, J. B. Harborne and B. L. Turner, eds) pp. 147–217. Academic Press, London, New York.

Stix, E. (1960). Pollenmorphologische Untersuchungen an Compositen. *Grana* **2,** 41–114.

Takahashi, M. (1980). On the development of the reticulate structure of *Hemerocallis* pollen (Liliaceae). *Grana* **19,** 3–6.

Thanikaimoni, G. (1975). Pollen morphological terms: Proposed definitions. *Proc. IVth Int. Palynol. Conf., Lucknow* **1,** 228–239.

Tomb, A. S. (1975). Pollen morphology in tribe Lactuceae (Compositae). *Grana* **15,** 93–98.

Van Campo, M. (1966). Pollen et phylogénie. Les Breviaxes. *Pollen Spores* **8,** 57–73.

Van Campo, M. (1976). Patterns of pollen morphological variation within taxa. *In* "The Evolutionary Significance of the Exine" (I. K. Ferguson and J. Muller, eds) pp. 125–138. Academic Press, London, New York.

Van Campo, M. (1978). La face interne de l'exine. *Rev. Palaeobot. Palynol.* **26,** 301–311.

Van Campo, M. and Sivak, J. (1972). Structure alvéolaire de l'ectexine des pollens à ballonets des Abietacées. *Pollen Spores* **14,** 115–141.

Wodehouse, R. P. (1935). "Pollen Grains". McGraw-Hill, New York.

Karyology and Genetics

9 | Chromosomal Evidence in Taxonomy

J. GREILHUBER

Institute of Botany, University of Vienna, Austria

Abstract: It is the objective of the present contribution to review technical approaches in plant chromosome analysis as far as they will be routinely applied by the cytotaxonomist (light microscopical, mainly non-molecular techniques), and to discuss the relevant structural and quantitative features of the karyotype with regard to their evolution and their use in taxonomy and phylogenetic research.

In addition to chromosome number and morphology of the conventionally stained karyotype, it will often be necessary to determine (1) the DNA content of a species, (2) the distribution and amount of constitutive heterochromatin by various banding techniques, (3) the cytochemical distinction of different heterochromatin types, mainly by fluorochromy with a set of specific dyes, (4) the location of transcriptionally active nucleolar organizers by silver-impregnation techniques. In favourable situations it will be possible (5) to identify genomic satellite DNA components and to relate them to heterochromatin bands by auto- and heterologous *in situ* hybridization.

Taxonomic applications of these methods are demonstrated in certain plant groups. It is possible to make some generalizations: (1) The genomic DNA content is usually fairly stable within a species (in the narrowest sense), although it can differ significantly between species. (2) Significant differences, occasionally even at the subspecific level, can be found in heterochromatin content. From our studies it appears safe to conclude that heterochromatin is an additive component of the genome and, if accumulated, increases the DNA content of a species. (3) The geographic stability of banding patterns can be very different from species to species. (4) There is evidence that the same heterochromatin type (as defined by its satellite DNA) can arise independently in related taxa. The possibility of parallelism should always be given serious consideration.

It is, therefore, recognized that the "new karyosystematics" has its impact mainly at the infrageneric level and, at any rate, in the "diffuse" zone, where it becomes difficult to separate taxa by traditional karyological and other methods.

Systematics Association Special Volume No. 25, "Current Concepts in Plant Taxonomy", edited by V. H. Heywood and D. M. Moore, 1984. Academic Press, London and Orlando.
ISBN 0 12 347060 9

INTRODUCTION

The contemporary taxonomist is familiar with the fact that the genomes of extant higher plant species (only these are focussed on here) are highly variant not only with regard to chromosome number and structure, but also genome size as expressed in pg DNA or number of base pairs. It is almost commonplace to state that this variation is largely unrelated to genetic or organismal complexity. Variation at the family or generic level is smaller, but sometimes still astonishing (Bennett and Smith 1976). A reasonable appraisal of the taxonomic value of chromosomal parameters depends on the knowledge about the nature and amount of chromosomal variation within and between populations, and between closely related but reproductively more or less isolated gene pools or major populations. In the past decade this subject has been treated comprehensively by Stebbins (1971), Jackson (1971), and K. Jones (1978). A contemporary account on the use of chromosome data in taxonomy was given by Moore (1978). Special topics such as chromosome evidence in angiosperm origin and phylogeny have been reviewed (Ehrendorfer 1976), and a comprehensive overview on polyploidy is available (Lewis 1980). Therefore, in the author's opinion, the bias towards recent methodical approaches in karyosystematics, which the reader will recognize, will be justified. In contrast (not in contradiction) to the classical approach, it is more widely appreciated now that chromosomes can accumulate and discharge DNA sequences of various quality keeping their linkage relationships, at least in principle, uninjured. The use of some of these changes as taxonomic indicators will be the subject of the following pages.

CHROMOSOME TECHNIQUES

It is not necessary to deal with traditional techniques of chromosome preparation. However, it seems appropriate to make some comments on banding, heterochromatin characterization, and DNA photometry. The chromosome banding in plants differs from banding in vertebrates in essential points, and there is still no generally accepted "banding philosophy". While DNA photometry is not so controversial, the most critical points for data reproducibility shall be mentioned.

C-banding reveals constitutive heterochromatin and consists of (*i*) depurination, (*ii*) denaturation, (*iii*) breaking the sugar-phosphate backbone (β-elimination), (*iv*) washing out DNA fragments, and (*v*) staining the remaining DNA, retained in regions of constitutive heterochromatin, with Giemsa (Holmquist 1979). In the past, even experts would have agreed that C-banding in plants is far from being problem-free. There is always the possibility that bands go undetected for technical reasons. A remarkable improvement seems to be the technique of Schwarzacher *et al.* (1980). *G-banding*, which is so characteristic for vertebrates, has not been reliably demonstrated in plants. Pachytene chromomere patterns can be correlated with mitotic G-banding in vertebrates, and it was calculated that the much higher ratio of contraction in mitotic vs. pachytene chromosomes in plants would be sufficient to explain the virtual absence of G-banding (Greilhuber 1977). *Hy-banding* (Greilhuber 1973, 1975) and *N-banding* (Funaki *et al.* 1975; Jewell 1981) involve relatively harsh treatments and reveal particular fractions of constitutive heterochromatin which seem to be correlated with base content (Hy-banding, Greilhuber 1975) or DNA sequence (N-banding, Jewell 1981). *Q-type-banding*, a class of banding obtained with quinacrine and some other dyes with similar binding and/or fluorescence specificity, is correlated with base composition and reveals, in addition to constitutive heterochromatin, a segmentation in euchromatin of vertebrate chromosomes, which correlates with G-banding (Comings 1978). As with G-banding, typical Q-banding has still not been demonstrated conclusively in plants. However, Holm (1976) and Kongsuwan and Smyth (1977) in *Lilium*, and Filion and Vosa (1980) in *Paris* demonstrated fluorescent banding virtually outside of heterochromatin. Kongsuwan and Smyth (1980) concluded from replication patterns that this sort of Q-banding is substantially different from vertebrate-type Q-banding. *Fluorescent banding* in plants offers the possibility of refined heterochromatin characterization by a number of established AT- or GC-specific fluorochromes such as quinacrine, Hoechst 33258, DAPI, or chromomycin, which can be employed in conjunction with fluorescent or nonfluorescent counterstains (Schweizer 1980, 1981). Since C-banding does not distinguish heterochromatin types, fluorochrome characterization is important for establishing heterochromatin homologies. *In situ hybridization* of satellite DNAs has been widely applied in animal karyosystematics. In plants however, apart from cereal crops, it has only been used on several *Scilla* species (Greilhuber *et al.* 1981; Deumling and Greilhuber 1982). The method (Pardue and Gall 1971) can be applied also in interspecific hybridization experiments and yield more conclusive evidence about interspecific sequence homologies of heterochromatic DNA. All these banding techniques can be applied, in principle, both to mitotic and meiotic chromosomes. *Silver impregnation techniques* as a method for the recognition of active nucleolar organizers have been established for plants by Schweizer and Ambros (1979). Silver impregnation of synaptonemal complexes in spread microsporocytes of plants, as pioneered by Gillies (1981), will probably have great importance in the analysis of homology relationships in hybrids. For *DNA-content determinations* the method of choice is still Feulgen-DNA-cytophotometry. Of paramount importance in karyosystematics is the simultaneous processing of an internal standard of known DNA amount (see Bennett and Smith 1976). Even highly standardized conditions without use of an internal standard have resulted in unreliable data (see Teoh and Rees 1976). To date, there has been no break-through of fluorometric techniques in plant cytophotometry (see, however, Geber and Hasibeder 1980).

HETEROCHROMATIN

(1) Heterochromatin variation

Constitutive heterochromatin (for brevity termed heterochromatin in his account), a genome component characterized by chromocentre formation, positive C-banding capacity and – at least most often – content of particular DNA sequences such as satellite DNA or highly repetitive DNA, is quantitatively highly variable even at specific or lower levels. The association of heterochromatin with nucleolar organizers, centromeres, and often telomeres suggests a biological function. The recent discovery of transcription of satellite DNA in lampbrush chromosomes (Varley *et al.* 1980) reinforces this assumption. However, there is agreement that the bulk of heterochromatin is genetically inert and cannot have a vital biological function, although there may well be biological significance in heterochromatin variation. It has been quantitatively shown in plants that heterochromatin is an additive component of the genome (Greilhuber and Speta 1976; Greilhuber 1982; Deumling and Greilhuber 1982) as is also quite obvious from Fig. 2. We may deduce that, according to the "nucleotype hypothesis" (Bennett 1972), heterochromatin changes would also change nuclear volume, cell cycle duration, cell size, and developmental speed. This topic needs much greater investigation (Nagl and Ehrendorfer 1974). Nagl (1974) concluded that annuals could develop faster through increasing heterochromatin content but this assumption seems inconclusive because (*i*) replication is slower in satellite DNA (Bostock *et al.* 1972) and (*ii*) heterochromatin is a burden to the genome due to its additive character. Amount and location of heterochromatin affects meiotic recombination which is certainly of adaptive importance (G. H. Jones 1978; Loidl 1979). Position effects of heterochromatin may also have significance.

To state that heterochromatin may have adaptive significance is not at variance with the idea that heterochromatin may be composed of "parasitic" or "selfish" DNA (Doolittle and Sapienza 1980; Orgel and Crick 1980). A certain amount of heterochromatin may indeed exist in a population merely because it is "selfish". On the other hand, perhaps the amount of heterochromatin or particular patterns are due to an equilibrium struck between accumulation of and natural selection against too much or ill-placed "selfish" sequences.

For a reasonable assessment of banding data in taxonomy, the amount of intraspecific heterochromatin variation is important. Completely stable karyotypes in sexually reproducing species are unlikely to occur. Little significant variation on a geographic scale has been reported in *Scilla vindobonensis*, *S. bifolia*, *S. mischtschenkoana* (Greilhuber and Speta 1977, 1978), *Adoxa* (Greilhuber 1979a), several *Allium* taxa, *Brimeura fastigiata* (Vosa 1976a, b, 1979), several *Hordeum* taxa (Linde-Laursen *et al.* 1980). Bennett *et al.* (1977) ascribed diagnostic value to C-banding patterns in *Secale* species. Between-population differences with unknown taxonomic implications were reported by Vosa (1976b) in *Allium pulchellum*. Abundant C-band polymorphism in *Leopoldia comosa* was found by Bentzer and Landström (1975), although this variation is well within the usual frame of variation we would expect from the more recent banding studies. As a model system for heterochromatin variation on a geographical scale, the intensively studied cold-sensitive H-segments of *Trillium* species are still the most suitable. Indeed, ample heteromorphism has been found both in the presence and size of H-segments, leading finally to somewhat complex interpretations of the presumed adaptive significance of geographically fixed or variable pattern combinations (Haga and Kurabayashi 1954, Kurabayashi 1958, 1963, Fukuda and Channell 1975, Fukuda and Grant 1980). At this point, in view of a possible nucleotypic role of heterochromatin, it would be interesting to know what the geographical or ecological distribution of heterochromatin amounts (as opposed to degrees of H-segment heterozygosities) would be in *Trillium*.

(2) Chromosome banding as a tool in taxonomy

Several C-banding studies have attempted to recognize ancestral genomes by heterochromatin markers in allopolyploids. One of the outstanding problems, the identification of the B-genome donor in *Triticum aestivum*, has still not been settled (Gill and Kimber 1974). *Hordeum roshevitzii* ($4x$) may be an allopolyploid with *H. roshevitzii* ($2x$) and a taxon near *H. californicum* as the ancestors (Linde-Laursen *et al.* 1980). Allopolyploidy was supposed in *Allium trifoliatum* subsp. *trifoliatum* ($3x$, $2n=21$) and *A. neapolitanum* ($4x$, $2n=28$) by Badr and Elkington (1977). In *A. neapolitanum* ($5x$, $2n=35$), Bruhns (1981) found 2:2:1 heteromorphism in two quintuples of identifiable chromosomes, stated predominant meiotic pairing of homomorphic chromosomes, and considered a genomic formula of AAA_1BB. It should be noted at this point that these heteromorphic proportions and meiotic pairing relationships could also occur on an autopolyploid basis in the presence of banding heteromorphism, particularly if selfing between unreduced gametes took place, and if meiotic recombination was suppressed or went undetected. There is evidence that structurally identical chromosomes show pairing preference over homologous but nonidentical chromosomes (Giraldez and Santos 1981). In the case of sexual reproduction, usually in orthoploids, selection for bivalent formation would reinforce pairing of homostructural chromosome pairs.

It is only fair to quote outstanding examples of allopolyploid recognition by heterochromatic markers from the pre-banding era. Haga (1956) traced the allohexaploid origin of *Trillium hagae* (6*x*) via triploid hybrids (*T. x hagae*, 3*x*) of *T. kamtschaticum* (2*x*) and *T. tschonoskii* (4*x*) by H-segment patterns. Teppner (1971a, b) provided evidence for the allotetraploid origin of *Onosma helveticum* agg., $2n=26$, 12L+14K, from *O. echioides* agg. ($2n=14$, 14K) and an *O. arenarium*-like ancestor (probably $2n=12$, 12L). The same ancestral genomes seem to be involved in the permanent triploid *O. arenarium*, $2n=20$, 12L+8K. The genomes differ markedly in size and allocyclic behaviour.

There is now sufficient information about heterochromatin variation in natural groups to avoid suggesting exaggerated taxonomic implications from chromosome banding. Its utility depends on the constancy of the banding profile or style as a character rather than on the stability of the detailed pattern. However, when marked and discontinuous presumptive infraspecific differences exist, the taxonomist should be prompted to treat the problem more thoroughly. A good example in this regard is the C-banding study of Kenton (1978) on diploid and tetraploid *Gibasis karwinskyana* (Commelinaceae) from different provenances in central Mexico. It was shown in this study that di- and tetraploids from the same area are chromosomally more similar than di- or tetraploids from different areas, and that the previous assumptions of one diploid and one tetraploid race have been incorrect. Incipient speciation in one diploid strain could be inferred.

Marks and Schweizer (1974) compared 6 *Anemone* species and *Hepatica nobilis* and were the first to demonstrate the impact that banding would have on plant systematics, because they found remarkable interspecific variation, and conditional agreement with taxonomic groupings. Overviews on genera such as *Allium* (Vosa 1976a, b; Badr and Elkington 1977), *Fritillaria* (La Cour 1978), and *Tulipa* (Blakey and Vosa 1981, 1982), have been given. All these studies showed that (*i*) a large amplitude of infrageneric variation in heterochromatin amount exists within each genus and (*ii*) chromosome banding results usually corresponded to existing taxonomic grouping, although sometimes disagreements or within-group heterogeneities were noted. To quote only a few examples, in the Saxatiles group of *Tulipa* subgen. Eriostemones, *T. urumiensis* poses problems concerning its heterochromatin pattern in relation to taxonomic grouping (Blakey and Vosa 1981). In *Fritillaria,* according to La Cour (1978), large differences exist in the amount and distribution of heterochromatin not only between groups but also within them, the divergence being extreme in the American *Liliarhiza* group, or within the European *Caucasica* group. In other words, the findings often negate the investigator's desire "to formulate a basic banding pattern characteristic of a given section" (Blakey and Vosa 1981) or other taxonomic unit. Since taxonomic units are presumably based on thorough morphological knowledge, cytologists are right not to change the taxonomists' system on a purely chromosomal basis. However, what improvement to the *system* could such studies provide?

The matter of fact is that a congruence of taxonomic grouping and grouping based on similarity in chromosome structure is not expected to occur generally even in an ideally perfect system. The problem is, of course, exactly the same as with other traits (Hennig 1966; Wiley 1981). I would like to emphasize, in light of recent heated discussions of this topic in botanical taxonomy, that it is a great advantage that chromosomal change is not necessarily, and often not, synchronous with change in other taxonomically "important" characters so that it contributes significantly to phylogeny reconstruction. I would like to exemplify this first in the *Scilla siberica* alliance.

(*a*) *The* Scilla siberica *alliance.* The *Scilla siberica* alliance currently embraces about 20 species (Speta 1980), and is characterized by a unique testa structure (papillae) and, at least as far as is currently known, by a bisporic embryo sac of the *Allium* type, both certainly apomorphies. The *Polygonum* type embryo-sac is probably ancestral in *Scilla* and occurs in the next related *S. messeniaca*-*S. bifolia*- and *S. hohenackeri*-alliance (Svoma 1981; Greilhuber 1982). Present systematic grouping is performed mainly according to seed types, which may be denoted as 1a (*S. bithynica* group), 1b (*S. mischtschenkoana* group and *S. melaina* group), 2a (*S. amoena* group), and 2b (*S. siberica* group). Type 1 and 2, seeds are black and yellow, respectively, types 1b and 2b carry nonhomologous elaiosomes, type a seeds none (type 2a seeds carry what apparently is a highly reduced elaisome). A helobial endosperm in *S. mischtschenkoana* was considered a symplesiomorphy with other Scilleae. The nuclear type in *S. siberica,*

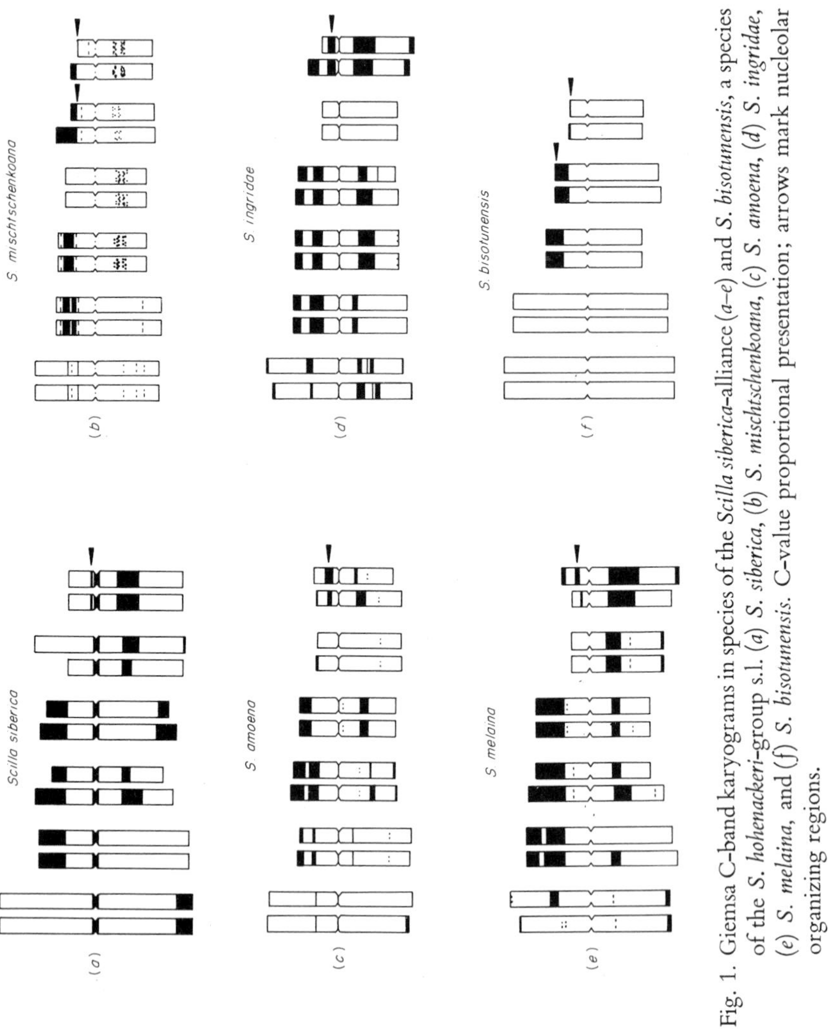

Fig. 1. Giemsa C-band karyograms in species of the *Scilla siberica*-alliance (*a–e*) and *S. bisotunensis*, a species of the *S. hohenackeri*-group s.l. (*a*) *S. siberica*, (*b*) *S. mischtschenkoana*, (*c*) *S. amoena*, (*d*) *S. ingridae*, (*e*) *S. melaina*, and (*f*) *S. bisotunensis*. C-value proportional presentation; arrows mark nucleolar organizing regions.

S. amoena and *S. ingridae* were considered a synapomorphy (Svoma 1981).

The data in *S. siberica*, *S. amoena*, *S. ingridae* and *S. mischtschenkoana* include satellite DNA sequencing and *in situ* hybridization (Deumling 1981; Deumling and Greilhuber 1982), so that I would like to consider these taxa first (Figs 1, 4 and 5). In terms of chromosomes, *S. amoena* and *S. ingridae* share most common traits. The basic karyotype structure, genome size, C-band arrangement is similar; fluorescence characters and satellite DNA sequence are the same. *Scilla siberica* shares with them the same satellite DNA sequence and a simliar karyotype, but there are also distinct differences: the heterochromatin fluorescence is somewhat different, C-band arrangement is less complex, there is a particular pericentromeric heterochromatin component, and the genome size is markedly higher. *Scilla mischtschenkoana* contrasts markedly with the foregoing taxa: 2 NOR-pairs instead of 1, particular clustered mode of C-band arrangement, particular fluorescence characters of the major band fraction, and a unique satellite DNA as judged from Hae III restriction enzyme digestion patterns (Deumling and Greilhuber 1982).

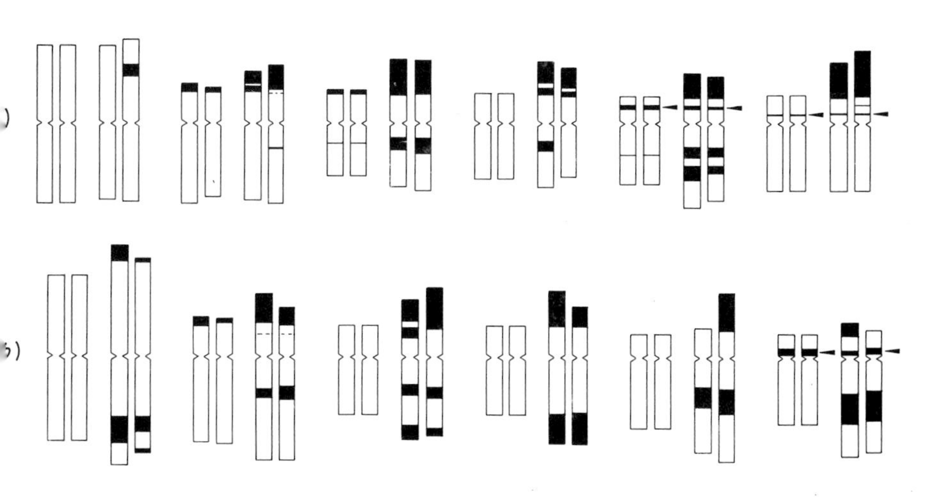

Fig. 2. DNA-value proportional pair by pair comparison in C-band karyograms of heterochromatin-poor and -rich sister taxa of the *Scilla siberica* alliance: (*a*) *S. bithynica* subsp. *radkae* (left), and *S. bithynica* subsp. *bithynica* (right), (*b*) *S. rosenii* (left) and *S. koenigii* (right).

These data seem in complete agreement with the systematics, since *S. mischtschenkoana*, *S. amoena* and *S. ingridae*, and *S. siberica* belong to separate groups. Moreover, the close affinity of *S. amoena* and *S. ingridae* would suggest a more recent common ancestor for these species than of either with *S. siberica*, while *S. mischtschenkoana* appears quite remotely related. A chemotaxonomic screening study in these species has shown that this is not necessarily correct: according to phenolic compound patterns *S. ingridae* and *S. mischtschenkoana*, and *S. amoena* and *S. siberica* should be grouped together (Harmer 1980). If parallelism is not involved, the dilemma can only be solved assuming the phylogeny in Fig. 3. The somewhat unexpected consequence is that the complex C-band pattern of *S. amoena* and *S. ingridae* is likely a plesiotypic condition, and the pattern in *S. siberica* due to simplification, with DNA increase and paracentromeric heterochromatin accumulation as a derived acquisition.

A new dimension was added to the problem when *interspecific in situ* cross-hybridization experiments were performed (Deumling and Greilhuber 1982). In accordance with the sequencing study, perfect cross hybridization was observed between *S. siberica*, *S. amoena* and *S. ingridae*, but not between any of these and *S. mischtschenkoana*. Quite unexpectedly, strong cross hybridization resulted with centromeric regions in *S. siberica* and satellite DNA of *S. mischtschenkoana* (Figs 4 and 5).

The most conflicting result of these experiments is that one and the same satellite DNA was accumulated in two related species in quite different chromosomal locations. This is a kind of similarity which is not easily explained by common descent of these species: (*i*) dislocation of these segments by common structural rearrangements is not likely for reasons of karyotype structure, and (*ii*) selective elimination of either centromeric or intercalary heterochromatin in a hypothetical ancestor group originally containing homologous heterochromatic sequences in both locations is not supported by the present evidence in the *S. siberica* alliance. As a further possibility, (*iii*) independent amplification of the same sequence in different chromosomal positions in different species should be considered. This kind of parallelism seems far from a reasonable probability when the genetical inertness of satellite DNA is considered, and when the choice for a sequence to be amplified would be by chance. However, in rodents Fry and Salser (1977) gained evidence for independent amplification of the same satellite DNA sequence in quite distantly related taxa separated by 40 million years of divergent evolution. According to their hypothesis, there is a limited set of potential sequences to be amplified, and an accompanying set of genes coding for recognition proteins associated with these sequences (library). Under conditions of selective pressure a certain sequence could be amplified, even in long separated taxa.

The sobering consequence of these considerations, from the viewpoint of taxonomy, is that independent amplification of one and the same sequence

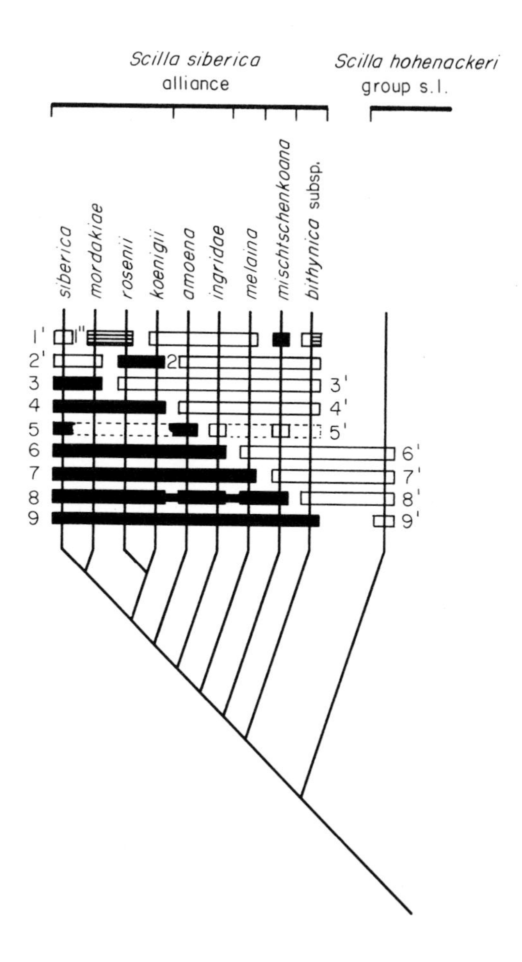

Fig. 3. Hypothesis of relationships in C-banded species of the *Scilla siberica* alliance. Outgroup: *Scilla hohenackeri* group s.l. Apomorphies black, plesiomorphies white, false synapomorphy (1″) hatched. 1–1′–1″ blocked banding – clustered banding – no or little banding, 2–2′ filaments broad – filaments narrow, 3–3′ heavy centromeric banding – little or no centromeric banding, 4–4′ "type 2b" elaiosomous seeds (see text) – other seed types, 5–5′ pattern of phenolic compounds simple – pattern complex (character distribution only incompletely known), 6–6′ seeds yellow – seeds black, 7–7′ one pair of NOR-chromosomes – two pairs, 8–8′ seeds primarily with elaiosomes (character transition indicated) – seeds primarily without, 9–9′ papillose testa – smooth testa. Top line indicates systematic grouping according to Speta (1980).

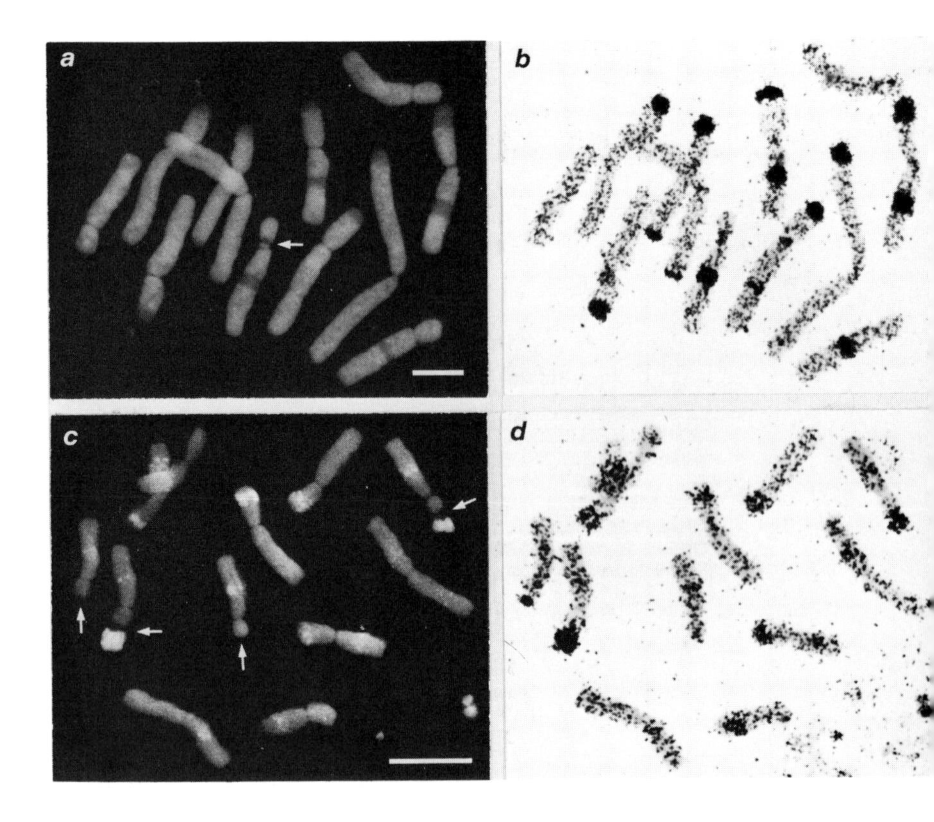

Fig. 4. Autologous *in situ* hybridization of satellite-DNA-cRNA onto fluorochrome stained chromosomes. (*a*) and (*b*) *Scilla siberica* "Spring Beauty" chromosomes (*a*, quinacrine staining), (*c*) and (*d*) *Scilla mischtschenkoana* (*c*, DAPI staining). Arrows mark NORs, bar represents 10 μm (after Deumling and Greilhuber 1982).

could also occur in *closely* related taxa in *similar* chromosomal positions. Ultimately, a distinction of heterochromatin identity by common ancestry or by parallelism could become very difficult.

I would even go further to consider a certain degree of probability for *independent* accumulation of heterochromatic sequences (homologous or not) according to *similar* patterns in different species. As is now widely acknowledged, C-bands are not restricted to centromeres but can occur at any chromosomal position. However, heterochromatin distribution is rarely random in a karyotype. Rather there is a regularity of patterns in the chromosome arms, as noted already by Heitz (1932) as the "principle of equilocal heterochromatin distribution". Recently, Greilhuber (1982) and Loidl (1981) have stated that

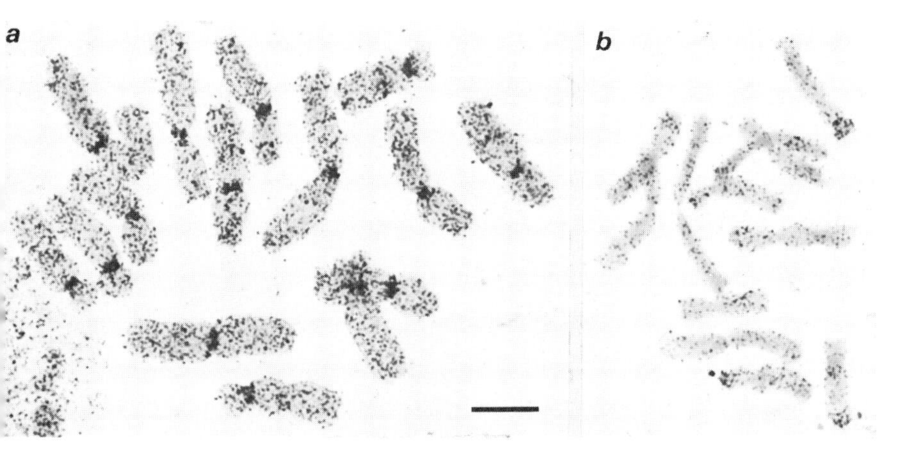

Fig. 5. Cross hybridization of satellite-DNA-cRNA in *Scilla siberica* and *S. mischtschenkoana*. (*a*) *Scilla siberica* chromosomes hybridized with *S. mischtschenkoana* satellite-DNA-cRNA, (*b*) *S. mischtschenkoana* chromosomes hybridized with *S. siberica* satellite-DNA-cRNA. Bar represents 10 μm (after Deumling and Greilhuber 1982).

banding patterns are frequently built up according to even more specific rules: (*i*) C-bands, which are supposed to be evolutionarily new, emerge in 2-armed chromosomes preferentially symmetrically banded (Greilhuber 1982). (*ii*) If intercalary bands occur, their position in the arms corresponds to the length of certain short arms in the karyotype; an unbanded region peripheral to the centromere corresponds to the length of the shortest chromosome arm in the karyotype (Loidl 1981).

The most plausible rationale behind these rules seems to be that heterochromatic sequences are mobile elements in a phylogenetic time scale, that they may be accumulated first at a telomere, and are then passed on from this site to other sites in spatial contiguity (see also Bennett *et al.* 1977), and so on. Furthermore, recent models of chromosome disposition indicate that in the interphase nucleus haploid genomes are arranged as chains, nonhomologous chromosome arms of similar size and, therefore, interphase volume being associated (Bennett 1982). This model would imply a tendency towards co-evolution of banding patterns in nonhomologous chromosome arms of similar size (Bennett 1982; Flavell 1982). This prediction is shown particularly well in *Scilla vindobonensis*, *S. voethorum* and *S. resslii* of the *S. bifolia* alliance (Greilhuber and Loidl 1983), which represent a likely case of evolutionary heterochromatin accumulation (see Fig. 6). Consequently the final C-band distribution may be deterimined in a given karyotype by the chromosome arrangement in the polarized nucleus. Chromosome disposition as well as the supposed determinant role of the shortest and other short chromosome arms in the creation of intercalary banding implies that the particular banding profile of a species can be ultimately derived from the external morphology of the karyotype itself. Karyotypes, which are similar due to common ancestry (or whatever reason), may tend to accumulate satellite DNA at similar loci independently. This, for instance, is probably the case in *S. siberica* and *S. mischtschenkoana*, which accumulated different sequences in highly similar positions, though in a different design (Fig. 1).

If, in addition to the taxa discussed above, we now add to our phylogenetic consideration *S. mordakiae* (Greilhuber and Speta 1978), *S. rosenii*, *S. koenigii*, *S. bithnica* subspp. (Fig. 2, see also Greilhuber 1982), and *S. melaina* (Fig. 1), the interpretation of banding patterns becomes a still more complex problem in terms of the evolutionary polarity of unbanded and heavily banded karyotypes. A parsimonious solution is presented in Fig. 3. As the out-group, the *S. hohenackeri* group s.l. with $x=5$ was chosen because it is the only group showing significant chromosomal similarities to the *S. siberica* alliance. This is particularly evident in species such as *S. bisotunensis* (Fig. 1f) or *S. furseorum* of the *S. bisotunensis* alliance *sensu* Speta (1980), which shows a relatively simple structural relationship to the karyotypes in the *S. siberica* alliance. It is hypothesized that black seeds without an elaiosome and two NOR-pairs are plesiomorphic, that seeds were modified in a transformation series 1a–1b–2a–2b (see above), and that 1 NOR pair as well as accumulated centromeric heterochromatin represent synapomorphies. Phytochemistry is considered as well. *S. rosenii* and *S. koenigii* show synapomorphies in flower structure. Heavy banding as an ancestral condition at least in *S. melaina*, *S. ingridae* and *S. amoena*, if not also in *S. bithynica* subsp. *bithynica*, promotes the assumption of symplesiomorphic heavy banding in *S. koenigii* and *S. siberica*, and heterochromatin paucity in *S. rosenii* and *S. mordakiae*, probably also in *S. bithynica* subsp. *radkae*, as a convergence. Any other hypothesis would create new parallelisms, including independent amplification of the same satellite DNA sequence in similar positions.

It is not the intention of this contribution to discuss the relation of phylogenetic analysis to taxonomy, but it is evident that only by phylogenetic analysis could the enormous resolution of banding and DNA data, as compared to conventional karyology, be properly (i.e. critically) used in taxonomy. This holds true also for more basic evolutionary studies, for instance, for the recognition and correct interpretation of so-called evolutionary trends or strategies, whether they have the character of transformation series or of parallelisms. It is appropriate to note that previous evolutionary considerations in the *Scilla siberica* alliance as well as in the *S. hohenackeri* group s.l. were based on in-group parsimony (Greilhuber and Speta 1978; Deumling and Greilhuber 1982) or the commonality principle (Greilhuber and Speta 1976), rather than on out-group comparison, and are therefore questionable.

(*b*) *Anacyclus*. In *Anacyclus* (Asteraceae, Anthemideae) Schweizer and Ehrendorfer (1976) performed a detailed C-banding study of 1 perennial and 5 annual taxa and considered many parameters, such as genome size, band numbers, amount of banding (terminal, centromeric, intercalary), in NOR-chromosomes and in the rest of the genome. Banding patterns were highly diagnostic for species relationships and discrimination of morphologically ill-separated taxa. The authors concluded, in short, that banding became stronger during the evolution from perennials to annuals. The cladistic analysis by Humphries (1981), which is based on a quite different data set, can be justified by the chromosome data (Ehrendorfer *et al.* 1977; Humphries 1981). By the out-group criterion, *within* the banded annuals the moderate banding appears ancestral to heavier banding, which in turn is ancestral to even stronger banding. However, the supposed evolutionary trend from light banding in perennials to enhanced banding in annuals rests on an intuitive assumption which cannot be derived from the data set. With the same justification one might assume that moderate banding is ancestral in *Anacyclus*, and light banding is one of the numerous apomorphies characteristic for the perennial taxon studied, *A. depressus*. Only if one of the two most basal annual taxa shared its banding style with *A. depressus* or, if not, with an out-group of *Anacyclus*, could the supposed trend from perennials to annuals be justified (see Humphries 1981).

EUCHROMATIN

Although heterochromatin is the most labile component in the genome and an *a priori* cause of variation in DNA content, euchromatin accounts for most of the dramatic differences in genome size which we observe among higher plants. In the following C-values are given as a measure of euchromatin variation, and it is assumed that they are not unduely biased by heterochromatin variation. Earlier reports on extensive DNA variation within taxa of coniferous trees have been due to methodical artefacts, as demonstrated by Teoh and Rees (1975). In angiosperms only a few investigations include large scale geographic screening of DNA constancy. Amounts are fairly stable in *Scilla bifolia* and *S. vindobonensis* (Greilhuber 1979), *Triticum monococcum* and *T. aestivum* (Furuta *et al.* 1978), and *Aira caryophyllacea* (Albers 1980). A distinction should be made between infra-specific variation at the level of subspecies or variety (which is still at the level of taxa) and continuous variation within, in the strictest sense, a homogeneous taxon, as was observed by Furuta *et al.* (1975) within and among *Aegilops squarrosa* varieties. Special consideration should be given to genome size variation in the annual inbreeders *Microseris bigelovii* and *M. douglasii* (Cichoriaceae), which have been measured with a method considered to be particularly accurate (Price *et al.* 1980, 1981a, b). In *M. bigelovii* no significant variation was found within populations, but populations differed significantly by up to 20%. In *M. douglasii*, in addition to between population variation, that within populations was significant. Eventually, high and low population levels could be correlated with mesic and xeric environments, respectively.

There are sound reasons for the assumption that, in each species, or even population, C-value adjustment is of adaptive importance quite independent of the informational content of DNA. Physical parameters of the genome (the nucleotype, Bennett 1972) such as the DNA amount and perhaps average base composition, amount of satellite DNA, indirectly influence the phenotype via nuclear and cell volume, surface-volume ratios, nuclear and cell cycle duration. A large genome means large nuclei, large cells and slow nuclear cycles. Minimum generation time in herbaceous plants is positively correlated with genome size (Bennett 1972). Price and Bachmann (1977) demonstrated that accelerated development can be achieved by different tactics, either by reducing genome size and shortening cell cycle duration, or by increasing genome size and increasing cell size. It is important to note that polyploidy, though multiplying genome and cell size, does not increase cell cycle duration (Bennett and Smith 1972; Verma and Lin 1979). The reason is very probably that the ratio of coding genes to inert DNA is not altered as is the case when non-genic DNA accumulates in a taxon at constant ploidy level. Gene numbers are not different in diploid related species with divergent DNA amounts (Roose and Gottlieb 1978).

The utility and importance of DNA values in taxonomy may perhaps best be demonstrated in the *Scilla bifolia* alliance, a group of plants with a constant basic number of $x=9$ and little evidence for major structural rearrangements (Greilhuber and Speta 1977; Greilhuber 1979b). Twenty-six species are recognized (Speta 1980). Most remarkable from the standpoint of traditional systematics is the incorporation of *Chionodoxa* Boiss. into this group as the series Chionodoxa (Boiss.) Speta (Speta 1976). The sectional grouping is preponderantly based on seed colour, which is yellow, grey, brown or black. C-values range from 4.2 to 9.4 pg (1C). Higher C-values are correlated with yellow seeds, intermediate values with grey or brown seeds, and intermediate to low values with black seeds. There is no parametric overlap between the taxonomic groups when euchromatin estimates rather than total DNA amounts are considered. The group as a whole appears to be monophyletic due to such supposed apomorphies as a predominantly bifoliate condition, a particular elaiosome type, and a nuclear endosperm. A hypothesis of phylogenetic relationships is presented in Fig. 6. *Scilla messeniaca* ($2n=18$) is the only taxon closely connected to the *S. bifolia* alliance, although it represents a morphological link to the *S. siberica* alliance too. By out-group comparison, almost free perigon, yellow seeds, *Polygonum* type embryo-sac, and very low heterochromatin amount are recognized as plesiotypic. All these traits are found in *S. kladnii*. Within the yellow seeded taxa, there is a heterochromatin-rich monophyletic group with Drusa I type embryo-sac, which exhibits a character transformation series from medium towards heavy banding and from high to low C-values. There is a conflict with the yellow seeded, tetrasporic *S. taurica*: if it roots basal to *S. vindobonensis* its lower DNA amount is a parallelism, if it roots basal to *S. decidua* the embryo-sac type shows parallel origin. *Scilla decidua* is the only grey-seeded species known. The seed colour is perhaps not a transition between yellow and brown, but more likely a unique derivative of a yellow condition. In terms of *S. kladnii*, smaller DNA amounts are apotypic to larger ones. A transformation series is recognized from 8.6 pg in *S. kladnii* to 4.2 pg in *S. nivalis*, and correlates nicely with the seed colour changes from yellow to brown to black seeds. In terms of *S. nivalis*, a connate perigon, white-eyed flower and Drusa I type embryo-sac are apotypic. Within the tetrasporic species, *S. nana* and *S. albescens* are characterized by synapotypic small flowers. In *S. luciliae* and *S. tmoli* the flower structure is plesiotypic with respect to *S. siehei*, but the exact position of *S. tmoli* is not clear. However, *S. tmoli* has accumulated heterochromatin with constant euchromatin amounts.

In the *S. bifolia* alliance DNA amounts have been a valuable source of phylogenetic and, consequently, also of taxonomic evidence. It appears now that the present classification in this group avoids polyphyly but not paraphyly. From a more general viewpoint, hypotheses regarding evolutionary trends could be discussed and tested more properly: (*i*) What accounts for the trend towards lower euchromatin amounts, which occurs even when heterochromatin is accumulated? In terms of nucleotypic adaptation, this trend may reflect a gradual environmental change during the history of the group, in which selection favoured suitable variants in peripheral stressed populations. (*ii*) Why

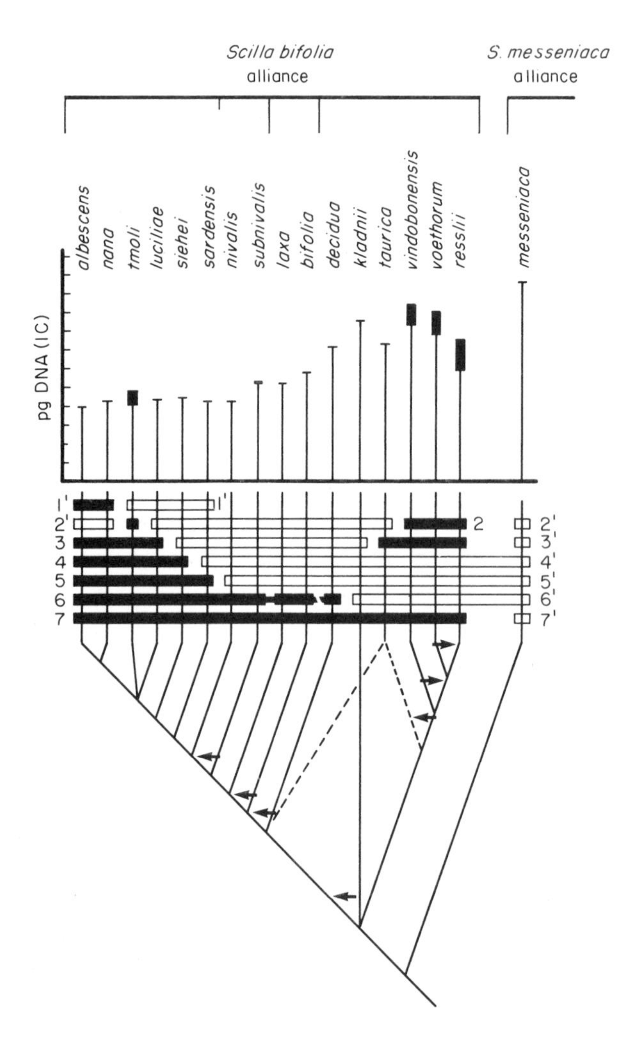

Fig. 6. Hypothesis of relationships in species of the *Scilla bifolia* alliance based on morphological criteria, C-banding and DNA contents. Outgroup: *S. messeniaca* (*S. messeniaca* alliance). Apomorphies black, plesiomorphies white. 1–1′ flowers small – flowers large, 2–2′ strong banding – little banding (note parallelism), 3–3′ *Drusa*-1-type embryo-sac – *Polygonum*-type embryo-sac (note parallelism), 4–4′ flower white-eyed – flower not eyed, 5–5′ perigon partly connate – perigon almost free, 6–6′ seeds not yellow (presumed character transition indicated) – seeds yellow, 7–7′ plants preponderantly bifoliate, nuclear endosperm type – plants plurifoliate, helobial endosperm type. Cladistic position in *S. taurica* ambiguous. Critical steps of banding increase and/or DNA reduction justifying clades in absence of clear morphological apomorphies indicated by arrows.

is heterochromatin accumulated, even when it creates what appears to be a nucleotypic contradiction in *S. vindobonensis*, *S. voethorum* and *S. resslii* (Fig. 6)? The currently most attractive hypothesis is that C-band accumulation restricts amount and location of meiotic recombination and therefore contributes to the maintenance of adaptively superior gene combinations. (*iii*) Did diversification in euchromatin amounts, once established, function as a reproductive barrier? The limited evidence available does not seem to support this hypothesis strongly (Gupta and Rees 1975; Price *et al.* 1981b). However, Jones *et al.* (1981) recently reported reduced fertility in chromosomally homomorphic hybrids of two subspecies of *Gibasis venustula* with significantly different C-values.

CONCLUSIONS

Considerable progress has been made in the past decade in plant chromosome technology, which karyosystematics is now starting to exploit. The methods for cytochemical dissection of the karyotype into eu- and heterochromatin, as well as a shift to quantitative karyology, have produced a new level of taxonomic resolution, and added significantly to karyomorphology and cytogenetics as the well established fundamentals of karyosystematics. Therefore, emphasis has been given in the present contribution to examples in which these approaches have been used for the clarification of specifically taxonomic problems. Evidently, the karyotype of higher plants exhibits an enormous degree of evolutionary plasticity, and events as improbable as the independent amplification of the same satellite DNA sequence may well be in its repertoire. Although it is obvious that the new approaches will contribute much to an improved systematics, it should be realized as well that only *phylogenetic analysis*, as opposed to traditional systematic, or pure phenetic, concepts, will eventually promote an understanding of chromosome evolution, and in particular of evolutionary trends or strategies.

ACKNOWLEDGEMENTS

I cordially thank Miss Janice Dawe, for correcting and typing the English manuscript. My thanks go also to Dr D. Schweizer for many discussions on this topic.

REFERENCES

Albers, F. (1980). Vergleichende Karyologie der Gräser-Subtriben *Aristaveninae* und *Airinae* (Poaceae-Aveneae). *Pl. Syst. Evol.* **136,** 137–167.

Badr, A. and Elkington, T. T. (1977). Variation of Giemsa C-band and fluorochrome banded karyotypes, and relationships in *Allium* subgen. *Molium*. *Pl. Syst. Evol.* **128,** 23–35.

Bennett, M. D. (1972). Nuclear DNA content and minimum generation time in herbaceous plants. *Proc. Roy. Soc. London.* (*B*) **181,** 109–135.

Bennett, M. D. (1982). Nucleotypic basis of the spatial ordering of chromosomes in eukaryotes and the implications of the order for genome evolution and phenotypic variation. *In* "Genome Evolution" (G. A. Dover and R. B. Flavell, eds) pp. 239–261. Academic Press, London, New York.

Bennett, M. D. and Smith, J. B. (1976). Nuclear DNA amounts in angiosperms. *Phil. Trans. Roy. Soc. Lond.* (*B*) **274,** 227–274.

Bennett, M. D., Gustafson, J. P. and Smith, J. B. (1977). Variation in nuclear DNA in the genus *Secale*. *Chromosoma* (*Berl.*) **61,** 149–176.

Bentzer, B. and Landstöm, T. (1975). Polymorphism in chromosomes of *Leopoldia comosa* (Liliaceae) revealed by Giemsa staining. *Hereditas* **80,** 219–232.

Blakey, D. H. and Vosa, C. G. (1981). Heterochromatin and chromosome variation in cultivated species of *Tulipa* subg. *Eriostemones* (Liliaceae). *Pl. Syst. Evol.* **139,** 47–55.

Blakey, D. H. and Vosa, C. G. (1982). Heterochromatin and chromosome variation in cultivated species of *Tulipa* subg. *Leiostemones* (Liliaceae). *Pl. Syst. Evol.* **139,** 163–178.

Bostock, C. J., Prescott, D. M. and Hatch, F. T. (1972). Timing of replication of the satellite and mainband DNAs in cells of the kangaroo rat (*Dipodomys ordii*). *Exp. Cell Res.* **74,** 487–495.

Bruhns, T. (1981). Cytological identification of the genomes in pentaploid *Allium neapolitanum* using Giemsa C-banding. *Pl. Syst. Evol.* **139,** 1–10.

Comings, D. E. (1978). Mechanisms of chromosome banding and implications for chromosome structure. *Ann. Rev. Genet.* **12,** 25–46.

Deumling, B. (1981). Sequence arrangement of a highly methylated satellite DNA of a plant, *Scilla*: A tandemly repeated inverted repeat. *Proc. Nat. Acad. Sci.* (*Wash.*) **78,** 338–342.

Deumling, B. and Greilhuber, J. (1982). Characterization of heterochromatin in different species of the *Scilla siberica* group (Liliaceae) by *in situ* hybridization of satellite DNAs and fluorochrome banding. *Chromosoma* (*Berl.*) **84,** 535–555.

Doolittle, W. F. and Sapienza, C. (1980). Selfish genes, the phenotype paradigm and genome evolution. *Nature* (*London*) **284,** 601–603.

Ehrendorfer, F. (1976). Evolutionary significance of chromosomal differentiation patterns in gymnosperms and primitive angiosperms. *In* "Origin and Early Evolution of Angiosperms" (C. B. Beck, ed.) pp. 220–240. Columbia University Press, New York, London.

Ehrendorfer, F., Schweizer, D., Greger, H. and Humphries, C. J. (1977). Chromosome banding and synthetic systematics in *Anacyclus* (Asteraceae-Anthemideae). *Taxon* **26,** 387–394.

Filion, W. G. and Vosa, C. G. (1980). Quinacrine fluorescence studies in *Paris polyphylla*. *Can. J. Genet. Cytol.* **22,** 417–420.

Flavell, R. B. (1982). Sequence amplification, deletion and rearrangement: Major sources of variation during species divergence. *In* "Genome Evolution" (G. A. Dover and R. B. Flavell, eds) pp. 301–323. Academic Press, London, New York.

Fry, K. and Salser, W. (1977). Nucleotide sequences of HS-α satellite DNA from kangaroo rat *Dipodomys ordii* and characterization of similar sequences in other rodents. *Cell* **12,** 1069–1084.

Fukuda, I. and Channell, R. B. (1975). Distribution and evolutionary significance of chromosome variation in *Trillium ovatum*. *Evolution* **29,** 257–266.

Fukuda, I. and Grant, W. F. (1980). Chromosome variation and evolution in *Trillium grandiflorum*. *Can. J. Genet. Cytol.* **22,** 81–91.

Funaki, K., Matsui, S. and Sasaki, M. (1975). Location of nucleolar organizers in animal and plant chromosomes by means of an improved N-banding technique. *Chromosoma (Berl.)* **49,** 357–370.

Furuta, Y., Nishikawa, K. and Haji, T. (1978). Uniformity of nuclear DNA content in *Triticum monococcum* L. *Jap. J. Genet.* **53,** 361–366.

Furuta, Y., Nishikawa, K. and Makino, T. (1975). Intraspecific variation of nuclear DNA content in *Aegilops squarrosa* L. *Jap. J. Genet.* **50,** 257–263.

Geber, G. and Hasibeder, G. (1980). Cytophotometrische Bestimmung von DNA-Mengen: Vergleich einer neuen DAPI-Fluoreszenzmethode mit Feulgen-Absorptions-photometrie. *Microscopica Acta, Suppl.* **4,** 31–35.

Gill, B. S. and Kimber, G. (1974). Giemsa C-banding and the evolution of wheat. *Proc. Nat. Acad. Sci. (Wash.)* **71,** 4086–4090.

Gillies, C. B. (1981). Electron microscopy of spread maize pachytene synaptonemal complexes. *Chromosoma (Berl.)* **83,** 575–591.

Giraldez, R. and Santos, J. L. (1981). Cytological evidence for preferences of identical over homologous but non-identical meiotic pairing. *Chromosoma (Berl.)* **82,** 447–451.

Greilhuber, J. (1973). Differential staining of plant chromosomes after hydrochloric acid treatments (Hy bands). *Öster. Bot. Z.* **122,** 333–351.

Greilhuber, J. (1975). Heterogeneity of heterochromatin in plants: Comparison of Hy- and C-bands in *Vicia faba*. *Pl. Syst. Evol.* **124,** 139–156.

Greilhuber, J. (1977). Why plant chromosomes do not show G-bands. *Theor. Appl. Genet.* **50,** 121–124.

Greilhuber, J. (1979a). C-band distribution, DNA content and base composition in *Adoxa moschatellina* (Adoxaceae), a plant with cold-sensitive chromosome segments. *Pl. Syst. Evol.* **131,** 243–259.

Greilhuber, J. (1979b). Evolutionary changes of DNA and heterochromatin amounts in the *Scilla bifolia* group (Liliaceae). *Pl. Syst. Evol., Suppl.* **2,** 263–280.

Greilhuber, J. (1982). Trends in the chromosome evolution of *Scilla* (Liliaceae). *Stapfia* **10,** 11–51.

Greilhuber, J. and Loidl, J. (1983). On regularities of C-banding patterns, and their possible cause. *In* "Kew Chromosome Conference. II" (P. Brandham and M. D. Bennett, eds) p. 344. George Allen and Unwin, London.

Greilhuber, J. and Speta, F. (1976). C-banded karyotypes in the *Scilla hohenackeri* group, *S. persica*, and *Puschkinia* (Liliaceae). *Pl. Syst. Evol.* **126,** 149–188.

Greilhuber, J. and Speta, F. (1977). Giemsa karyotypes and their evolutionary significance in *Scilla bifolia, S. drunensis,* and *S. vindobonensis* (Liliaceae). *Pl. Syst. Evol.* **127,** 171–190.

Greilhuber, J., and Speta, F. (1978). Quantitative analyses of C-banded karyotypes, and systematics in the cultivated species of the *Scilla siberica* group (Liliaceae). *Pl. Syst. Evol.* **129,** 63–109.

Greilhuber, J. Deumling, B. and Speta, F. (1981). Evolutionary aspects of chromosome banding, heterochromatin, satellite DNA, and genome size in *Scilla* (Liliaceae). *Ber. dt. Bot. Ges.* **94,** 249–266.

Gupta, P. K. and Rees, H. (1975). Tolerance of *Lolium* hybrids to quantitative variation in nuclear DNA. *Nature (London)* **257,** 587–588.

Haga, T. (1956). Genome and polyploidy in the genus *Trillium*. VI. Hybridisation and speciation by chromosome doubling in nature. *Heredity* **10,** 85–98.

Haga, T. and Kurabayashi, M. (1954). Genome and polyploidy in the genus *Trillium*. V. Chromosomal variation in natural populations of *Trillium kamtschaticum* Pall. *Mem. Fac. Sci. Kyushu Univ., E (Biol.)* **1(4),** 159–185.

Harmer, M. (1980). Beitrag zur Chemotaxonomie der Gattung *Scilla* s.l. M.Sc. Thesis, Vienna.

Heitz, E. (1932). Die Herkunft der Chromozentren. *Planta* **18,** 571–636.

Hennig, W. (1966). "Phylogenetic Systematics". University of Illinois Press, Urbana, Ill.

Holm, P. B. (1976). The C and Q banding patterns of the chromosomes of *Lilium longiflorum* (Thunb.). *Carlsberg Res. Comm.* **41,** 217–224.

Holmquist, G. (1979). The mechanism of C-banding: depurination and β-elimination. *Chromosoma (Berl.)* **72,** 203–224.

Humphries, C. J. (1981). Cytogenetic and cladistic studies in *Anacyclus* (Compositae: Anthemideae). *Nordic J. Bot.* **1,** 83–96.

Jackson, R. C. (1971). The karyotype in systematics. *Ann. Rev. Ecol. Syst.* **2,** 327–368.

Jewell, D. C. (1981). Recognition of two types of positive staining chromosomal material by manipulation of critical steps in the N-banding technique. *Stain Techn.* **56,** 227–234.

Jones, G. H. (1978). Giemsa C-banding of rye meiotic chromosomes and the nature of "terminal" chiasmata. *Chromosoma (Berl.)* **66,** 45–57.

Jones, K. (1978). Aspects of chromosome evolution in higher plants. *Adv. Bot. Res.* **6,** 119–194.

Jones, K., Bhattarai, S. and Hunt, D. (1981). Contributions to the cytotaxonomy of the Commelinaceae. *Gibasis linearis* and its allies. *Bot. J. Linn. Soc.* **83,** 141–156.

Kenton, A. (1978). Giemsa C-banding in *Gibasis* (Commelinaceae). *Chromosoma (Berl.)* **65,** 309–324.

Kongsuwan, K. and Smyth, D. R. (1977). Q-bands in *Lilium* and their relationships to C-banded heterochromatin. *Chromosoma (Berl.)* **60,** 169–178.

Kongsuwan, K. and Smyth, D. R. (1980). Late labelled regions in relation to Q- and C-bands in chromosomes of *Lilium longiflorum* and *L. pardalinum*. *Chromosoma (Berl.)* **76,** 151–164.

Kurabayashi, M. (1958). Evolution and variation in Japanese species of *Trillium*. *Evolution* **12,** 286–310.

Kurabayashi, M. (1963). Karyotype differentiation in *Trillium sessile* and *T. ovatum* in the western United States. *Evolution* **17,** 296–306.

La Cour, L. F. (1978). The constitutive heterochromatin in chromosomes of *Fritillaria* sp., as revealed by Giemsa banding. *Phil. Trans. Roy. London. (B)* **285,** 61–71.

Lewis, W. H. (ed.) (1980). "Polyploidy: Biological Relevance". Plenum Press, New York, London.

Linde-Laursen, I., Bothmer, R. v. and Jacobsen, J. (1980). Giemsa C-banding in Asiatic taxa of *Hordeum* section Stenostachys with notes on chromosome morphology. *Hereditas* **93,** 235–254.

Loidl, J. (1979). C-band proximity and absence of terminalisation in *Allium flavum* (Liliaceae). *Chromosoma (Berl.)* **73,** 45–51.

Loidl, J. (1981). Das Heterochromatin einiger *Allium*-Arten: Cytochemische Charakterisierung und cytogenetische Aspekte. Thesis, Vienna.

Marks, G. E. and Schweizer, D. (1974). Giemsa banding: Karyotype differences in some species of *Anemone* and in *Hepatica nobilis*. *Chromosoma (Berl.)* **44,** 405–416.

Moore, D. M. (1978). The chromosomes and plant taxonomy. *In* "Essays in Plant Taxonomy" (H. E. Street, ed.) pp. 39–56. Adademic Press, London, New York.

Nagl, W. (1974). Role of heterochromatin in the control of cell cycle duration. *Nature (London)* **249,** 53–54.

Nagl, W. and Ehrendorfer, F. (1974). DNA content, heterochromatin, mitotic index, and growth in perennial and annual *Anthemideae* (Asteraceae). *Pl. Syst. Evol.* **123,** 35–54.

Orgel, L. E. and Crick, F. H. C. (1980). Selfish DNA: the ultimate parasite. *Nature (London)* **284,** 604–607.

Pardue, M. L. and Gall, J. G. (1971). Chromosomal localization of mouse satellite DNA. *Science* **168,** 1356–1358.

Price, H. J. and Bachmann, K. (1976). Mitotic cycle time and DNA content in annual and perennial Microseridinae (Compositae, Cichoriaceae). *Pl. Syst. Evol.* **126,** 323–330.

Price, H. J., Bachmann, K., Chambers, K. L. and Riggs, J. (1980). Detection of intraspecific variation in nuclear DNA content in *Microseris douglasii*. *Bot. Gaz.* **141,** 195–198.

Price, H. J., Chambers, K. L. and Bachmann, K. (1981a). Genome size variation in diploid *Microseris bigelovii* (Asteraceae). *Bot. Gaz.* **142,** 156–159.

Price, H. J., Chambers, K. L. and Bachmann, K. (1981b). Geographic and ecological distribution of genomic DNA content variation in *Microseris douglasii* (Asteraceae). *Bot. Gaz.* **142,** 415–426.

Roose, M. L. and Gottlieb, L. D. (1978). Stability of structural gene number in diploid species with different amounts of nuclear DNA and different chromosome numbers. *Heredity* **40,** 159–163.

Schwarzacher, T., Ambros, P. and Schweizer, D. (1980). Application of Giemsa banding to orchid karyotype analysis. *Pl. Syst. Evol.* **134,** 293–297.

Schweizer, D. (1980). Fluorescent chromosome banding in plants: applications, mechanisms, and implications for chromosome structure. *In* "The Plant Genome" (D. R. Davies and R. A. Hopwood, eds.) pp. 61–72. *Proc. 4th John Innes Symp.*, Norwich 1979. John Innes Charity, Norwich.

Schweizer, D. (1981). Counterstain enhanced chromosome banding. *Hum. Genet.* **57,** 1–14.

Schweizer, D. and Ambros, P. (1979). Analysis of nucleolus organizer regions (NORs) in mitotic and polytene chromosomes of *Phaseolus coccineus* by silver staining and Giemsa C-banding. *Pl. Syst. Evol.* **132,** 27–51.

Schweizer, D. and Ehrendorfer, F. (1976). Giemsa banded karyotypes, systematics and evolution in *Anacyclus* (Asteraceae-Anthemideae). *Pl. Syst. Evol.* **126,** 107–148.

Speta, F. (1976). Über *Chionodoxa* Boiss., ihre Gliederung und Zugehörigkeit zu *Scilla* L. *Naturk. Jahrb. Stadt Linz* **21,** 9–79.

Speta, F. (1980). Die frühjahrsblühenden *Scilla*-Arten des östlichen Mittelmeerraumes. *Naturk. Jahrb. Stadt Linz* **25,** 19–198.

Stebbins, G. L. (1971). "Chromosomal Evolution in Higher Plants". Edward Arnold, London.

Svoma, E. (1981). Zur systematischen Embryologie der Gattung *Scilla* L. (Liliaceae). *Stapfia* **9,** 1–124.

Teoh, S. B. and Rees, H. (1976). Nuclear DNA amounts in populatios of *Picea* and *Pinus* species. *Heredity* **36,** 123–137.

Teppner, H. (1971a). Cytosystematik, bimodale Chromosomensätze und permanente Anorthoploidie bei *Onosma* (Boraginaceae). *Öster. Bot. Z.* **119,** 196–233.

Teppner, H. (1971b). Cytosystematische Studien an *Onosma* (Boraginaceae). Die Formenkreise von *O. echioides, O. helveticum* und *O. arenarium. Ber. dt. Bot. Ges.* **84,** 691–696.

Varley, J. M., Macgregor, H. C. and Erba, H. P. (1980). Satellite DNA is transcribed on lampbrush chromosomes. *Nature* (*London*) 283, 686–688.

Verma, R. S. and Lin, M. S. (1979). The duration of DNA synthetic (S) period in *Zea mays*: a genetic control. *Theor. Appl. Genet.* **54,** 277–282.

Vosa, C. G. (1976a). Heterochromatin patterns in *Allium*: 1. The relationship between the species of the *cepa* group and its allies. *Heredity* **36,** 383–393.

Vosa, C. G. (1976b). Heterochromatin patterns in *Allium*: 2. Heterochromatin variation in the species of the *paniculatum* group. *Chromosoma* (*Berl.*) **57,** 119–133.

Vosa, C. G. (1979). Heterochromatic banding patterns in the chromosomes of *Brimeura* (Liliaceae). *Pl. Syst. Evol.* **132,** 141–148.

Wiley, E. O. (1981). "Phylogenetics. The Theory and Practice of Phylogenetic Systematics". John Wiley, Chichester, Brisbane.

10 Variation, Polymorphism and Gene-flow within Species

Q. O. N. KAY

Department of Botany and Microbiology, University College of Swansea, Wales

Abstract: Many factors, some independent of one another but others interlinked to varying degrees, affect the level and nature of intraspecific genetic variation. The major factors that may be involved include habitat diversity, population size, structure and density, the nature and intensity of selection and the relative importance of stochastic processes, the breeding system, the type of pollen or seed dispersal and consequent patterns of gene-flow, the existence of barriers to gene-flow, interspecific hybridization and other interspecific interactions. Theoretical analyses of the roles of these processes are now complemented by an increasing body of observations of their effects in the field.

Intraspecific patterns of variation range from essentially continuous to sharply discontinuous types, both within populations and over the geographical or ecological range of a species. Most sympatric discontinuities are a consequence of genetic polymorphism, while allopatric discontinuities commonly result from barriers to gene-flow. A genetically variable species is most likely to be taxonomically subdivided if its pattern of variation shows clear discontinuities in easily observed morphological characters, especially when the forms are also geographically or ecologically distinct. These taxonomically recognized forms, whether they are sympatric or allopatric, may differ in only one or a few alleles or may be genotypically more widely separated from one another. In either case the existence of named forms is often a pointer to an evolutionarily interesting situation. At the same time the absence of taxonomically recognized variants is not necessarily an indication of genetic uniformity within a species, but may be a consequence of the absence of convenient discontinuities or easily recognizable diagnostic characters within a species that is in fact genetically variable.

Systematics Association Special Volume No. 25, "Current Concepts in Plant Taxonomy", edited by V. H. Heywood and D. M. Moore, 1984. Academic Press, London and Orlando.
ISBN 0 12 347060 9

INTRODUCTION

There are several possible approaches to the problem of variation at and below the species level. Although the best approach would probably be a broadly integrative one, in practice evolutionary studies tend to be fragmented into specialist fields which may differ markedly in their scope and sometimes have little contact with one another. The classical taxonomist, for example, may be primarily concerned to seek patterns, discontinuities and diagnostic characters that can be used to divide the range of variation into recognizable named taxa, and to reconcile these taxa and their levels in the taxonomic hierarchy with those that have been named within the group in the past. The difficulties that the taxonomist may encounter and the nomenclatural devices that have been used to classify variation around the species level have been reviewed by Stace (1978). The experimental taxonomist or genecologist may try to achieve a broader evolutionary understanding of the origins and inter-relationships of what he may initially regard as a group of populations by a range of techniques including comparative cultivation and studies of cytology and breeding relationships. Both classical and experimental taxonomists will draw on an increasing range of biochemical and micromorphological data, often provided by other specialists, and they will usually share a common vocabulary and the same broad objectives.

The approach of the geneticist is often rather different. The evolutionary or biochemical geneticist may investigate a group of plants if it appears to be particularly suitable for the study of the evolutionary processes in which he is interested, but may not be concerned with the overall pattern of variation. This is also true of the mathematical population geneticist, who is likely to study a group of organisms only if it has characteristics that appear to be particularly suitable for mathematical analysis or are related to the results of mathematical modelling.

At another level, the organic chemist or the comparative biochemist may study chemical or biochemical variation within a group of plants, using specialized techniques that can provide valuable new evidence for taxonomic and evolutionary studies; the comparative micromorphologist, using scanning electron microscope (SEM) or transmission electron microscope (TEM) techniques to reveal new characters, has a similar role, and like the chemist may work in the context of a wider study of the group. In contrast to these taxonomic and evolutionary approaches that emphasize diversity and variation, classical physiological and biochemical studies usually stressed unity rather than diversity and were extremely limited both in the range of organisms that were studied and in awareness of the types of variation that may exist within and among species. Physiological and biochemical genecologists have corrected this imbalance, and the patterns of variation and adaptation in physiological and biochemical characters that may exist in species that include populations exposed to salt stress, heavy metal toxicity and other limiting factors are increasingly well understood (Bradshaw 1971).

The interface between taxonomy and evolutionary studies is broad and active, and any attempt to discuss current work on intraspecific variation must necessarily be very selective. In this chapter the main emphasis has been placed on intraspecific variation in floral characters that may affect gene-flow and the overall pattern of variation as a result of interactions with pollinators, but some other related topics have also been considered.

DIOECY

Dioecy is of great evolutionary and ecological interest and provides the most extreme example of sharp genetic and morphological discontinuity without breakdown of genetic continuity and gene-flow within a species of flowering plant. Although the existence of dioecy within a species is often used as a taxonomic character, the male and female plants are not recognized as separate taxa unless their relationship to one another has not been discovered. Nevertheless they may differ from one another in fitness, abundance, size and ecological characteristics (Putwain and Harper 1972; Bawa *et al.* 1982) and may behave in some respects as different competing species. The selective processes that are thought to be involved in the origin and maintenance of dioecy and subdioecy have recently been reviewed and discussed by Bawa (1980) and Ross (1982).

Dioecy is widespread among angiosperms, and dioecious taxa are often closely related to taxa with hermaphrodite flowers. It is clear that dioecy has evolved independently from hermaphroditism on many occasions in different groups of flowering plants. Dioecy may arise directly from hermaphroditism, or indirectly through the intermediate stages of gynodioecy, androdioecy or monoecy (Bawa 1980). It has often been assumed that the evolution of dioecy or of the various intermediate stages that may lead to dioecy has resulted from selective presure for outcrossing, but many other factors may also have been involved.

Dioecious, subdioecious, gynodioecious and androdioecious populations all show a special type of polymorphism in which the population consists of two or more forms that differ in the allocation of resources during the reproductive cycle, and interact competitively in different ways: male plants compete to produce and disperse pollen, female plants compete to produce and disperse seed, and all forms compete among themselves and with the other conspecific forms at the vegetative level. The seasonal phasing of growth may differ in the different sexes (Gross and Soule 1981) and these and other differences between the sexes may result in more efficient use of the available resources, as was first suggested by Darwin (1877). The relative fitness of the different forms and of different genotypically controlled strategies of resource allocation will vary, however, depending on the relative importance of seed reproduction compared with vegetative reproduction. Thus the proportions of the sexes, or of unisexual and hermaphrodite plants, may be expected to vary in different populations, in different microhabitats and from year to year, and this is well-known to be the case in a number of examples that have been studied. Various types of stabilizing selection may act to control the ratios.

For the taxonomist interested in variation and evolution within species, several aspects of dioecy, subdioecy and gyno- or androdioecy provide opportunities for investigation. Dioecy and partial unisexuality are often overlooked, especially if the degree of morphological differentiation is small, and in cases where dioecy has been described the extent to which the species is in fact clearly or uniformly dioecious is often uncertain (Lloyd and Myall 1976; Ross 1982). Thus careful observation of floral structure and of pollen or seed production in the field may provide valuable new data. Field observations may also provide valuable data on the proportions of the different forms in natural populations and on the extent to which they are aggregated or spatially separated from one another (Barrett and Helenurm 1981). At another level, it is known that the great majority of dioecious angiosperms are zoophilous (Bawa 1980), not anemophilous as has often been incorrectly assumed. The pollination biology of zoophilous dioecious species, in which male and female flowers may differ widely in morphology, number and resource provision, is very poorly understood in the majority of cases, and the responses of seed dispersers and predators to the different resources provided by male and female plants are also largely unknown (Bawa 1980).

Investigation of pollinator behaviour in dioecious species may help to clarify the ways in which pollinators may respond to gynodioecious and androdioecious species and to comparable types of intraspecific floral variation in hermaphrodite species. For example, a study of pollinator discrimination in the dioecious insect-pollinated perennial herbs *Cirsium arvense* and *Silene dioica* using marked individual pollinators (Kay 1982) showed that some of the most important insect visitors discriminated very strongly between the sexes. In a mixed population of spaced plants of male and female *Cirsium arvense* honeybees (*Apis mellifera*) and some bumblebees (*Bombus* spp.) and wasps (*Vespula* spp.) showed strong or very strong preferences for female capitula. The number of visits to male capitula by nectar-seeking *Apis mellifera* usually appeared to be well below the minimum necessary for effective pollination; one individual, for example, made only 26 visits to male capitula but 773 visits to female capitula in a cumulative total of 799 observed visits (the percentage of male capitula in the population ranged from 24% to 38% during the period of observation). In a mixed population of spaced plants of male and female *Silene dioica* some nectar-seeking *Bombus* individuals showed equally strong preferences for female flowers; a *B. lapidarius* queen, for example, visited male flowers only 33 times but female flowers 216 times in a total of 249 observed visits (female flowers comprised only 18% of the population) while pollen-seeking *Apis mellifera* discriminated almost unerringly between male and female flowers of *S. dioica* in the same population, with only 5 visits to female flowers being seen in a total of 803 observed visits. Thus pollinator discrimination is likely to be a substantial selective disadvantage of dioecy in these dioecious species and may have comparably strong effects on gene-flow in other species with similar types of floral variation.

FLORAL SCENT

Variation in floral scent may be an important factor in gene-flow and differentiation at and below the species level. Butler (1951), Manning (1956) and Klostehalfen *et al.* (1978) have shown that under some conditions scent may be a more important determinant of floral choice by honeybees than colour and other factors, and scent may also determine the behaviour of other pollinators to an equal or greater extent. If pollinators discriminate between different floral scents changes in the scents, or in the range of pollinators that visit a species that varies in floral scent, may produce barriers to gene-flow, which may lead to further genetic differentiation. Intraspecific differences in floral scent are not easily observed or studied in the field and most of the few cases that have been studied also involve variation in other floral characters. For example, Loper and Waller (1970) have reported that bees discriminate between different races of *Medicago sativa* (alfalfa, lucerne) on the basis of differences in scent. Flower colour also varies in *M. sativa*, and several observers have reported that pollinators discriminate between different flower colour forms (Pedersen and Todd 1949; Clement 1965; Pedersen 1967) while Kauffeld and Sorenson (1971) studied the relationships between pollinator preferences and flower colour, scent and nectar production in this cultivated species.

Galen and Kevan (1980) investigated intraspecific floral polymorphism in both scent and colour in natural populations of *Polemonium viscosum* in Colorado. They reported that two floral scent morphs, "sweet" and "skunky" occurred; 62% of *P. viscosum* plants had sweet-scented flowers at higher altitudes, but only 39% at lower altitudes. Bumblebees preferentially visited sweet-scented flowers, but *Hyles lineata* (a sphingid moth) did not. The colour morphs also differed in altitudinal frequency, but no differences in their attractiveness to bumblebees were observed.

Among orchids and other groups in which pollinator specificity is particularly important, differences in floral scent sometimes appear to be the chief factor that maintains isolation between related species. Hills (1972) stated that differences in scent constituted the main barriers against interspecific hybridization in the orchid genus *Catasetum*; a change in only one of 39 compounds found in the floral scent could produce a new barrier. Cases in which pollinators are attracted to flowers by pheromonal scents, as in *Ophrys* (Kullenberg 1956, 1973) or by scents resembling those of specialized food-sources, as in *Arum conophalloides*, which apparently emits a scent resembling that of the skin of the mammals on which the blood-sucking female midges that are its chief pollinators feed (Knoll 1926) also demonstrate the potential importance of floral scent as a character that may be subject to intraspecific variation and polymorphism that is not easily apparent to the human observer, but may be of great evolutionary and ecological importance. It is still difficult to observe and quantify variation in floral scent but taxonomists and genecologists should be aware of its importance and should investigate it when it is practicable to do so.

ULTRAVIOLET FLORAL CHARACTERS

The ultraviolet patterning and absorption characteristics of the flower are even more cryptic than scent to the human observer, but are of great importance to many pollinators, and variation in ultraviolet characters may have a similar or greater influence on gene-flow and differentiation at and below the species level. Photography with suitable filters now provides a simple and fairly rapid technique for determination of the ultraviolet characteristics of flowers, at least at a qualitative level (Hill 1977; Kevan 1978; Kay 1979; Silberglied 1979) but it is very rare for taxonomists to consider ultraviolet floral characters and even floral biologists quite commonly fail to investigate them. Many pollinating insects, including bees (Daumer 1956, 1958; Labhart 1974), butterflies (Struwe 1972a, 1972b), moths (Höglund *et al.* 1973) and syrphid flies (Bishop and Chung 1972) have been shown to have essentially trichromatic colour vision with the near ultraviolet as one of their primary colours. The ultraviolet pattern of a flower is important for pollinator orientation (Manning 1956; Daumer 1958) and may also play an important role in discrimination between otherwise similar forms or species (Ornduff and Mosquin 1970).

Preliminary surveys of the ultraviolet characteristics of the flowers of a fairly wide range of species have been made by Kugler (1963, 1966) and others (see Silberglied 1979). These have shown that different species within some genera, e.g. *Potentilla*, may be distinguished from one another by differences in ultraviolet patterns. There is however a frustrating absence of published information about the ultraviolet floral characteristics of the great majority of species. Intraspecific variation in ultraviolet floral characteristics has only been studied in a handful of cases, but is of particular interest both for its possible role in the origin of interspecific differentiation and because of the possibility that intraspecific polymorphisms in ultraviolet components may parallel visible flower colour polymorphisms, and may also exist in species in which flower colour polymorphisms are not apparent to the human eye. Intraspecific variation in ultraviolet floral characters has been described by Ornduff and Mosquin (1970) in the *Nymphoides indica* complex, in which the pantropical white-flowered form has uniformly ultraviolet-absorbing flowers while the discontinuously distributed yellow-flowered forms show differing patterns of ultraviolet reflection in different areas. White and yellow-flowered forms both occur on the Isle of Pines near Cuba but it was not known whether both forms occurred in the same populations and were thus truly sympatric.

Although in some cases ultraviolet floral patterns are identical to the patterns that are visible to the human eye, and are sometimes produced by the same pigments that produce the visible pattern, in general ultraviolet absorption and especially ultraviolet patterns appear to be produced by flavone, flavonol or anthochlor pigments that are chemically distinct from the visible anthocyanin and carotenoid pigments and are either virtually colourless to the human eye, as in many Brassicaceae, or have yellow colours that are masked by yellow UV-transmitting carotenoid pigments present in the same flower, as in many Asteraceae and Fabaceae (Harborne 1972; Harborne and Smith 1978; Daoud 1980; Kay *et al.* 1981). The ultraviolet pigments may also be perceived by pollinators in a different fashion to visible pigments, and ultraviolet markings are often simpler, cruder in outline and cover a greater relative area of the flower than visible markings. Thus there are both biochemical and functional reasons for supposing that intra- and interspecific variation and differentiation in ultraviolet floral characters may differ in scope and nature from variation in visible floral characters.

visible
uv
a
b
c
d
e
f
g
h

From the limited data that are available, it is possible to draw some tentative conclusions about the types of differentiation in ultraviolet floral characters that occur in groups of related species. There are many cases where clear and unambiguous differentiation in the ultraviolet contrasts with a lack of differentiation in the visible range of the spectrum. Thus among the yellow-flowered species of the genus *Potentilla* some species have completely ultraviolet-absorbing flowers and others have flowers with patterns of ultraviolet reflectance and absorbance that often differ from species to species, although some species have similar ultraviolet patterns; the white-flowered species in the genus appear to be uniformly ultraviolet-absorbing (Kugler 1963). A similar pattern of differentiation has been observed in *Ranunculus* and the Brassicaceae-Brassiceae (Kugler 1963; Kay unpublished) and in the *Nymphoides indica* complex (see above). In the genus *Lotus* and in many other apparently uniformly yellow-flowered groups in the Fabaceae I have observed a surprising range of different patterns of ultraviolet absorption and reflection (Figs 1 and 2) and in this family, as in the yellow-flowered members of the Asteraceae, patterns of ultraviolet absorption may change as the flower or capitulum matures.

There are little or no data on intrapopulation variation in ultraviolet floral characters. In *Raphanus raphanistrum*, which has conspicuous visible petal polymorphisms both in lamina colour (ultraviolet-reflecting yellow or white) and in the presence or absence of anthocyanin veining (Kay 1976, 1978) the visible colour differences are not associated with significant differences in ultraviolet characters. The related species *R. sativus*, with which *R. raphanistrum* sometimes hybridizes, has white or anthocyanin-flushed petals which absorb ultraviolet light, and this character might reasonably be expected to pass into some populations of *R. raphanistrum* and perhaps to be maintained as a polymorphism by the same mechanisms that are responsible for maintaining the visible petal colour polymorphisms, but surveys of several large populations of *R. raphanistrum* in southern England and Wales for ultraviolet petal polymorphisms (Kay unpublished) have shown that plants with ultraviolet-absorbing petals only

Fig. 1. Visible and ultraviolet floral patterns in yellow-flowered *Lotus* species (Fabaceae). The left-hand photograph of each pair shows the pattern visible to the human eye, while the right-hand photograph shows the pattern in ultraviolet light of *c.* 330–400 nm, which can be seen by many insect pollinators. The photographs were taken in the field using electronic flash and Ilford FP4 film, with a Schott UG1 filter for the ultraviolet exposures (Kay 1979). (*a, b*) *Lotus edulis* L., Spain; (*c, d*) *L. ornithopodioides* L., Crete; (*e, f*) *L. uliginosus* Schkuhr, Britain; (*g, h*) *L. pedunculatus* Cav., Spain.

Fig. 2. Visible and ultraviolet floral patterns in yellow-flowered members of the Fabaceae, (see Fig. 1). (*a*, *b*) *Lathyrus annuus* L., Spain; (*c*, *d*) *Pultenaea linophylla* Schrad. E. Australia; (*e*, *f*) *Ononis viscosa* L., Italy; (*g*, *h*) *Scorpiurus vermiculatus* L., Spain.

occur at very low frequencies (less than 1 in 500 in the populations that have been surveyed so far) and that ultraviolet absorption is associated with some morphological abnormalities in the petals, suggesting that the character is not derived from *R. sativus* but is the result of a mutation. The white-flowered forms of *R. raphanistrum* are however most unusual in having ultraviolet-reflecting white petals; as was first reported by Richtmyer (1923) and Lutz (1924) most white-flowered plants have ultraviolet-absorbing petals, in which ultraviolet absorption is usually due to the presence of flavones or flavonols (Roller 1956) in solution in the cells of the petal epidermis (Kay *et al.* 1981). Rare mutant or recombinant forms in which the flowers were ultraviolet-reflecting as a consequence of the absence of flavone or flavonol pigmentation might thus be expected to occur in species which normally have ultraviolet-absorbing white flowers, in the same way in which rare albino-flowered forms lacking normal anthocyanin pigmentation occur in many species with red, purple or blue anthocyanin-pigmented flowers. These might reasonably be expected to become established as polymorphic forms in some species if the polymorphism is favoured by selection, in the same way as flower-colour polymorphisms which occur in some species with anthocyanin pigmentation.

FLORAL COLOUR

In contrast to variation in scent or in ultraviolet floral characters, variation in vislble flower colour is easily observed and has long been recognized at the taxonomic level. Differences in visible flower colour are often used as aids to the identification of species, and commonly contribute to the syndromes of floral differentiation which are thought to reduce gene-flow between related species in the field. At the level of intraspecific variation, flower colour variants can be observed in most species and are of much evolutionary interest. While they normally occur at very low frequencies and then probably result from mutations or recombinations that are at a selective disadvantage, flower colour polymorphisms, in which two or more forms occur sympatrically at relatively high frequencies, are not uncommon. Geographic variation in flower colour, with clinal changes over a geographic range, is also fairly frequent.

The traditional taxonomic treatment of flower colour variation tends to be unsatisfactory and often serves only to draw attention to its occurrence. Several factors have contributed to this unsatisfactory situation, among them the unavoidable reliance on herbarium specimens, in which flower colour may not be preserved, during the pioneer stage of taxonomy; the difficulty of accurately describing and comparing colours (Kevan 1978) and the consequent confusion of different forms, linked to the practical difficulties of unravelling synonymy and priority in infraspecific taxa; and, most frustrating for the evolutionary biologist, the all-too-common failure to give any information about the relative frequencies of different named forms and the extent to which they are clearly distinct from one another in the field. Flower colour variation in cultivated species is generally treated in a more satisfactory fashion. In most groups of cultivated plants the colours of the flowers of named cultivars will have been reasonably accurately described by reference to colour standards, and they are sometimes chemically characterized in terms of specific pigments or combinations of pigments, while the probable or certain origin of the cultivars is known in many cases. In wild species a comparable level of knowledge exists only in a relatively small number of groups that have been biosystematically studied. Accurate and consistent descriptions of flower colour forms and their relative frequencies are of great value and should whenever possible be included in local and regional Floras.

Several studies of the factors that may affect morph frequencies in species that show flower colour polymorphism have been published recently, and earlier work has been reviewed by Kay (1978). In a study of *Platystemon californicus*, a self-incompatible annual in which five different flower colour morphs predominate in different regions, Hannan (1981) found no evidence that selection by pollinators was involved, and suggested that the genes controlling flower colour might be linked with as yet undetected morphologically or physiologically adaptive characters. In a preliminary study of altitudinal variation in both floral colour (light blue, blue-purple or purple) and floral scent ("sweet" or "skunky") in *Polemonium viscosum*, Galen and Kevan (1980) reported that bumblebees discriminated between the scent forms but did not discriminate between the colour forms. They observed ralatively few visits (a total of 20 visits to flowers of known scent and 159 visits to flowers of known colour by bumblebees) and their conclusions must be regarded as very tentative, but their work is of particular interest as one of the few studies of the adaptive value of scent variants (see above). In another study of a montane species of the western United States, Miller (1981) investigated the factors that may affect geographic and locally polymorphic variation in flower colour in *Aquilegia caerulea*, a perennial herb in which flower colour varies from white to blue, with monomorphic white-flowered populations occurring in the northern and western parts of its range and monomorphic deep blue flowers at its south-eastern limits, with intermediate or mixed-colour populations between these extremes. He considered that the occurrence of flower colour variation in the species might be related to variations in the abundance and behaviour of sphingid moths, especially *Hyles lineata*. Galen and Kevan (1980) reported that this sphingid was a major visitor to polymorphic *Polemonium viscosum*, but did not give any data on its behaviour with respect to the colour forms although they considered that it did not discriminate between the scent forms of *P. viscosum*.

Waser and Price (1981) studied the factors that may affect the reproductive success of the pale or white-flowered ("albino") plants of *Delphinium nelsonii* which occur at low frequencies (about 0.1%) in normal blue-flowered populations in Colorado. They found that although the pale or white-flowered plants were vegetatively vigorous and of full potential fertility hummingbird and bumblebee pollinators discriminated against them with the result that they had a lower seed-set per plant than the blue-flowered plants; their low frequency in the population thus appeared to be maintained by an equilibrium between their production by mutation and their reproductive disadvantage *vis-à-vis* the blue plants. In another study of the factors that may affect the frequency of rare flower colour forms, Kay (1982) investigated the behaviour of pollinating insects in natural and artificial populations of the perennial herbs *Cirsium arvense* and *Succisa pratensis*, marking individual pollinators where possible. In both species the flowers are normally bluish-purple but pale or white-flowered individuals occur at low frequencies in natural populations. In the artificial populations white-flowered individuals and normal bluish-purple-flowered individuals, grown from seeds collected from the same wild populations, were mixed at random with the white-flowered plants at relatively high frequencies. Kay found that most butterflies, especially *Pieris* spp., discriminated moderately to very strongly against the white forms in both species; for example, in a total of 1192 observed visits by male and female *Pieris brassicae* to flowerheads of *Succisa pratensis*, 1190 were to normal bluish-purple heads and only 2 to white heads although white flower-heads formed 17%–20% of the population during the period of observation. In contrast to this it was found that bees (*Apis mellifera* and *Bombus* spp.) varied from individual to individual in the extent to which they discriminated between the colour forms; some individuals, especially of *A. mellifera*, discriminated quite strongly, usually in favour of the white form, but others failed to discriminate and some individuals changed or lost their preferences on subsequent days. It thus seemed possible that in natural populations discrimination by butterflies might be a factor that consistently reduced the fitness of the white morphs, but discrimination by bees might under some circumstances increase their fitness or maintain their frequency in the population. Mogford (1974, 1978) reported discrimination by bees between flower colour forms in natural polymorphic populations of *Cirsium palustre*, which shows a complex pattern of microgeographic and altitudinal variation in the frequencies of white, pale purple and purple flower colour morphs, with white morphs predominating in some populations and pale purple or purple morphs in others. Polymorphic populations do not however appear to occur in *Cirsium arvense* and *Succisa pratensis*. *Cirsium arvense* is dioecious, and pollinators generally discriminated more strongly between males and females than between the colour forms (see above).

In the self-incompatible weedy annual *Chrysanthemum coronarium*, which shows a complex pattern of variation in flower colour in the Mediterranean region, with white-rayed, cream-rayed and yellow-rayed morphs replacing one another geographically in some areas but occurring sympatrically as apparently balanced polymorphic populations in others, Kay (1982) studied pollinator behaviour in both experimental and natural populations. In an experimental population in Britain, marked individuals of the solitary bee *Colletes fodiens* showed a limited degree of discrimination, some preferring yellow-rayed plants but most preferring white-rayed plants (all forms show uniform and complete ultraviolet absorption) while the butterfly *Maniola jurtina* showed a strong preference for the yellow-rayed form. Most observations were made in a spontaneous population in Crete, where *C. coronarium* is abundant and extensively polymorphic, with white-, cream- and yellow-rayed forms occurring sympatrically at differing frequencies. Bees (*Apis mellifera*, *Colletes* spp. and *Xylocopa* sp.) either preferred the majority yellow-rayed form or showed no preference; their preferences were relatively weak and quantitative rather than qualitative. In contrast to this the butterfly *Pieris rapae* showed very strong individual preferences, with some individuals preferring yellow-rayed plants and others preferring white-rayed plants in the same polymorphic population; for example, in a population of *C. coronarium* with 64.0% of yellow-rayed plants, one *P. rapae* individual made 107 successive visits to yellow-rayed capitula and no visits to white-rayed capitula, while another individual made 55 visits to white-rayed capitula and only one visit to a yellow-rayed capitulum in a total of 56 visits. The *P. rapae* individuals ranged widely through the population, visiting only one or two capitula on each plant, and often flying several metres between visits; the white- and yellow-rayed plants were mixed randomly in the population and did not differ in size or habit. *P. rapae* is already known to discriminate strongly between the ultraviolet reflecting yellow- and white-flowered forms of *Raphanus raphanistrum*, but shows a consistent preference for the yellow-flowered form of that species (Kay 1976).

Although relatively few examples have been investigated, it is clear that pollinator discrimination is likely to be a significant factor in many cases of intraspecific flower colour variation. Several cases have however been described in which flower colour variation appears to be correlated with other factors, and no differential pollination has been observed. In addition to the case of *Platystemon californicus* which has already been mentioned, examples include *Anemone coronaria* (Horovitz 1976) and *Viola calaminaria* (Kakes and Everards 1976) in which flower colour variation is correlated with edaphic factors, *Eschscholzia californica* in which it is correlated with moisture availability (Frias *et al.* 1975) and *Epacris impressa*, in which it is correlated with the degree of shading of the habitat (Stace and Fripp 1977a, b). Nevertheless the particular interest of flower colour variation lies in its relationship to pollinator behaviour and the ways in which it may thus affect gene-flow, speciation and the barriers between species (Levin and Kerster 1967, Levin 1969) and the maintenance of special types of polymorphism within species (Kay 1978) as well as providing information about the factors controlling the behaviour of different types of pollinator and the nature of floral adaptations to particular groups of pollinators.

VARIATION AND GENE FLOW

Other types of intraspecific and intrapopulation variation range from those that are fairly clearly linked to adaptive interpretations, like the classical syndromes of characters involved in "ecotypic" adaptation to extreme habitats and in heavy metal tolerance (Bradshaw 1971) to those that are difficult to interpret in simple adaptive terms although they may provide examples of clear-cut polymorphisms linked to habitat differences, like the seed-coat and hairiness polymorphisms of *Spergula arvensis* (New 1958, 1978) and may also, like many examples of isozyme variation (Gottlieb 1976, 1977; Schoen 1982a, b) provide evidence of the levels of gene-flow and differentiation in a species or among related species.

The patterns of intraspecific variation that can be observed in plant species depend to a large extent on the interplay of gene-flow, selection and stochastic processes. The ways in which gene-flow may take place in seed plants, and the resulting patterns of variation, have been reviewed and discussed by Levin and Kerster (1974) and there have been many discussions of the plant/animal relationships involved in pollination systems (Gilbert and Raven 1975; Faegri and van der Pijl 1979) and seed dispersal systems (Wheelwright and Orians 1982). The apparently inefficient and localized nature of the gene-flow resulting from pollen and seed dispersal has been a contributory factor in the questioning of the classical concept of the biological species as a uniform, cohesive unit linked by free intraspecific gene-flow. This concept has been strongly criticized by Ehrlich and Raven (1969) who suggest that it is more realistic to take the population rather than the species as the unit of evolution. Ehrlich and Raven's ideas have been fully supported by Levin (1979), who states that "plant species lack reality, cohesion, independence and simple evolutionary or ecological roles."

If this view is accepted it follows that the population is the most natural unit of classification in many groups of plants, and that the species should often be regarded as an arbitrary but convenient taxon, comparable in many ways with the section or genus. Many taxonomists, especially perhaps those who are familiar with geographic patterns of variation in well-sampled species-rich Floras, will accept that there is much to be said for this point of view. Levin suggests that the classical delimitation of species may depend on the ignoring of what is deemed to be minor variation in order to establish a species grouping at an appropriate level of similarity; arbitrary groups of populations at a higher level of similarity would be regarded as races or subspecies, while groupings at a lower level of similarity would comprise genera. Nevertheless there are good reasons for supposing that many plant species are in fact biological species, unified by free gene-flow and separated from related species by breeding barriers, while others that may be interlinked by geographic or eoclogical clines with other species behave as biological species at the local or regional level. The distinction between intra- and interspecific variation may thus be partially valid, but one should perhaps accept that they often intergrade in plants, and that intrapopulation or intraspecific variation in characters that affect gene-flow is particularly likely to intergrade with interspecific differentiation in the same characters.

REFERENCES

Barrett, S. C. H. and Helenurm, K. (1981). Floral sex ratios and life history in *Aralia nudicaulis* (Araliaceae). *Evolution* **35,** 752–762.

Bawa, K. S. (1980). Evolution of dioecy in flowering plants. *Ann. Rev. Ecol. Syst.* **11,** 15–39.

Bawa, K. S., Keegan, C. R. and Voss, R. H. (1982). Sexual dimorphism in *Aralia nudicaulis* L. (Araliaceae). *Evolution* **36,** 371–378.

Bishop, L. G. and Chung, D. W. (1972). Convergence of visual sensory capabilities in a pair of Batesian mimics. *J. Insect Physiol.* **18,** 1501–1508.

Bradshaw, A. D. (1971). Plant evolution in extreme environments. *In* "Ecological Genetics and Evolution" (R. Creed, ed.) pp. 20–50. Blackwell, Oxford.

Butler, C. G. (1951). The importance of perfume in the discovery of food by the worker honey-bee (*Apis mellifera* L.). *Proc. Roy. Soc. Lond. (B) Biol. Sci.* **138,** 403–413.

Clement, W. M. (1965). Flower color, a factor in attractiveness of alfalfa clones for honeybees. *Crop Sci.* **5,** 267–268.

Darwin, C. (1877). "The Different Forms of Flowers on Plants of the Same Species". John Murray, London.

Daoud, H. S. (1980). "The Roles of Petal Pigments and Structure in Producing Ultra-Violet Colours and Patterns in Yellow and White Flowers". M. Sc. thesis, University College of Swansea, Wales.

Daumer, K. (1956). Reizmetrische Untersuchung des Farbensehens der Bienen. *Z. vgl. Physiol.* **38,** 413–478.

Daumer, K. (1958). Blumenfarben, wie sie die Bienen sehen. *Z. vgl. Physiol.* **41,** 49–110.

Ehrlich, P. R. and Raven, P. H. (1969). Differentiation of populations. *Science, N.Y.* **165,** 1228–1232.

Faegri, K. and van der Pijl, L. (1979). "Principles of Pollination Ecology" (3rd edn). Pergamon, Oxford.

Frias, D., Godoy, R., Iturra, P., Koref-Santibanez, S., Navarro, J., Pachecho, N. and Stebbins, G. L. (1975). Polymorphism and geographic variation of flower color in Chilean populations of *Eschscholzia californica. Pl. Syst. Evol.* **123,** 185–198.

Galen, C. and Kevan, P. G. (1980). Scent and color, floral polymorphisms, and pollination biology in *Polemonium viscosum* Nutt. *Amer. Midl. Nat.* **104,** 281–289.

Gilbert, L. E. and Raven, P. H. (eds) (1975). "Coevolution of Animals and Plants". University of Texas Press, Austin.

Gottlieb, L. D. (1976). Biochemical consequences of speciation in plants. *In* "Molecular Evolution" (F. J. Ayala, ed.) pp. 123–140. Sinauer Assoc., Sunderland, Mass.

Gottlieb, L. D. (1977). Electrophoretic evidence and plant systematics. *Ann. Missouri bot. Gard.* **64,** 161–180.

Gross, K. L. and Soule, J. D. (1981). Differences in biomass allocation to reproductive and vegetative structures of a dioecious perennial herb. *Silene alba* (Miller) Krause. *Amer. J. Bot.* **68,** 801–807.

Hannan, G. L. (1981). Flower color polymorphism and pollination biology of *Plastystemon californicus* Benth. (Papaveraceae). *Amer. J. Bot.* **68,** 233–243.

Harborne, J. B. (1972). Evolution and function of flavonoids in plants. *Recent Adv. Phytochem.* **4,** 107–141.

Harborne, J. B. and Smith, D. M. (1978). Anthochlors and other flavonoids as honey guides in the Compositae. *Biochem. Syst. Ecol.* **6,** 287–291.

Hill, R. J. (1977). Technical note: ultraviolet reflectance-absorbance photography; an easy, inexpensive research tool. *Brittonia* **29,** 382–390.

Hills, H. G. (1972). Floral fragrances and isolating mechanisms in the genus *Catasetum* (Orchidaceae). *Biotropica* **4,** 61–76.

Höglund, G., Hamdorf, K. and Rosner, G. (1973). Trichromatic visual system in an insect and its sensitivity control by blue light. *J. comp. Physiol.* **86,** 265–279.

Horovitz, A. (1976). Edaphic factors and flower colour distribution in the Anemoneae (Ranunculaceae). *Pl. Syst. Evol.* **126,** 239–242.

Jones, C. E. and Buchmann, S. L. (1974). Ultraviolet floral patterns as functional orientation cues in hymenopterous pollination systems. *Anim. Behav.* **22,** 481–485.

Kakes, P. and Everards, K. (1976). Genecological investigations on zinc plants. I. Genetics of flower colour in crosses between *Viola calaminaria* Lej. and its subspecies *westfalica* (Lej.) Ernst. *Acta Bot. Neerl.* **25,** 31–40.

Kauffeld, N. M. and Sorensen, E. L. (1971). Interrelations of honeybee preferences of alfalfa clones and flower color, aroma, nectar volume and sugar concentration. *Kanss. agric. Exp. Stn. Res. Publ.* **163,** 1–14.

Kay, Q. O. N. (1976). Preferential pollination of yellow-flowered morphs of *Raphanus raphanistrum* by *Pieris* and *Eristalis* spp. *Nature, London* **261,** 230–232.

Kay, Q. O. N. (1978). The role of preferential and assortative pollination in the maintenance of flower colour polymorphisms. *In* "The Pollination of Flowers by Insects" (A. J. Richards, ed.) pp. 175–190. Academic Press, London, New York.

Kay, Q. O. N. (1979). Ultraviolet photography of the colours and patterns of flowers. *Watsonia* **12,** 339–340.

Kay, Q. O. N. (1982). Intraspecific discrimination by pollinators and its role in evolution. *In* "Pollination and Evolution" (J. A. Armstrong, J. M. Powell and A. J. Richards, eds) pp. 9–28. Royal Botanic Gardens, Sydney.

Kay, Q. O. N., Daoud, H. S. and Stirton, C. H. (1981). Pigment distribution, light reflection and cell structure in petals. *Bot. J. Linn. Soc.* **83,** 57–84.

Kevan, P. G. (1978). Floral coloration, its colorimetric analysis and significance in anthecology. *In* "The Pollination of Flowers by Insects" (A. J. Richards, ed.) pp. 51–78. Academic Press, London, New York.

Klostehalfen, S., Fischer, W. and Bitterman, M. E. (1978). Modification of attention in honeybees. *Science, N.Y.* **201,** 1241–1243.

Knoll, F. (1926). Die Arum-Blütenstände und ihre Besucher (Insekten und Blumen. IV). *Abh. zool.-bot. Ges. Wien* **12,** 379–481.

Kugler, H. (1963). UV-Musterungen auf Blüten und ihr Zustandekommen. *Planta* **59,** 296–329.

Kugler, H. (1966). UV-Male auf Blüten. *Ber. dt. Bot. Ges.* **79,** 57–70.

Kullenberg, B. (1956). On the scents and colours of *Ophrys* flowers and their specific pollinators among the aculeate Hymenoptera. *Svensk bot. Tidskr.* **50,** 25–46.

Kullenberg, B. (1973). Field experiments with chemical sexual attractants on aculeate hymenoptera males. II. *Zoon Suppl.* **1,** 1–151.

Labhart, T. (1974). Behavioral analysis of light intensity discrimination and spectral sensitivity in the honey bee, *Apis mellifera*. *J. comp. Physiol.* **95,** 203–216.

Levin, D. A. (1969). The effect of corolla color and outline on interspecific pollen flow in *Phlox*. *Evolution* **23,** 444–455.

Levin, D. A. (1979). The nature of plant species. *Science, N.Y.* **204,** 381–384.

Levin, D. A. and Kerster, H. W. (1967). Natural selection for reproductive isolation in *Phlox*. *Evolution* **21,** 679–687.

Levin, D. A. and Kerster, H. W. (1974). Gene flow in seed plants. *Evol. Biol.* **7,** 139–220.

Lloyd, D. G. and Myall, A. J. (1976). Sexual dimorphism in *Cirsium arvense* (L.) Scop. *Ann. Bot.* **40,** 115–123.

Loper, G. M. and Waller, G. D. (1970). Alfalfa flower aroma and flower selection among honeybees. *Crop Sci.* **10,** 66–68.

Lutz, F. E. (1924). Apparently non-selective characters and combinations of characters, including a study of ultraviolet in relation to the flower visiting habits of insects. *Ann. N.Y. Acad. Sci.* **29,** 181–283.

Manning, A. (1956). Some aspects of the foraging behaviour of bumblebees. *Behaviour* **9,** 164–201.

Miller, R. B. (1981). Hawkmoths and the geographic patterns of floral variation in *Aquilegia caerulea*. *Evolution* **35,** 763-774.

Mogford, D. J. (1974). Flower colour polymorphism in *Cirsium palustre*. 2. Pollination. *Heredity* **33,** 257–263.

Mogford, D. J. (1978). Pollination and flower colour polymorphism, with special reference to *Cirsium palustre*. *In* "The Pollination of Flowers by Insects" (A. J. Richards, ed.) pp. 191–199. Academic Press, London, New York.

New, J. K. (1958). A population study of *Spergula arvensis*. I. Two clines and their significance. *Ann. Bot.* **22,** 457–477.

New, J. K. (1978). Change and stability of clines in *Spergula arvensis* L. (corn spurrey) after 20 years. *Watsonia* **12,** 137–143.

Ornduff, R. and Mosquin, T. (1970). Variation in the spectral qualities of flowers in the *Nymphoides indica* complex (Menyanthaceae) and its possible adaptive significance. *Canad. J. Bot.* **48,** 603–605.

Pedersen, M. W. (1967). Cross pollination studies involving three purple-flowered alfalfa, one white-flowered line, and two pollinator species. *Crop Sci.* **7,** 59–62.

Pedersen, M. W. and Todd, F. E. (1949). Selection and tripping in alfalfa clones by nectar collecting honey bees. *Agron. J.* **41,** 247–249.

Putwain, P. D. and Harper, J. L. (1972). Studies in the dynamics of plant populations. V. Mechanisms governing the sex ratio in *Rumex acetosa* and *R. acetosella*. *J. Ecol.* **60,** 113–129.

Richtmyer, F. K. (1923). The reflection of ultraviolet by flowers. *J. opt. Soc. Am.* **7,** 151–168.

Roller, K. (1956). Über Flavonoide in weissen Blumenblättern. *Z. Bot.* **44,** 477–500.

Ross, M. D. (1982). Five evolutionary pathways to subdioecy. *Amer. Naturalist* **119,** 297–318.

Schoen, D. J. (1982a). The breeding system of *Gilia achilleifolia*; variation in floral characteristics and outcrossing rate. *Evolution* **36,** 352–360.

Schoen, D. J. (1982b). Genetic variation and the breeding system of *Gilia achilleifolia*. *Evolution* **36,** 361–370.

Silberglied, R. E. (1979). Communication in the ultraviolet. *Ann. Rev. Ecol. Syst.* **10,** 373–398.

Stace, C. A. (1978). Breeding systems, variation patterns and species delimitation. *In* "Essays in Plant Taxonomy" (H. E. Street, ed.) pp. 57–78. Academic Press, London, New York.

Stace, H. M. and Fripp, Y. J. (1977). Raciation in *Epacris impressa*. I. Corolla colour and corolla length. *Austral. J. Bot.* **25,** 299–314.

Struwe, G. (1972a). Spectral sensitivity of the compound eye in butterflies (*Heliconius*). *J. comp. Physiol.* **79,** 191–196.

Struwe, G. (1972b). Spectral sensitivity of single photoreceptors in the compound eye of a tropical butterfly (*Heliconius numata*) *J. comp. Physiol.* **79,** 197–201.

Waser, N. M. and Price, M. V. (1981). Pollinator choice and stabilizing selection for flower color in *Delphinium nelsonii*. *Evolution* **35,** 376–390.

Wheelwright, N. T. and Orians, G. H. (1982). Seed dispersal by animals; contrasts with pollen dispersal, problems of terminology, and constraints on coevolution. *Amer. Naturalist* **119,** 402–413.

Ecology and Geography

11 Infraspecific Variation and its Taxonomic Implications

R. W. SNAYDON

Agricultural Botany Department, University of Reading, England

Abstract: The magnitude and pattern of infraspecific variation within any species depends upon the definition of the species. Apparent infraspecific variation will be increased by broad definition of the species, and diminished by narrow definition. The criteria for defining species (e.g. reproductive isolation or phenetic disjunction) are therefore important in determining the apparent magnitude and pattern of infraspecific variation. Conversely, the pattern of infraspecific variation is the basis for defining species in phenetic analyses; it may also be used to infer reproductive barriers. As a result, there is a danger of circular argument concerning the nature of infraspecific variation.

The special case of hybrids is considered. In some cases individuals or populations may not be truly hybrid, but may be self-perpetuating intermediates within a larger bimodal population. Greater care is needed in defining hybrids.

Sampling within and between populations is considered, and a plea made for more objective and more extensive sampling. Such sampling is essential if patterns of infraspecific variation are to be defined, and the limits of species firmly drawn. Recent examples of extensive sampling are considered, and the taxonomic implications discussed.

New techniques (e.g. isozyme analysis) and widespread sampling (e.g. gene bank collections) have led to renewed interest in infraspecific variation, both within and between populations. The results throw some doubt on previously held views concerning the relative importance of such factors as breeding system, migration and environmental variation in determining the magnitude and pattern of infraspecific variation.

Most studies of infraspecific variation have been undertaken for specific purposes, usually by non-taxonomists. As a result, the available taxonomic data are fragmentary. There is need for a more comprehensive, integrated and taxonomically oriented study of variation within species and species complexes. Some difficulties in the taxonomic treatment of that variation are discussed.

Systematics Association Special Volume No. 25, "Current Concepts in Plant Taxonomy", edited by V. H. Heywood and D. M. Moore, 1984. Academic Press, London and Orlando.
ISBN 0 12 347060 9

INTRODUCTION

Infraspecific variation in plants poses particular problems for taxonomists. Stace (1976) has pointed out that, as a result of these difficulties, taxonomic interest has shifted away from the infraspecific level in the last 50 years. This has led to a gradual decline in the use of the lower levels of the taxonomic hierarchy (Stace 1976). Probably the most important consequence of this taxonomic neglect of infraspecific variation is, as Burtt (1970) and Stace (1976) have stressed, that variation which is not named tends not to be recognized.

There are several practical and theoretical reasons why infraspecific variation poses problems for taxonomists. At its simplest, and most practical, there are difficulties in collecting, storing and retrieving sufficient herbarium material to represent adequately the variation which exists within each species. The question of sampling, in particular, will be considered later. There are also difficulties in collecting, storing and retrieving relevant data from available plant material, though these difficulties can usually be solved with available computer techniques. Finally, and ultimately most important, there are the theoretical difficulties of classifying plants on the basis of this infraspecific variation. This seems to have been the stumbling block on which most studies have foundered. One response to the difficulty has been to attempt to simplify infraspecific classification by restricting it to a single level – the subspecies (Burtt 1970; Raven 1974). These approaches will be considered later.

Perhaps the single most important difficulty faced by the taxonomist in handling infraspecific variation is the ambiguity in what constitutes taxonomic units. There is certainly an implicit assumption, and usually an explicit assumption, that the species is the "basic unit of formal taxonomy" (Davis and Heywood 1963). However, the operational taxonomic units are individual plants or, more often, herbarium sheets. As a result, the variation that exists within and between populations is interposed between the practical or observed units of taxonomy (i.e. individual plants) and the recognized theoretical units (i.e. species). The historical roots of this problem lie in typology, remnants of which still exist not only in taxonomy (e.g. type-specimens) but also in ecology (Antonovics 1976).

As a result of these various difficulties, infraspecific variation has been neglected by taxonomists. Most studies of infraspecific variation have been made by population biologists (e.g. population geneticists and ecological geneticists), who are usually interested in some specific aspect of variation (e.g. cytological or isozyme variation), and often in a particular context (e.g. a particular species, and a particular region or habitat). The resulting data are therefore fragmentary and often of limited use to the taxonomist.

In this paper I shall review some of the rather fragmentary data on infraspecific variation that have accumulated during the last few decades, and consider some of the implications that this has for taxonomy.

THE NATURE OF VARIATION

Initially, the most important advance that was made in the study of infraspecific variation was to distinguish between genetically determined variation and environmentally induced, phenotypic variation (Snaydon 1973). Most subsequent studies have attempted to eliminate environmental variation, either by growing plants under uniform conditions, or by measuring characters believed to be phenotypically stable (e.g. chromosome number and morphology, or isozyme complement). Not all studies have satisfactorily eliminated environmental effects, but these effects will not be considered further, except in so far as they have implications for the field identification of infraspecific variation.

The emphasis in studies of infraspecific variation has gradually shifted, over the last 50 years, largely reflecting concurrent developments in botany as a whole. Most of the early studies, from the pioneer experiments by Turesson and the classic studies by Clausen, Keck and Hiesey, concentrated on morphological variation. These have been reviewed by Heslop-Harrison (1964). Attention then shifted towards cytological variation, though the shift was not as dramatic as that in animal studies, where the work of Dobzhansky was particularly influential. Physiological variation next attracted attention, particularly the variation related to climatic and soil factors. Epstein and Jefferies (1964), Hiesey and Milner (1965) and Anotnovics *et al.* (1971) have reviewed these studies. More recently attention has shifted to biochemical variation (Johnson 1973; Hedrick *et al.* 1976; Brown 1979).

This trend in interest, from individuals to molecules and from genome to gene, has had important repercussions for taxonomists, since it is a trend from conspicuous morphological variation to cryptic variation (cytogenetic, physiological and biochemical). There has also been a parallel trend from qualitative to quantitative studies. As a result, many of the data that have been collected in the past few decades have been less attractive to taxonomists. The piecemeal way in which the data have accumulated has also reduced its value for taxonomic purposes, though it could be argued that all these data should be grist for the taxonomic mill. Although cryptic variation cannot be used for purposes of field identification, it can be used for purposes of classification (see below), and discriminant analysis can then be used to define conspicuous markers for the groupings so defined.

THE PATTERNS OF VARIATION

(1) The genetic pattern

The pattern of variation within species has been viewed differently at various times during the past 50 years. The general trend has been towards recognition of variation within progressively smaller groupings. Initially, variation within species was barely recognized: species were regarded as discrete and uniform groupings. Early genecological studies then demonstrated appreciable variation within species (Heslop-Harrison 1964). Many of those studies also showed evidence of variation within populations, but this tended to receive much less attention, and was largely regarded as random "noise" that had escaped selection. In recent years, within-population variation has received increasing attention, and studies have been made of the morphological, cytological and biochemical variation between individuals within populations. This variation has been increasingly seen as an adaptive feature of populations (Johnson 1973; Snaydon 1973; Bryant 1974; Soulé 1976; Nevo 1978) rather than as random or residual variation.

More recently, heterozygosity within individuals has attracted attention, with increasing interest in isozyme analysis. Surprisingly high levels of heterozygosity have been found (Johnson 1973; Hedrick *et al.* 1976; Brown 1979; Hamrick 1979) even in inbreeding species. These findings, and especially the high incidence of heterozygosity in inbreeders (Brown 1979), indicate selective advantages. There may be direct heterozygous advantage (Hedrick *et al.* 1976) or the advantage may be indirect, through the action of some kind of frequency-dependent selection (Ayala and Campbell 1974).

(2) The geographical pattern

Most of the early studies of the geographical pattern of variation were carried out on modest scales, both in terms of sampling intensity and the area covered, because of the limited amount of material that could be studied morphologically under controlled conditions. The recent developments in isozyme analysis (see above), and the large investment in collections of gene bank material (Frankel and Hawkes 1975), have provided opportunities for much wider study. For example, Jain *et al.* (1975) studied morphological variation in 3000 accessions of

Triticum turgidum, while Tolbert *et al.* (1979) studied 17 000 accession of *Hordeum vulgare*; both used the Shannon-Weaver index of diversity to measure overall genetic diversity within populations. Both studies showed complex geographical patterns of variation, with significant differences in diversity between countries and regions, but relatively weak clinal patterns overall. The pattern of variation within these crop species will have been influenced by human activity, such as migration, trade and artificial selection. No comparable extensive study has yet been made in any wild species. However, fairly extensive collections have been made of semi-natural pasture species, with up to 250 accessions, which have been studied morphologically (e.g. Edye *et al.* 1970; Burt *et al.* 1980; Williams *et al.* 1980). Geographical patterns of variation have been detected in some of those studies. Geographical patterns in isozyme frequency have also been found in natural populations (Johnson 1973; Nevo 1978) though, as Nevo (1978) has pointed out, there is not enough information to draw firm conclusions on the pattern of large-scale geographical variation.

(3) Analysis of patterns

Seen from the taxonomic viewpoint of classification at the species level, all variation within species, both between and within populations, is equally important. The integrity of species and their boundaries must be seen in the context of the total variation within the species. Valid definition of species and their boundaries depends on adequate, valid sampling of infraspecific variation (Snaydon 1973); this is especially important when species are defined phenetically. Few, if any, studies have yet been carried out in sufficient detail to satisfy these criteria.

When infraspecific variation is considered in its own right, it is essential to recognize the distinction between variation within populations and between populations. Seen from the viewpoint of population genetics and ecological genetics, different processes may determine the magnitude of interpopulation and intrapopulation variation. Several authors (e.g. Slobodkin and Rapoport 1974; Jain 1979) have commented that between-population variation, within-population variation, and within-genotype variation (i.e. heterozygosity and phenotypic flexibility) represent different adaptive strategies in response to different selection processes (e.g. spatial and temporal variation of the environment, seral or climax conditions). Variation between populations should be an effective strategy when environmental conditions vary spatially on a large scale. Variation between plants within populations should be an effective strategy when environmental variation in time is fairly short term (1–5 times the life cycle) or where spatial variation is fairly small-scale (1–10 times the size of the plant). Variation within genotypes, whether genetic, i.e. heterozygosity, or environmentally induced, i.e. phenotypic plasticity (Bradshaw 1965), would be effective where the environmental variation occurs within the life span of an individual or within its spatial range.

Seen from the viewpoint of classification of infraspecific variation, it would be possible to treat variation between and within pouplations either separately or together. Treated separately, it is possible to use multivariate techniques, such as canonical variates analysis (Jeffers 1978) which maximize the discrimination between populations. Although this may be valuable for some purposes, it could be argued that, for other purposes, variation between and within populations should be treated equivalently. However, if classification were carried out in this way, it seems likely that each population would contain individuals of several infraspecific taxa, so a complex and unworkable classification would result. Although there are no *a priori* reasons for treating variation between populations differently from that within populations, it might prove more practicable, therefore, to treat them separately. However, I know of no case where this has been tested.

MAGNITUDE OF VARIATION

The magnitude of infraspecific variation has rarely been measured quantitatively, mainly because there has rarely been adequate sampling of that variation. I know of no case where infraspecific variation within different species has been validly compared, though some comparisons (e.g. Babbel and Selander 1975) have been made on restricted samples. As a result, there is little sound information on which to draw conclusions. Evaluation is made more difficult by the numerous measures of variability that have been proposed and used (e.g. Lewontin 1966; Agnew 1968; Adams 1970; Coaldrake 1971; Latter 1973; Nei 1973; Brown 1979). These various methods may not give similar results (e.g. Davies 1977).

Sampling poses particular problems in measuring infraspecific variation. The geographical (and probably ecological) extent of sampling will affect the apparent magnitude of variation within species (Jain 1975; Stace 1976). Since the extent and intensity of sampling usually varies from study to study and species to species, it is extremely difficult to make valid comparisons of the variation within different species.

Various factors are said to affect the magnitude of infraspecific variation (Zangerl *et al.* 1977), these factors include: (*i*) the geographical and ecological range of the species (Van Valen 1965; Babbel and Selander 1975); (*ii*) geographical location (Agnew 1968; Soulé 1973); (*iii*) environmental heterogeneity (Hedrick *et al.* 1976); (*iv*) breeding system (Wright 1951); (*v*) the evolutionary age of the species (Soulé 1976; Hamrick 1979); (*vi*) plant longevity (Hamrick 1979). These factors may differ in importance, and may even have opposite effects, depending on whether variation is considered within or between populations. For example, inbreeding might be expected to reduce variation within populations, but increase variation between populations (Wright 1951).

The available evidence is, in most cases, conflicting and it is difficult to draw valid generalizations. Problems arise not only because of sampling differences (see above) but also because of the different forms of variability that have been measured (e.g. morphological cytological, and biochemical). These measures may be poorly correlated or even negatively correlated (Soulé 1973); the various indices of variation may also be poorly correlated (see above). In addition, it is usually difficult to disentangle the effects of the various factors. However, it seems that breeding system (Allard *et al.* 1968, but see Jain 1975, and Brown 1979) and plant longevity (Hamrick 1979) are relatively unimportant factors. On the other hand, environmental heterogeneity appears to be a very potent factor in determining the magnitude of variation both within and between populations (Levins 1968: Johnson 1973; Snaydon 1973; Hedrick *et al.* 1976; Brown 1979). Similarly, on a larger scale, the geographical and ecological range of the species appears to be an important factor determining the magnitude of variation between populations (Van Valen 1965; Babbel and Selander 1974; Nevo 1978; Brown 1979). However, it could equally be argued that the geographical and ecological range of the species is determined by the magnitude of genetic variation within it.

The apparent magnitude of variation within species will, of course, depend heavily on how the species is delimited. It is not the purpose of this chapter to consider the vexed question of what constitutes a species and how species should be defined, though I must declare my support for a phenetic, rather than a genetic or phylogenetic, approach. Species limits differ depending on whether phenetic disjunction or reproductive isolation are used as criteria (Solbrig 1967). As a result, the magnitude and pattern of variation within a species will depend on the criteria used to define the species.

INTERSPECIFIC HYBRIDS AND GENE-FLOW

Closely related to the question of species limits is the question of gene-flow between species, and hence interspecific hybrids. If there is appreciable gene-flow between species then, at the least, we may expect genetic variation within the recognized limits of those species to be increased. At the other extreme, hybridization may be so common that the species are not discrete, either phenetically or reproductively, and must be regarded as a single species, so increasing still further the extent of infraspecific variation. To what extent do these theoretical effects occur in the field? How common are interspecific hybrids, to what extent does hybridization lead to gene flow between species, and what proportion of individuals are intermediate?

Wagner (1967), Ornduff (1969), Heiser (1973), and Raven (1976) have evaluated the status and consequences of interspecific hybridization. I must confess that these conflicting reviews, the abundant scattered literature, and my own experiences in working on the ecological genetics of pasture species, leave me confused even about the frequency of interspecific hybridization, and even more confused about its consequences in determining the magnitude and pattern of infraspecific variation. One of the greatest problems is in satisfactorily identifying hybrids and hybrid swarms. In far too many studies, the presence of interspecific hybrids has been inferred merely from the presence of intermediate phenotypes in the field. In many cases, no attempt has been made to objectively sample the population and carry out a comprehensive multifactorial analysis of the data obtained, nor has there been an attempt to produce artificial hybrids for comparison.

One of the most important reasons for the conflicting views of interspecific hybridization are the apparent differences in the frequency of hybridization among plants with different breeding systems and life cycle strategies (Raven 1976). Early views of hybridization were coloured by the fact that short-lived, herbaceous species tended to be studied more commonly, and these species tend to be less subject to hybridization than long-lived woody species (Raven 1976). It also seems likely that early views were coloured by the current conception of species as being discrete and immutable. It is certainly apparent that, where biologists have more recently studied new areas (e.g. North America, Australia and New Zealand), large numbers of hybrids have been recognized (Raven 1976). More recent re-evaluations of long-studied areas have also shown widespread hybridization (Stace 1975).

While, on the one hand, the frequency of interspecific hybrids may have been underestimated, for reasons just considered, there is reason to suspect that some individuals and populations described as hybrids and hybrid swarms may not be hybrids at all. I suspect that some of these so-called interspecific hybrids, such as those in *Ranunculus* (Briggs 1962), in *Quercus* (Forde and Faris 1962) and in *Cercocarpus* (Brayton and Mooney 1966), may be intermediate populations in morphologically and ecologically diverse species. This, of course, leads back to the definition of species. Regardless of whether species are defined on the basis of reproductive isolation or phenetic disjunction, if intermediates occur commonly between two "species", they cannot be considered true species, since they will be neither reproductively isolated nor phenetically disjunct. The critical question then is "at what point does the frequency of intermediates become so great that the two groups can no longer be considered separate species?" This is a question that most population biologists would, I think thankfully, leave to taxonomists.

With this uncertainty as to what constitutes an interspecific hybrid, and how frequently such hybrids occur, it is difficult to evaluate the contribution of interspecific hybridization to infraspecific variation. In the absence of adequate and satisfactory data, it is tempting to approach the problem from a more theoretical viewpoint, and to consider the likely patterns of variation in response to known environmental patterns, and when constrained by certain reproductive systems. Viewed in this way, it is tempting to rephrase the question posed by Hutchinson (1968) and ask "Why are species necessary?"

THE ROLE OF SPECIATION

Several considerations point to the fact that the reproductive barriers, and morphological and physiological disjunction, associated with speciation may often be unnecessary, and can even be disadvantageous, from an adaptive viewpoint (Ehrlich and Raven 1969). Certainly, many plant species have exceptionally wide ecological and geographical distributions. Where such species have been studied, large amounts of ecotypic variation have usually been found. Such patterns are not really unexpected, at least with hindsight. Firstly, most spatial variation in environmental conditions is clinal, or consists of intergrading mosaics, rather than being distinct and abruptly changing. As a result, if populations and species are to fit their environment, there is a need for clinal variation rather than discrete variation. Such clinal patterns of variation are certainly widespread at the infraspecific level (see above). Secondly, environmental conditions are constantly changing in time, both randomly and cyclically, with frequencies ranging from centuries to minutes; these patterns of environmental variation, depending on frequency, tend to favour widespread intrapopulation variation and intra-individual variation, rather than uniform discrete groupings; again, this pattern is observed at the infraspecific level (see above). Set against this apparent requirement for clinal variation and intrapopulation variation, it has usually been assumed that some form of reproductive isolation is needed to prevent uncontrollable intermixing of the infraspecific variation. Most recent studies, however, have shown that the need for such devices has been greatly overestimated, since the magnitude of gene-flow is much less than was previously thought, while the magnitude of selection is often greater (Jain and Bradshaw 1966; Snaydon 1973; Levin and Kerster 1974). As a result, population divergence can occur over distances of less than 1 m (Snaydon and Davies 1976). In view of this, mechanisms of reproductive isolation between species seem largely unnecessary.

There is an increasing realization that such isolating barriers that exist between species may occur as a result of, rather than be an essential prerequisite for, differentiation (Lewis 1966; Ehrlich and Raven 1969; Schwarz 1974; Zouros 1974; Raven 1976; Van Valen 1976). It would seem therefore that the role of reproductive isolation in maintaining genetic diversity has been overemphasized in the past. Similarly, it could be argued that the role of species and speciation has probably been overemphasized. This is most apparent when considered from an ecological viewpoint. I have argued elsewhere (Snaydon 1973) that, because of large amount of ecotypic variation within many species, and the ecological similarity of many taxonomically diverse species, the species has only limited use in ecology. It is obviously unrealistic to expect a special purpose classification for ecological purposes, but perhaps a little of the attention, effort and enthusiasm being currently lavished on phylogeny might be devoted to the more practical and pragmatic problem of classification to meet present needs of field botanists. This would mean shifting attention from the higher levels of the hierarchy to the infraspecific level, and from floral characters to vegetative characters and to physiological and biochemical attributes (Snaydon 1973).

INFRASPECIFIC CLASSIFICATION

Although infraspecific variation has attracted little taxonomic attention recently (see above), there has been a surprisingly large number of attempts to classify and ordinate infraspecific variation. The studies have been undertaken for a variety of purposes. Some of the studies (e.g. Edye *et al.* 1970; Burt *et al.* 1971, 1980; Burt and Williams 1979; Robinson *et al.* 1980; Williams *et al.* 1980) are part of programmes to evaluate collections of pasture species for agriculture. Other studies (e.g. Lefèbvre 1974; Riggins *et al.* 1977; Johnstone and Hallam 1978; Prentice 1979, 1980; Dijk and Deldon 1981) seem to have been carried out primarily from an ecological genetic viewpoint, while a few studies (e.g. Drury and Randal 1969; Arroyo 1973; Clayton 1976; Duncan 1980) seem to have been carried out specifically for taxonomic purposes. The studies differ widely not only in the purposes for which they were carried out but, partly as a result, also in the characters used and the numerical methods applied. Morphological characters have been most commonly studied. Taxonomic studies were largely restricted to floral characters, while ecological genetic studies tended to concentrate on vegetative characters, and agricultural studies included agronomic characters as well as other vegetative characters. Only a few studies (e.g. Robinson *et al.* 1980; Dijke and Deldon 1981) have used isozyme complement.

Various methods of classification and ordination were used, ranging from cluster analyses to cladistic methods (Duncan and Baum 1981); in some cases several techniques were compared. No very strong pattern emerges, but classificatory methods seem to be used more commonly, and discontinuities more emphasized, in taxonomic studies, while ordination techniques and clinal patterns seem more emphasized in ecological genetic studies.

Most of the studies that have been carried out indicate that infraspecific variation is continuous, though those numerical methods which are based on sequential dichotomy tend to conceal this. Variation that was previously considered discrete is often found to be continuous, when further sampling is carried out (Fisher 1967). Too little attention has been given to sampling infraspecific variation, except perhaps in the context of the conservation of genetic resources (Frankel and Hawkes 1975).

The pattern of variation demonstrated by classification is also heavily dependent upon the attributes measured. Duncan and Baum (1981) have briefly considered the number and choice of characters, but surprisingly few studies have made direct comparisons of the effect of using different groups of attributes, e.g. floral vs. vegetative, morphological vs. physiological or biochemical. However, Robinson *et al.* (1980) found a remarkably good correlation between classifications based on isozyme complement and on morphological data. It would be interesting to know whether this is a common phenomenon.

So far, I have carefully avoided the vexed question of whether classification of infraspecific variation is desirable. I agree with Burtt (1970) and Stace (1976) that variation which is not described and, if possible, classified and named may not be recognized. This has important implications in taxonomy, where there is a danger that species may be regarded as monotypic, or at least highly discrete. There are more important implications in functional biology (ecology, physiology and biochemistry), where a monotypic view of species can be even more misleading (Antonovics 1976).

The question of whether infraspecific variation should be classified should, of course, be considered in the light of the uses to which that classification will be put. It is possible to envisage genetic, phylogenetic, ecological, biochemical or purely taxonomic classifications, and even sub-groupings of some of these. In the past it has proved fairly easy to subdivide species taxonomically into subspecies, on the basis of simple morphological characters. In some cases these subdivisions paralleled geographical, or even ecological distribution. However, those subspecies seem to have gained little favour with functional biologists or geneticists, presumably because they served little purpose. Would a more "natural" classification, or special purpose classification, prove to be more popular or useful? Does the fault lie less in the classification than in the users? Little attempt seems to have been made to evaluate currently available classifications, and even less to compare different classifications. This needs to be done in the context of the various purposes of classification (Warburton 1967), i.e. (*i*) ease of identification, (*ii*) convenience for communicating ideas, (*iii*) retrieval of information, and (*iv*) predictive value; I do not regard the other two purposes (demonstration of phylogenetic relationships and intellectual satisfaction) as relevant in the present context.

CONCLUSIONS

A welter of information on infraspecific variation has accumulated in the last few decades. Much of the evidence is apparently conflicting, but several general conclusions can, I think, be drawn from the numerous and diverse studies. (1) Most of the evidence indicates that infraspecific variation is usually continuous, rather than discrete, and usually involves many attributes, morphological, physiological and biochemical. Such complex patterns of variation can only be satisfactorily analysed by multivariate techniques. (2) Infraspecific variations exists at three levels: between populations, between individuals within populations, and within individuals. Variation at each of these levels has increasingly been seen as adaptive, rather than random or residual. (3) If the complex patterns of infraspecific variation are to be adequately described, and perhaps classified, they must first be adequately sampled; more widespread and objective methods of sampling will be needed than those normally used previously. (4) If infraspecific variation is to be *usefully* classified, the classification must be based on a wide range of attributes, morphological (vegetative and floral), physiological and biochemical, rather than on single, simple characters. However, after carrying out such a classification, it should be possible to use discriminant analysis to define simple morphological characters which can be used to distinguish between the groupings in the field.

It would be totally unrealistic to think that comprehensive studies of infraspecific variation will be carried out on more than a very small proportion of species. What is particularly needed, at the moment, is a series of comparative studies using contrasting species, i.e. species differing in extent of ecological or geographical distribution, in breeding system, or in taxonomic integrity. Such comparative studies should help to clarify the present rather confused state of knowledge on the magnitude and pattern of infraspecific variation, and the factors which determine that variation.

REFERENCES

Adams, R. P. (1970). Contour mapping and differential systematics of geographical variation. *Syst. Zool.* **19,** 385–390.

Agnew, A. D. Q. (1968). Variation and selection in an isolated series of populations of *Lysimachia volkensii. Evolution* **22,** 228–236.

Allard, R. W., Jain, S. K. and Workman, P. L. (1968). The genetics of inbreeding populations. *Adv. Gen.* **14,** 55–131.

Antonovics, J. (1976). The input from population genetics: "The New Ecological Genetics". *Syst. Bot.* **1,** 233–245.

Antonovics, J., Bradshaw, A. D. and Turner, R. G. (1971). Heavy metal tolerance in plants. *Adv. Ecol. Res.* **7,** 1–85.

Arroyo, M. T. K. (1973). A taximetric study of infraspecific variation in autogamous *Limnanthes floccosa. Brittonia* **25,** 177–191.

Ayala, F. J. and Campbell, C. (1974). Frequency-dependent selection. *Ann. Rev. Ecol. Syst.* **5,** 115–138.

Babbel, G. R. and Selander, R. K. (1975). Genetic variability in edaphically restricted and widespread plant species. *Evolution* **28,** 619–630.

Bradshaw, A. D. (1965). Evolutionary significance of phenotypic plasticity in plants. *Adv. Gen.* **13,** 115–155.

Brayton, R. and Mooney, H. A. (1966). Population variability of *Cercocarpus* in the White Mountains of California as related to habitat. *Evolution* **20,** 382–391.

Briggs, B. G. (1962). Interspecific hybridization in the *Ranunculus lappaceus* group. *Evolution* **16,** 372–390.

Brown, A. H. D. (1979). Enzyme polymorphism in plant populations. *Theoretical Population Biology* **15,** 1–42.

Bryant, E. H. (1974). On the adaptive significance of enzyme polymorphism in relation to environmental variation. *Amer. Naturalist* **108,** 1–19.

Burt, R. L. and Williams, W. T. (1979). Strategy of evaluation of a collection of tropical herbaceous legumes from Brazil and Venezuela. III The use of ordination techniques in evaluation. *Agro-ecosystems* **5,** 135–146.

Burt, R. L., Edye, L. A., Williams, W. T., Grof, B. and Nicholson, C. H. L. (1971). Numerical analysis of variation patterns in the genus *Stylosanthes* as an aid to plant introduction and assessment. *Austral. J. Agri. Res.* **22,** 737–757.

Burt, R. L., Williams, W. T., Gillard, P. and Pengelly, B. C. (1980). Variation within and between some perennial *Urachloa* species. *Austral. J. Bot.* **28,** 343–356.

Burtt, B. L. (1970). Infraspecific categories in flowering plants. *Biol. J. Linn. Soc.* **2,** 233–238.

Clayton, W. D. (1976). Some discriminant functions of *Hyparrhenia. Kew Bull.* **30,** 511–520.

Coaldrake, J. E. (1971). Variation in some floral, seed and growth characteristics of *Acacia harphophylla. Austral. J. Bot.* **19,** 335–352.

Davies, T. M. (1977). Rapid population differentiation in *Anthoxanthum odoratum.* Ph.D. thesis, University of Reading.

Davis, P. H. and Heywood, V. H. (1963). "Principles of Angiosperm Taxonomy". Oliver and Boyd, Edinburgh.

Dijk, H. van and Delden, W. van (1981). Genetic variability in *Plantago* species in relation to their ecology. I. Genetic analysis of allozyme variation in *P. major. Theor. Appl. Gen.* **60,** 285–290.

Drury, D. G. and Randal, J. M. (1969). A numerical study of the variation in the New Zealand *Erecatities arguta-scaberula* complex. *New Zealand J. Bot.* **7,** 56–75.

Duncan, T. (1980). A Cladistic analysis of the *Ranunculus hispidus* complex. *Taxon* **29,** 441–454.

Duncan, T. and Baum, B. R. (1981). Numerical phenetics: its uses in botanical systematics. *Ann. Rev. Ecol. Syst.* **12,** 387–404.

Edye, L. A., Williams, W. T. and Pritchard, A. J. (1970). A numerical analysis of variation patterns in Australian introductions of *Glycine wightii* (*G. javanica*). *Austral. J. Agri. Res.* **21,** 57–69.

Ehrlich, P. R. and Raven, P. H. (1969). Differentiation of populations. *Science, N.Y.* **165,** 1228–1232.

Epstein, E. and Jefferies, R. L. (1964). The genetic basis of selective ion transport in plants. *Ann. Rev. Pl. Physiol.* **15,** 165–184.

Fisher, F. J. F. (1967). The role of geographical and ecological studies in taxonomy. *In* "Modern Methods in Plant Taxonomy" (V. H. Heywood, ed.) pp. 241–259. Academic Press, London, New York.

Forde, M. B. and Faris, D. G. (1962). Effects of introgression on the serpentine endemism of *Quercus durata. Evolution* **16,** 338–347.

Frankel, O. H. and Hawkes, J. G. (1975). "Crop Genetic Resources for Today and Tomorrow". Cambridge University Press, Cambridge.

Hamrick, J. L. (1979). Genetic variation and longevity. *In* "Topics in Plant Population Biology". (O. T. Solbrig, S. Jain, G. B. Johnson and P. H. Raven, eds) pp. 84–107. Columbia University Press, New York.

Hedrick, P. W., Ginevan, M. E. and Ewing, E. P. (1976). Genetic polymorphism in heterogenous environments. *Ann. Rev. Ecol. Syst.* **7,** 1–32.

Heiser, C. B. (1973). Introgression re-examined. *Bot. Rev.* **39,** 347–366.

Heslop-Harrison, J. (1964). Forty years of genecology. *Adv. Ecol. Res.* **2,** 159–247.

Hiesey, W. M. and Milner, H. W. (1965). Physiology of ecological races and species. *Ann. Rev. Pl. Physiol.* **16,** 203–216.

Hutchinson, G. E. (1968). When are species necessary? *In* "Population Biology and Evolution" (R. C. Lewontin, ed.) pp. 177–186. Syracuse University Press, Syracuse, New York.

Jain, S. K. (1975). Population structure and the effect of breeding system. *In* "Crop

Genetic Resources for Today and Tomorrow" (O. H. Franklel and J. G. Hawkes, eds). Cambridge University Press, Cambridge.
Jain, S. K. (1979). Adaptive strategies: polymorphism, plasticity and homeostasis. *In* "Topics in Plant Biology" (O. T. Solbrig, S. Jain, G. B. Johnson, and P. H. Raven, eds) pp. 160–187. Columbia University Press, New York.
Jain, S. K. and Bradshaw, A. D. (1966). Evolutionary divergence among adjacent plant populations. I. The evidence and its theoretical analysis. *Heredity* **21,** 407–441.
Jain, S. K., Qualset, C. O., Bhatt, G. M. and Wu, K. K. (1975). Geographical patterns of phenotypic diversity in a world collection of durum wheats. *Crop Science* **15,** 700–704.
Jeffers, J. N. R. (1978). "An Introduction to Systems Analysis". Edward Arnold, London.
Johnson, G. B. (1973). Enzyme polymorphism and biosystematics; the hypothesis of selection neutrality. *Ann. Rev. Ecol. Syst.* **4,** 93–116.
Johnstone, P. C. and Hallam, N. D. (1978). Clinal variation in *Eucalyptus incrassata. Proc. Roy. Soc. Victoria* **91,** 193–206.
Latter, B. D. H. (1973). The estimation of genetic divergence between populations based on gene frequency data. *Amer. J. Hum. Gen.* **25,** 247–261.
Lefèbvre, C. (1974). Population variation and taxonomy in *Armeria martima* with special reference to heavy-metal-tolerant populations. *New Phytol.* **73,** 209–219.
Levin, D. A. and Kerster, H. (1974). Gene flow in seed plants. *Evol. Biol.* **7,** 139–220.
Levins, R. (1968). "Evolution in Changing Environments". Princeton University Press, Princeton, N.J.
Lewis, H. (1966). Speciation in flowering plants. *Science, N.Y.* **152,** 167–172.
Lewontin, R. C. (1966). On the measurement of relative variability. *Syst. Zool.* **15,** 141–142.
Nei, M. (1973). Analysis of gene diversity in subdivided populations. *Proc. Nat. Acad. Sci. (Wash.)* **70,** 3321–3323.
Nevo, E. (1978). Genetic variation in natural populations: patterns and theory. *Theor. Pop. Biol.* **13,** 121–177.
Ornduff, R. (1969). The systematics of populations in plants. *In* "Systematic Biology" (F. Blair, ed.) pp. 104–131. National Academy of Science, Washington, D.C.
Prentice, H. C. (1979). Numerical analysis of infraspecific variation in European *Silene alba* and *S. dioica. Bot. J. Linn. Soc.* **78,** 181–212.
Prentice, H. C. (1980). Variation in *Silene dioica*: numerical analysis of populations from Scotland. Watsonia **13,** 11–26.
Raven, P. H. (1974). Nomenclatural proposals to the Leningrad congress. *Taxon* **23,** 828–831.
Raven, P. H. (1976). Systematics and plant population biology. *Syst. Bot.* **1,** 284–316.
Riggins, R., Pimental, R. A. and Walters, D. R. (1977). Morphometrics of *Lupinus nana.* I. Variation in natural populations. *Syst. Bot.* **2,** 317–326.
Robinson, P. J., Burt, R. L. and Williams, W. T. (1980). Network analysis of genetic resources data. II. The use of isozyme data in elucidating geographical relationships. *Agro-ecosystems* **6,** 111–118.
Schwarz, S. S. (1974). Intraspecific variability and species formation. *Trans. 1st Theriol. Congr.* **2,** 136–139.
Slobodkin, L. B. and Rapoport, A. (1974). An optimal strategy of evolution. *Quart. Rev. Biol.* **49,** 181–200.

Snaydon, R. W. (1973). Ecological factors, genetic variation and speciation in plants. *In* "Taxonomy and Ecology" (V. H. Heywood, ed.) pp. 1–29. Academic Press, London, New York.

Snaydon, R. W. and Davies, M. S. (1976). Rapid population differentiation in a mosaic environment. IV. Populations at sharp boundaries. *Heredity* **37,** 9–25.

Solbrig, O. T. (1967). Fertility, sterility and the species problem. *In* "Modern Methods in Plant Taxonomy" (V. H. Heywood, ed.) pp. 77–96. Academic Press, London, New York.

Soulé, M. (1973). The epistasis cycle: a theory of marginal populations. *Ann. Rev. Ecol. Syst.* **4,** 165–187.

Soulé, M. (1976). Allozyme variation: its determination in space and time. *In* "Molecular Evolution" (F. J. Ayala, ed.) pp. 60–77. Sinaur Assoc., Sunderland, Mass.

Stace, C. A. (1975). "Hybridization and the Flora of the British Isles". Academic Press, London, New York.

Stace, C. A. (1976). The study of infraspecific variation. *Curr. Adv. Pl. Sci.* **23,** 513–523.

Tolbert, D. M., Qualset, C. O., Jain, S. K. and Craddock, J. C. (1979). A diversity analysis of a world collection of barley. *Crop Sci.* **19,** 789–794.

Van Valen, L. (1965). Morphological variation and the width of the ecological niche. *Amer. Naturalist* **94,** 377–390.

Van Valen, L. (1976). Ecological species, multispecies and oaks. *Taxon* **25,** 233–239.

Wagner, W. H. (1967). Hybridization, taxonomy and evolution. *In* "Modern Methods in Plant Taxonomy" (V. H. Heywood, ed.) pp. 113–136. Academic Press, London, New York.

Warburton, F. E. (1967). The purposes of classifications. *Syst. Zool.* **16,** 241–245.

Williams, W. T., Burt, R. L., Pengelly, B. C. and Robinson, P. J. (1980). Network analysis of genetical resources data. I. Geographical relationships. *Agro-ecosystems* **6,** 99–109.

Wright, S. (1951). The Genetic structure of populations. *Ann. Eugen.* **15,** 323–354.

Zangerl, A. R., Pickett, S. T. A. and Bazzaz, F. A. (1977). Some hypotheses on variation in plant populations and an experimental approach. *The Biologist* **59,** 113–122.

Zouros, E. (1974). Genic differentiation associated with the early stages of speciation in the *mulleri* subgroup of *Drosophila*. *Evolution* **27,** 601–621.

12 | Taxonomy and Geography

D. M. MOORE

Department of Botany, University of Reading, England

Abstract: The interplay between taxonomy and geography has a long and distinguished history. Indeed, the morpho-geographical approach to taxonomy, still important today, was employed well before Darwin and Wallace formally recognized the role of spatial isolation in speciation, while Hooker's enthusiastic acceptance of the theory of evolution showed its importance to phytogeographers as well as to taxonomists. The techniques and concepts of biosystematics, chemical taxonomy, numerical taxonomy, cladistics, geophysics, ecology and population genetics, for example, have all contributed to underlining the interdependence of taxonomy and geography in dealing with the problems of systematizing and understanding the diversity and distribution of the world's plants (and animals). This paper attempts to assess the current situation.

INTRODUCTION

For now we see through a glass, darkly; but then face to face; now I know in part; but then shall I know even as also I am known (Corinthians XIII: 12)

Whatever other interpretations may be placed upon it, the above part of one of Paul's epistles seems to be particularly apposite in considering the relationships between taxonomy and geography. As a traveller he would have been aware, like all voyagers, that the further apart regions are the more their Floras differ, because of the greater dissimilarities in the taxa within each genus, family and order that they share. Over the centuries the explanations for this phenomenon have been variously based upon divine creation in separate areas, evolution and dispersal of organisms over a stable world, and rafting of organisms in an ever-changing combination of continents. Each viewpoint has had, and has, its

---

Systematics Association Special Volume No. 25, "Current Concepts in Plant Taxonomy", edited by V. H. Heywood and D. M. Moore, 1984. Academic Press, London and Orlando.
ISBN 0 12 347060 9

zealots. Paul would have recognized the role of human prejudices in interpreting observations.

Twenty years ago the Systematics Association published *Taxonomy and Geography* (Nichols 1962), at a time when the techniques and concepts of neo-Darwinism influenced much taxonomic work – biosystematics was the fashion – and continental drift, though invoked by some biologists, was denied by most geophysicists. Today, evolutionary processes continue to be discussed: "The history of evolution is not one of stately unfolding, but a story of homeostatic equilibria disturbed only rarely . . . by rapid and episodic events of speciation" (Eldredge and Gould 1972), while plate tectonics is respectable and taxonomy has been enriched by the work of, for example, cytogeneticists, ecologists, chemists and computer experts. Furthermore, the current vogue for cladistics has led to the emergence of the rather curiously named vicariance biogeography – continuing evidence of the long history of the interplay between taxonomy and geography. This relationship was recognized at the formalization of modern taxonomy by Linnaeus, who included geographical data with his descriptions, by Darwin and Wallace, who explained the importance of spatial isolation in evolution, enthusiastically supported by the great phytogeographer Joseph Hooker, and has subsequently been aided by our continually increasing knowledge of evolving plants and this evolving planet. Why, then, should biblical references to partial knowledge and obscured vision be so appropriate to our present understanding of the relationships between taxonomy and geography?

It is because, although we have many data and concepts that can be generally accepted, we still lack many vital pieces of information about the evolving world and its plants, while the abounding theories on plants and their distribution cannot be tested experimentally. This paper attempts to assess current views on the interdependence of taxonomy and geography in systematizing and understanding the diversity and distribution of the world's plants against the background of changing vogues in taxonomy and geography.

THE CHANGING FACE OF TAXONOMY

(1) Morphology

The early observations that, in general, plants are more dissimilar the further apart they occur was, as noted above, formally recognized by Linnaeus. Darwin's explanation of the role in speciation of what we now call spatial isolation had an immediate impact on the theory, but not the practice, of taxonomy. Nevertheless, the morpho-geographical approach to taxonomy has had two principal practical consequences. Firstly, the major infraspecific category, the subspecies (with which the term variety has often been used synonymously), is normally used in the sense of Du Rietz (1930) to comprise a geographically related morphological mode in the variation pattern within a species. Secondly, the larger the distance separating two populations the greater the *a priori* assumption that they would belong to different taxa. Thus, for example, Dusén (1900) was virtually preconditioned to recognize a single collection of *Koenigia* from Tierra del Fuego as a new species, *K. fuegina*, because it was found about 13 000 km from the nearest occurrence of *K. islandica* L., with which it is clearly conspecific (Moore 1972). With this phenomenon may be linked what might be called "political speciation", whereby restricted views of variation patterns seem to have overemphasized the features of local populations, and any large scale regional revision contains revealing cases such as the following from "Flora Europaea" (Tutin *et al.* 1976), in which *Leontodon pyrenaicus* Gouan has both *L. cantabricus* Widder and *L. helveticus* Mérat included as synonyms.

The symposium (Nichols 1962) already referred to earlier, concluded with a "geographer's postscript" (Stamp 1962) in which the importance of accurately mapping taxa was emphasized. As Walters (1957) has documented, from the early nineteenth century distribution maps had progressed from the relatively crude indications of the limits of taxa to computer-based maps, such as those initiated for the British Flora (Perring and Walters 1962), which have now become such a prominent part of the taxonomic-geographic scene (e.g. Soper 1966; Adams 1974), though still for too few groups and too few parts of the world, while even 3-dimensional maps are now increasingly available (Tomlinson 1972).

Concomitant with this accurate mapping, however, is the need for the highest level of taxonomic accuracy. Over much of the world this still means that we must try to be sure that when, for example, we talk about a species having a north-temperate distribution, the Eurasiatic populations really do belong to the same taxon as those occurring in North America. However, it should be remembered that by circumscribing taxa we are effectively recognizing "packages" of variation and, from the morphological point of view at least, more detailed understanding of geographical variation patterns nowadays leans heavily on an accurate representation of the distributions of character-combinations. This seems to rest on two apparently disparate foundations – the cladistic view, which determines by some means what are primitive and advanced characters, and the neo-Darwinist view, which is, I believe, at least in part concerned with trying to find out how such differences can be recognized and explained within the complex fabric of demonstrable evolutionary change.

A good example of the former approach is that of Humphries (1981) who used "primitive" and "advanced" states of 22 characters to erect a phylogenetic scheme for *Nothofagus* and then compared this with the sequential fragmentation of Gondwanaland. Interestingly, this "analytical" study gave results agreeing with modern "narrative" studies (e.g. Moore 1972), which utilize data on the dispersal abilities of species, their apparent taxonomic relationships and the fossil record, supporting the important role of continental drift in determining the present distribution of the genus. The neo-Darwinist approach to studying the geography of character-combinations is expressed in the techniques and concepts of biosystematics, which will be considered next.

(2) Biosystematics

The use of cytogenetical principles and methods to assist our understanding of variation patterns and in deciding upon their taxonomic treatment – an approach which came to be known as biosystematics – was arguably the dominant force in taxonomy for the three decades or so up to about the middle 1960s. Those aspects of the subject dealing with the adaptation of populations to local environmental conditions, which effectively comprise genecology, basically formulated from the initial studies of Turesson in the 1920s, are treated by Snaydon elsewhere in this volume (Chapter 11). However, the confirmation of genetically based clines on a continental scale, such as those of Böcher (1949) in *Prunella vulgaris*, led to considerations as to how geographical differentiation and speciation may develop in spite of apparently continuous gene-flow (e.g. Endler 1977), thus counterbalancing the Darwinian view of the overwhelming importance of strictly allopatric speciation.

(*a*) *Morphology*. The theoretical and experimental basis provided by such studies undoubtedly gave some impetus to mapping the distributions of characters and character-combinations which, although their cytogenetical basis might be unknown, could yet be explained in terms of neo-Darwinian processes, in which gene-flow and the mechanisms modifying, controlling or preventing it are paramount. It has long been known that many genera show one or more centres of diversity, where most of their component species occur, and these have on occasions been rather tenuously regarded as centres of origin, with decreasing numbers of species towards the distributional limits of the genus; isoflors have often been used to document such situations. Similarly, mapping character-combinations frequently shows the same sort of pattern, with "centres of variability" grading outwards into areas of lower variation. In many cases these data seem to accord with ideas of biotype depletion during migration (e.g. Böcher 1972). In other cases, however, the areas where character-combinations display the greatest variability are towards a taxon's distributional limit, where it is hybridizing with a formerly allopatric taxon, often giving introgression on a regional basis, as was shown by Ehrendorfer (1958), for example, in the *Galium canum-graecum* complex of the east Mediterranean region and Anatolia. There are still relatively few attempts to map by means of computers the geographical distribution of character-combinations (but see, e.g. Adams and Turner 1970). This must be one of the more useful modern approaches to the morpho-geographical aspect of taxonomy, in the same way that combinations of taxa are now most readily handled by using computer programmes to delimit the phytochoria of floristic plant geography (Birks 1976).

(*b*) *Chromosomes*. Amongst the most prominent tools in biosystematics has been the use of chromosomes to chart some of the variation (which indeed they are to a considerable extent involved in producing and controlling) that it is the basic concern of the taxonomist to describe and understand. Indeed, it is probably fair to say that only data on chromosomes, or rather one aspect of them, their number, are available on anything like a sufficient scale to give some sort of a general view of cytogenetics in relation to taxonomy and geography, and even information on chromosome numbers ("cytogeography") lags far behind that which we have on morphological variation in this context. Nevertheless, it *was* data on chromosome numbers which drew attention to the geographical importance of polyploids, in Floras (e.g. Hagerup 1931; Löve and Löve 1949, 1971; Favarger 1967) and within individual genera such as *Eremophila* (Barlow 1971) and species such as *Themeda australis* (Hayman 1960), *Eriophyllum lanatum* (Mooring 1975) and *Symphytum officinale* (Gadella and Kliphuis 1972).

These indications of the incidence of polyploids in colonizing situations provided some of the earliest demonstrations of the impact of "modern" tools of taxonomy in understanding the dynamics of geographical variation (e.g. Darlington 1963). Since, under inbreeding, polyploids move more slowly towards homozygosity than diploids, their presence in the more dispersed marginal populations where in breeding may be more likely could thus be explained. Furthermore, since most natural polyploids are thought to be at least partly allopolyploid they will have levels of heterozygosity, and perhaps heterosis, commensurate with their hybridity – features of value to colonizers.

The value of chromosomal variation in charting the migrations of plant species and genera has also been increasingly shown by studies of structural heterozygosity, which itself promotes those features of polyploids already referred to.

Thus, for example, *Paeonia* (Walters 1942) and *Oenothera* (Cleland 1960) in North America, *Isotoma* (James 1970; Beltran and James 1974) in Australia and *Haworthia reinwardtii* var. *chalumnensis* (Brandham 1974) in South Africa show geographical patterns of chromosomal structural differences which help in understanding the routes by which these taxa achieved their present distributions. A particularly apposite example on a world scale is provided by *Epilobium*. The almost universal occurrence of $2n=36$ throughout this genus, with no indication of the interchanges prevalent in some other parts of the Onagraceae (Kurabayashi *et al*. 1962), led to it being considered chromosomally uninformative. However, Seavey and Raven (1977a, b, c, 1978) showed that chromosomal arrangements, differing by one or two interchanges, identify species from primarily North America (AA), Eurasia, Africa and Australasia (BB), and the circumboreal Alpinae (CC) – chromosomal taxonomic and geographical variation writ large! Recently, Solomons (1981) has studied in detail the difficult *Epilobium* species in South America. Apart from aliens, these have almost all been shown to possess either the AA or the BB chromosomal arrangements. One species, the Fuegian endemic *E. conjungens* Skottsb., shows many morphological features found in the creeping species of Australia and New Zealand. The similarities have been considered by Raven and Raven (1976) to result from convergent evolution and by Skottsberg (1906) and Solomons (1982) to reflect amphi-Antarctic affinities; as soon as information on the chromosomal arrangement (which may be predicted as BB) of *E. conjungens* becomes available its taxonomic and geographical status will be clarified beyond reasonable doubt.

(*c*) *Breeding Systems*. As Ornduff (1969), among others, has pointed out, studies of breeding systems occupy a prominent place in biosystematic investigations. This is not surprising, given the important role of the relationship between inbreeding and outbreeding in shaping the variation patterns which it is the aim of the taxonomist to describe and understand.

In the last twenty years or so many authors have pointed out that there is often a correlation between the breeding systems of plant populations and their distribution. Many pertinent references are given by Lloyd (1980), and essentially their message may be summarized as follows: populations at the margin of a species' range are likely to be more widely separated from each other than those in central regions, and more likely to suffer fluctuations in size because they are also ecologically marginal, conditions which may also tend to a paucity of pollinators. These factors are, of course, particularly important in areas colonized by long-distance dispersal, such as oceanic islands. Reference has already been made (p. 223) to chromosomal mechanisms militating against the homozygosity resulting from the higher frequency of inbreeding found in marginal populations. What other indicators are available from modern taxonomic studies?

The most widespread approach to the problem has been to examine the mechanical features controlling the breeding system. Thus, plants apparently capable of self-pollination, with the pollen and stigma of a flower maturing at the same time, may be considered self-fertilizing, and often this has been demonstrated experimentally, as has the converse mechanical and fertilizing syndrome. However, most experimental studies are still on far too limited a geographical scale, particularly where self-incompatibility systems are involved, since these can require detailed and arduous programmes to unravel. Baker (1959), however, demonstrated that such large-scale studies can be carried out on heteromorphic sporophytic self-incompatibility systems. In most of these the system of any population can be determined by examining styles, anthers, stigmas and pollen on herbarium sheets, so long as some experimental background is also obtained. Such studies over the years have demonstrated that groups with heteromorphic self-incompatibility almost always show a strong tendency to homomorphy in marginal, isolated populations. Similarly, dioecism is detectable morphologically and has been shown to be low in the floras of oceanic islands. The well-known correlation between apomixis and polyploidy is also significant in relation to the colonizing abilities of taxa.

However, the efficacy of self- and cross-fertilization in giving increased homo- or heterozygosity is very difficult to test on a large enough scale to be of much help in interpreting the geographical distribution of taxa. One approach which shows some promise in this respect is the use of isozymes and allozymes, detected by gel electrophoresis, to indicate the amount of homo- and heterozygosity at the gene loci determining the enzymes under study. By these means geographical patterns of heterozygosity have been found in a number of taxa. A good example is that of *Lycopersicon pimpinellifolium*, which Rick *et al.* (1977) showed to have its greatest variability in NW Peru, with a decrease both to the south and northwards into Ecuador. Within this general trend they were able to discern a number of component trends, termed simple regional clines, single-peaked clines, double-peaked clines, etc., which reflected the different geographical patterns of the alleles at the various gene-loci studied. The increasing popularity of such studies suggests that these may be the most significant way in which knowledge of differences in breeding systems and their consequence may be utilized in understanding geographical distributions in the future, but many technical and interpretative aspects of this approach still have to be clarified.

(3) Phytochemistry

As Heywood (this volume, p. 8) commented in his introduction, the rise of chemotaxonomy has been one of the noteworthy phenomena since "Modern Methods in Plant Taxonomy" (Heywood 1968), the forerunner of this symposium, was published. Even at that time it had become clear that there was frequently a correlation between the chemical constituents of plants and their geographical distribution. Such a correlation was early demonstrated, for example, in the flavonoids of Eurasiatic *Pyrus* (Challice and Williams 1968), Mediterranean *Crocus* (Bate-Smith 1968) and Australian–South American *Eucryphia* (Bate-Smith *et al.* 1967), the essential oils of *Monarda* (Scora 1967), and the terpenes of *Abies* (Zavarin *et al.* 1970), *Pinus* (Mirov 1967) and *Bursera microphylla* (Mooney and Emboden 1968). This last indicated that affinities between the populations in northern Mexico were in a north-south direction rather than over the much shorter east-west distance across the Gulf of California.

Even now there are still too few examples of studies, such as that on *Bursera*, involving detailed sampling of species or species-groups in relation to what Gottlieb (1980) has termed "ecogeographic phytochemistry" – "a novel approach to micromolecular systematics with an exceptionally high predictive value". One difficulty, no doubt, is that in many groups no clear chemogeographic pattern can be detected, perhaps because of local environmental effects, which seemed to be important in an abortive study that my colleague Professor Harborne and I undertook on the flavonoids of the widespread, geographically interesting species, *Phleum alpinum* L. Nevertheless, the power of this approach has been demonstrated. A careful study of the phenolic constituents of *Sophora* (Markham and Godley 1972) clarified the relationships of *S. microphylla* Aiton in New Zealand, Gough Island in mid-Atlantic and Chile, while a study of flavonoids in *Empetrum* (Moore *et al.* 1970) was similarly informative. This showed the overlap in flavonoid profiles between the Northern Hemisphere *E. nigrum* and the temperate South American *E. rubrum*. Furthermore, comparison of flavonoid profiles within *E. rubrum* indicated a correlation with actual and historical vegetation zones, while Similarity Index Values calculated for the flavonoid content of all populations provided strong clues to the migration routes taken by this species; the Tristan da Cunha populations, almost certainly derived by trans-oceanic dispersal from South America, were not dioecious as elsewhere in its range (see also Section 2c) and had a markedly depauperate flavonoid content. Interestingly, in a survey of flavonoids in *Ribes* Bate-Smith (1976) showed that the southernmost species in the Americas, *R.*

magellanicum, was similarly depauperate; such examples may reflect the limited synthetic ability in small marginal or isolated populations.

Proteins, much-discussed in claims and counterclaims about their importance in systematics and phylogenetic studies (e.g. Jensen 1981), have not so far yielded the same quantity of geographically useful information as micro-molecular substances. Nevertheless, analyses of non-protein amino acids in *Acacia* (Bell and Evans 1978) drew attention to palaeogeographic links between Australia and the Mascarene Islands, while a computer-generated "evolutionary tree" based on the composition of Fraction 1 proteins of *Nicotiana* (Chen *et al.* 1976) indicated that the genus could have radiated from an evolutionary centre in what is now South America to North America, Africa and Australia when the ancient supercontinents of Gondwanaland and Laurasia were still extant.

Gel electrophoresis has provided a powerful tool for isozyme and allozyme assays which has been enthusiastically adopted by population geneticists and biosystematists, although too often the data have been handled uncritically and both techniques and interpretation need further refinement (Hurka 1980). However, studies such as those of Rick and Fobes (1975) and Rick *et al.* (1977) on *Lycopersicon cheesmanii* and *L. pimpinellifolium* in western South America demonstrate the potential power of thorough investigations of isozyme and allozyme profiles in populations throughout a species' range for understanding some of the genetic factors involved in the attainment of geographical distributions by taxa. As Jensen (1981), among others, has observed, the stability of seed storage proteins should make them particularly useful in phylogenetic studies, and we may perhaps look forward to their future contribution to charting and understanding the geography of taxa.

One intriguing link between chemotaxonomy and geography involves the known positions of many compounds in biosynthetic pathways. Is a plant which accumulates a metabolite produced early in the pathway more "primitive" than one which accumulates a biosynthetically more derived substance? Harborne *et al.* (1976) and Harborne (1977), for example, have discussed this problem and pointed out the uncertainties involved in determining what constitutes a primitive or an advanced feature. However, Gottlieb (1978) and his co-workers (e.g. Gomes and Gottlieb 1980) have suggested a set of principles which can help in solving this problem and he has provided various examples involving primary precursors and secondary metabolites in alkaloid biosynthesis. Of particular interest here is his use of quinolizidine alkaloids as systematic markers in the Leguminosae subfam. Papilionoideae (Salatino and Gottlieb 1980). Based on the position of these alkaloids in what Gottlieb terms biogenetic maps, and their occurrence in genera of the Papilionoideae, Gottlieb (1980) has derived a very convincing map of the world indicating postulated migration routes of the subfamily which provides a firm basis for testing by any other techniques which are or become available.

THE CHANGING FACE OF GEOGRAPHY

Just as taxonomists have had to cope with the incorporation of varying kinds of information, described above, and the changing concepts about the origins of the taxa they study, ranging from special creation to the continuous flux of evolution, so have they had to set these against views on the geographical background of a world which changed from being immutable to one in which the present pattern of land and sea is but one in an everchanging kaleidoscope resulting from plate tectonics operating over aeons. When the known world was considered to have come into being in a relatively few days and the diversity of organisms was due to special creation taxonomy and geography faced no problems, plants and animals were there because they were there. The theory of evolution by natural selection was certainly assisted by the widespread understanding by the early nineteenth century of the chequered history of the earth. Darwin was aware, for example, of Forbes' (1846) recognition of the role of the glaciations in determining the distributions of plants and animals at higher latitudes, the gradually changing conditions of glacial advances and retreats according well with his ideas on the gradual selection of adaptive features which gave geographically related variation.

Until about twenty years ago most discussions of plant taxonomy and geography were dominated by neo-Darwinist views of plants dispersing by various means over the world. Whilst these views took account of the relatively well-known history of the land-masses, involving glaciations, marine-transgressions and retreats, mountain orogenesis, etc., they had to be set against a background of continents whose spatial relationships to each other and the intervening oceans had changed little through the ages. Of course, phytogeographers had long pointed to intercontinental floristic links which demanded explanation. Wegener (1929) developed ideas on continental drift which satisfied many biologists but which received little or no support from geophysicists.

Whilst some phytogeographers, such as Croizat (e.g. 1952), essentially ignored the geophysicists in interpreting the distributions of plants, the majority did not. Eustatic changes in sea-level, whose importance had been recognized since Forbes' (1846) publications, allowed periods of migration within many archipelagos and between continents and adjacent islands, but floristic links across the oceans could not be so explained. Long-distance dispersal (see e.g. Ridley 1930; Carlquist 1967) was used by some to explain these links, while others invoked submerged land-bridges (e.g. Van Steenis 1962) or such foundered continents as "Pacifica" (Melville 1966). However, in the 1960s the burgeoning evidence for continental drift was firmly supported by palaeomagnetic studies of the rocks of the ocean floors and plate tectonics became an accepted facet of the world's geography (e.g. Irving 1964).

With the acceptance of plate tectonics, taxonomists had an increasingly well-documented geographical background against which to set their studies on the diversity and distributions of plants. The long-known morphological affinities across the world's oceans in groups such as the Proteaceae (Johnson and Briggs 1963) were further supplemented and refined by data from other taxonomic disciplines. Thus, for example, the amino acids of seeds in *Acacia* (Bell and Evans 1978) assisted in assessing relationships between austral members of this genus, as have multidisciplinary studies of the amphi-Antarctic Restionaceae. The flavonoid chemistry of this family showed that the Australasian members are more closely related to each other than they are to the South African representatives (Harborne 1979), thus supporting anatomical data (Cutler 1972) indicating that Australasian and South African species, hitherto considered to be congeneric, belong to different genera. Similarly, within the classical austral family Podocarpaceae recent studies of ovule-development (Quinn 1982) and flavonoid chemistry (Quinn and Gadek 1981) have suggested that the only South American species of the genus *Dacrydium*, *D. fonkii*, should be separated as *Lepidothamnus* Phil., together with two New Zealand species, thus moving the relationships across the southern oceans from the family to the genus level.

There is little doubt that the general acceptance of plate tectonics has been of considerable importance to the interpretative aspects of the subject area encompassed by taxonomy and geography. However, there are indications that the zealots have become involved in this area with attractive siren-calls. As Nelson and Platnick (1981), in their stimulating book considering systematics and biogeography, said "But it *is* possible that the world is as orderly with respect to the distribution of organisms as it is with respect to their characters, and that nature, as Croizat maintained, 'forever repeats', in the distribution of group after group, the same pattern of area relationships, caused by the same events".

I came of age when major discontinuities in the geographical distributions of plant taxa were the result of *either* land-bridges *or* long distance dispersal. Nowadays it seems that there is a move to consider them to be the result of *either* plate tectonics *or* dispersal of some kind. This thesis *is* possible. But until I am sure that the rather similar disjunct distributions, between Australasia and South America, of the well-diversified tree genus *Nothofagus* (Van Steenis 1972), with a fossil history back at least to the Cretaceous break-up of Gondwanaland,

and of *Ranunculus biternatus* (Moore 1972), which now occurs on intermediate areas over the Southern Ocean completely glaciated some 15 000 years ago, result from similar events, I remain sceptical. The invocation of different evolutionary rates by some workers is not overly helpful in this instance.

WHITHER NOW?

Assuming that there is no major change in geophysical theory, and assuming that there will be increasingly exact information on the timing of the various events which brought the world's continents to their present positions and conditions, the onus for further progress in understanding the geography of plants lies with botanists. There is a great need for accurate mapping of distributions, which is itself inextricably intertwined with the requirement for accurate taxonomy at all levels of the hierarchy.

Accurate taxonomy will be aided by refining and employing the various tools available, and developing others. However, it is well known that in many instances different lines of evidence – morphology, anatomy, chemistry and cytogenetics, for example – do not accord in delimiting taxa, and this has also been shown to be true in relation to geographical distribution. Thus, for example, Lewis and Bloom (1972) showed that morphological discontinuities marked one zone of contact between *Clarkia speciosa* Lewis & Lewis and *C. nitens* Lewis & Lewis, while cytological features marked another discontinuity some 100 km distant. Similarly, a group of taxa, including species of *Blennospermum*, *Chamissonia*, *Clarkia*, *Lasthenia* and *Phacelia*, showing essentially the same amphitropical disjunct distribution between temperate western North and South America, revealed markedly dissimilar degrees of morphological, cytological and genic differentiation (Moore and Raven 1970); Endler (1977: 7–8) has drawn attention to comparable phenomena in animals.

We need to know whether differences such as those just mentioned result from dissimilar rates of evolution of the various features, which may be suspected from evolutionary taxonomic studies, or from distinct phases of migration, or some combination of several such factors. What is certain is that attaching overwhelming importance to one kind of taxonomic information, to one method for assessing the data or to one kind of geographical event is premature. Given the diverse ways in which plants evolve and extend their ranges, is it not more reasonable to keep assessing the geographical variation of plants, fossil as well as contemporary, with the increasingly refined tools available and thus asking the plants, in relation to the currently known geographical parameters, whence they come and how?

REFERENCES

Adams, R. F. (1974). Computer graphic plotting and mapping of data in systematics. *Taxon* **23,** 53–70.

Adams, R. F. and Turner, B. L. (1970). Chemosystematic and numerical studies in natural populations of *Juniperus ashei* Buch. *Taxon* **19,** 728–751.

Baker, H. G. (1959). The contribution of autecological and genecological studies to our knowledge of the past migrations of plants. *Amer. Naturalist* **93,** 255–270.

Barlow, B. A. (1971). Cytogeography of the genus *Eremophila. Austral. J. Bot.* **19,** 295–310.

Bate-Smith, E. C. (1968). The phenolic constituents of plants and their taxonomic significance. II. Monocotyledons. *J. Linn. Soc.* (*Bot.*) **60,** 325–356.

Bate-Smith, E. C. (1976). Chemistry and taxonomy of *Ribes. Biochem. Syst. Ecol.* **4,** 13–23.

Bate-Smith, E. C., Davenport, S. M. and Harborne, J. B. (1967). Comparative biochemistry of flavonoids. III. A correlation between chemistry and plant geography in the genus *Eucryphia. Phytochemistry* **6,** 1407–1413.

Bell, E. A. and Evans, C. S. (1978). Biochemical evidence of a former link between Australia and the Mascarene Islands. *Nature* (*London*) **273,** 295–296.

Beltran, I. C. and James, S. H. (1974). Complex hybridity in *Isotoma petraea*. IV. Heterosis in interpopulational hybrids. *Austral. J. Bot.* **22,** 251–264.

Birks, H. J. B. (1976). Distribution of European pteridophytes: a numerical analysis. *New Phytol.* **77,** 257–287.

Böcher, T. W. (1949). Racial divergence in *Prunella vulgaris* in relation to habitat and climate. *New Phytol.* **48,** 285–314.

Böcher, T. W. (1972). Evolutionary problems in the Arctic flora. *In* "Taxonomy, Phytogeography and Evolution" (D. H. Valentine, ed.) pp. 101–113. Academic Press, London, New York.

Brandham, P. E. (1974). Interchange and inversion polymorphism among populations of *Haworthia reinwardtii* var. *chalumnensis. Chromosoma* (*Berl.*) **47,** 85–108.

Carlquist, S. (1967). The biota of long-distance dispersal. V. Plant dispersal to Pacific Islands. *Bull. Torrey Bot. Club* **94,** 129–162.

Challice, J. S. and Williams, A. H. (1968). Phenolic compounds of the genus *Pyrus*. II. A chemotaxonomic survey. *Phytochemistry* **7,** 1781–1801.

Chen, K., Johal, S. and Wildman, S. G. (1976). Role of chloroplast and nuclear DNA genes during evolution of Fraction I protein. *In* "Genetics and Biogenesis of Chloroplasts and Mitochondria" (T. Bücher, W. Neupert, W. Sebald and S. Werner, eds) pp. 3–11. North Holland-Elsevier, Amsterdam.

Cleland, R. E. (1960). A case history of evolution. *Proc. Indiana Acad. Sci.* **69,** 51–64.

Croizat, L. (1952). "Manual of Phytogeography". Junk, The Hague.

Cutler, D. F. (1972). Vicarious species of Restionaceae in Africa, Australia and South America. *In* "Taxonomy, Phytogeography and Evolution" (D. H. Valentine, ed.) pp. 73–83. Academic Press, London, New York.

Darlington, C. D. (1973). "Chromosome Botany and the Origins of Cultivated Plants" (3rd edn). Allen & Unwin, London.

Du Rietz, G. E. (1930). The fundamental units of biological taxonomy. *Bot. Tidskr.* **24,** 333–428.

Dusén, P. (1900). Die Gefässpflanzen der Magellandsländer. *Svenska Exped. Magellansl.* **3(5),** 77–266.

Ehrendorfer, F. (1958). Ein Variabilitätszentrum als "fossiler" Hybrid-Komplex: Der ostmediterrane *Galium graecum* L. – *G. canum* Req. Formenkreis. Eine Monographie. (Zur Phylogenie der Gattung *Galium*, VI.) *Öster. Bot. Z.* **105,** 229–279.

Eldredge, N. and Gould, S. J. (1972). Punctuated equilibria: an alternative to phyletic gradualism. *In* "Models in Paleobiology" (T. J. M. Schopf, ed.) pp. 82–115. W. H. Freeman, San Francisco.

Endler, J. A. (1977). "Geographic Variation, Speciation and Clines". Princeton University Press, Princeton, N.J.

Favarger, C. (1967). Cytologie et distribution des plantes. *Biol. Rev.* **42,** 163–206.

Forbes, E. (1846). On the connection between the distribution of the existing fauna and flora of the British Isles and the geographical changes which have affected their area. *Mem. Geol. Surv. Gr. Brit.* **1,** 336–432.

Gadella, T. W. J. and Kliphuis, E. (1972). Cytotaxonomic studies in the genus *Symphytum* IV. Cytogeographic investigations in *Symphytum officinale* L. *Acta Bot. Neerl.* **21,** 169–173.

Gomes, C. M. R. and Gottlieb, O. R. (1980). Alkaloid evolution and Angiosperm systematics. *Biochem. Syst. Ecol.* **8,** 81–87.

Gottlieb, O. R. (1978). Biochemical systematics based on secondary metabolites: principles and methods. *Rev. Latinoamer. Quim.* **9,** 138–147.

Gottlieb, O. R. (1980). Micromolecular systematics: principles and practice. *In* "Chemosystematics: Principles and Practice" (F. A. Bisby, J. G. Vaughan and C. A. Wright, eds) pp. 329–352. Academic Press, London, New York.

Hagerup, O. (1931). Über Polyploidie in Beziehung zu Klima, Ökologie und Phylogenie. *Hereditas* **16,** 19–40.

Harborne, J. B. (1977). Flavonoids and the evolution of the Angiosperms. *Biochem. Syst. Ecol.* **5,** 7–22.

Harborne, J. B. (1979). Correlations between flavonoid chemistry, anatomy and geography in the Restionaceae. *Phytochemistry* **18,** 1323–1327.

Harborne, J. B., Heywood, V. H. and King, L. (1976). Evolution of yellow flavonols in flowers of Anthemideae. *Biochem. Syst. Ecol.* **4,** 1–4.

Hayman, D. L. (1960). The distribution and cytology of the chromosome races of *Themeda australis* in South Australia. *Austral. J. Bot.* **8,** 58–68.

Heywood, V. H. (1968). "Modern Methods in Plant Taxonomy". Academic Press, London, New York.

Hurka, H. (1980). Enzymes as a taxonomic tool: a botanist's view. *In* "Chemosystematics: Principles and Practice" (F. A. Bisby, J. G. Vaughan and C. A. Wright, eds) pp. 103–121. Academic Press, London, New York.

Humphries, C. J. (1981). Biogeographical methods and the southern beeches. *In* "The Evolving Biosphere" (P. H. Forey, ed.) pp. 283–297. British Museum (Nat. Hist.), London.

Irving, E. (1964). "Palaeomagnetism and its Application to Geological and Geophysical Problems". John Wiley, London, New York.

James, S. H. (1970). Complex hybridity in *Isotoma petraea*. II. Components and operation of a possible evolutionary mechanism. *Heredity* **25,** 53–77.

Jensen, U. (1981). Proteins in plant evolution and systematics. *Fortschr. Botanik* **43,** 344–369.

Johnson, L. A. S. and Briggs, B. G. (1963). Evolution in the Proteaceae. *Austral. J. Bot.* **11,** 21–61.

Kurabayashi, M., Lewis, H. and Raven, P. H. (1962). A comparative study of mitosis in the Onagraceae. *Amer. J. Bot.* **49(9),** 1003–1026.

Lewis, H. and Bloom, W. L. (1972). Interchanges and interpopulational gene exchange in *Clarkia speciosa*. *Chromosomes Today* **3,** 268–284.

Lloyd, D. G. (1980). Demographic factors and mating patterns in angiosperms. *In* "Demography and Evolution in Plant Populations" (O. T. Solbrig, ed.) pp. 67–88. Blackwell, London.

Löve, A. and Löve, D. (1949). The geobotanical significance of polyploidy. I. Polyploidy and latitude. *Port. Acta. Biol.* (*A*) R. B. Goldschmidt, Vol. 273–352.

Löve, A. and Löve, D. (1971). Polyploidie et géobotanique. *Nat. Can.* **98,** 469–494.

Markham, K. R. and Godley, E. J. (1972). Chemotaxonomic studies in *Sophora*. I. An evaluation of *Sophora microphylla* Ait. *New Zealand J. Bot.* **10,** 627–640.

Melville, R. (1966). Continental drift, Mesozoic continents and the migrations of the angiosperms. *Nature* (*London*) **211,** 116–120.

Mirov, N. T. (1967). "The Genus *Pinus*". Ronald Press, New York.

Mooney, H. A. and Emboden, W. A. (1968). The relationships of terpene composition, morphology and distribution of populations of *Bursera microphylla* (Burseraceae). *Brittonia* **20,** 44–51.

Moore, D. M. (1972). Connections between cool temperate floras, with particular reference to southern South America. *In* "Taxonomy, Phytogeography and Evolution" (D. H. Valentine, ed.) pp. 115–138. Academic Press, London, New York.

Moore, D. M. and Raven, P. H. (1970). Cytogenetics, distribution and amphitropical affinities of South American *Camissonia* (Onagraceae). *Evolution* **24,** 816–823.

Moore, D. M., Harborne, J. B. and Williams, C. A. (1970). Chemotaxonomy, variation and geographical distribution of the Empetraceae. *Bot. J. Linn. Soc.* **63,** 277–293.

Mooring, J. S. (1975). A cytogeographic study of *Eriophyllum lanatum* (Compositae, Helenieae). *Amer. J. Bot.* **62,** 1027–1037.

Morton, J. K. (1961). The incidence of polyploidy in a tropical flora. *Rec. Adv. Bot.* **1,** 900–903.

Nelson, G. and Platnick, N. (1981). "Systematics and Biogeography". Columbia University Press, New York.

Nichols, D. (ed.) (1962). "Taxonomy and Geography". Systematics Association, London.

Ornduff, R. (1969). Reproductive biology in relation to systematics. *Taxon* **13,** 121–133.

Perring, F. H. and Walters, S. M. (1962). "Atlas of the British Flora". Bot. Soc. Brit. Isles, London.

Quinn, C. J. (1982). Taxonomy of *Dacrydium* Sol. ex Lamb. emend. de Laub. (Podocarpaceae). *Austral. J. Bot.* **30,** 311–320.

Quinn, C. J. and Gadek, P. (1981). Biflavones in *Dacrydium* sensu lato. *Phytochemistry* **20,** 677–681.

Raven, P. H. and Raven, T. E. (1976). The genus *Epilobium* (Onagraceae) in Australasia: a

systematic and evolutionary study. *New Zealand Dept. Sci. Industr. Res. Bull* **216,** 1–321.
Rick, C. M. and Fobes, J. F. (1975). Allozymes of Galapagos tomatoes: polymorphism, geographic distribution and affinities. *Evolution* **29,** 443–457.
Rick, C. M., Fobes, J. F. and Holle, M. (1977). Genetic variation in *Lycopersicon pimpinellifolium*: evidence of evolutionary change in mating systems. *Plant Syst. Evol.* **127,** 139–170.
Ridley, H. N. (1930). "The Dispersal of Plants Throughout the World". Reeve, Ashford.
Salatino, A. and Gottlieb, O. R. (1980). Quinolizidine alkaloids as systematic markers of the Papilionoideae. *Biochem. Syst. Evol.* **8,** 133–146.
Scora, R. W. (1967). Study of the essential oils of the genus *Monarda* (Labiatae). *Amer. J. Bot.* **54,** 446–452.
Seavey, S. R. and Raven, P. H. (1977a). Chromosomal evolution in *Epilobium* sect. *Epilobium* (Onagraceae). *Pl. Syst. Evol.* **127,** 107–119.
Seavey, S. R. and Raven, P. H. (1977b). Chromosomal evolution in *Epilobium* sect. *Epilobium* (Onagraceae), II. *Pl. Syst. Evol.* **128,** 195–200.
Seavey, S. R. and Raven, P. H. (1977c). Chromosomal differentiation and the sources of the South American species of *Epilobium* (Onagraceae). *J. Biogeog.* **4,** 55–59.
Seavey, S. R. and Raven, P. H. (1978). Chromosomal evolution in *Epilobium* sect. *Epilobium* (Onagraceae), III. *Pl. Syst. Evol.* **130,** 79–83.
Skottsberg, C. J. F. (1906). Zur Flora des Feuerlandes. *Wiss. Ergebn. Schwed. Südpolarexped.* **4,** 1–41.
Solomons, J. C. (1982). The systematics and evolution of *Epilobium* (Onagraceae) in South America. *Ann. Missouri Bot. Gard.* **69,** 239–335.
Soper, J. H. (1966). Machine-plotting of phytogeographical data. *Canad. Geographer* **10,** 15–26.
Stamp, L. D. (1962). A geographer's postscript. *In* "Taxonomy and Geography" (D. Nichols, ed.) pp. 153–158. Systematics Association, London.
Tomlinson, R. F. (1972). "Geographical data handling". 2GU Commission of Geographical Data Sensing and Processing, Ottawa.
Tutin, T. G., Heywood, V. H., Burges, N. A., Moore, D. M., Valentine, D. H., Walters, S. M. and Webb, D. A. (eds.) (1976). "Flora Europaea" Vol. 4. Cambridge University Press, Cambridge.
Van Steenis, C. G. G. J. (1962). The land-bridge theory in botany with particular reference to tropical plants. *Blumea* **11(2),** 235–372.
Van Steenis, C. G. G. J. (1972). *Nothofagus,* key genus to plant geography. *In* "Taxonomy, Phytogeography and Evolution" (D. H. Valentine, ed.) pp. 275–288. Academic Press, London, New York.
Walters, J. L. (1942). Distribution of structural hybrids in *Paeonia californica. Amer. J. Bot.* **39,** 270–275.
Walters, S. M. (1957). Distribution maps of plants – an historical survey. *In* "Progress in the Study of the British Flora" (J. E. Lousley, ed.) pp. 89–95. Bunde, Arbroath.
Wegener, A. (1929). "Die Entstehung der Kontinente und Ozeane". Friedr. Vieweg & Sohn, Braunschweig.
Zavarin, E., Snajberk, K., Reicher, T. and Tsien, E. (1970). On the geographic variability of the monoterpenes from the cortical blister oleoresin of *Abies lasiocarpa. Phytochemistry* **9,** 377–395.

Chemistry, Taxonomy and Systematics

13 | Chemical Data in Practical Taxonomy

J. B. HARBORNE

Department of Botany, University of Reading, England

Abstract: The most widespread application of chemical characters in plant taxonomy has been in the resolution of problems where the critical biological data are ambiguous. Some examples will be given of such successful applications at the species, generic and family levels among the flowering plants. Chemical data may also be called in to solve problems of plant identification, for example with medicinal plants where only certain organs (e.g. the roots or leaves) are available for testing. Probably the most important practical application of chemical data is in the identification of cultivars (e.g. of cereal crops) by gel electrophoresis of the seed proteins. Such techniques are also widely used for plant debris identification in forensic science. One other area where chemical data may have the edge over morphological or cytological criteria is in the recognition of hybrid plants and the documentation of their parental origin.

INTRODUCTION

Considering that chemical data have only been applied to taxonomic problems on any scale since the early 1960s, the development of chemosystematics as part of the modern biosystematic scene has been remarkably rapid. Chemosystematics is now widely recognized as an essential ingredient of the taxonomic process. This is true with animals (Wright 1974; Ferguson 1980) as well as with plants (Bisby *et al.* 1980; Stace 1980).

From the viewpoint of practical taxonomy, some of the most important developments have been in solving problems of bacterial identification and classification. Chemical and enzymic tests are a routine part of the process of microbial identification (Skinner and Lovelock 1979). In addition, pryolysis

Systematics Association Special Volume No. 25, "Current Concepts in Plant Taxonomy", edited by V. H. Heywood and D. M. Moore, 1984. Academic Press, London and Orlando.
ISBN 0 12 347060 9

gas chromatography is being developed as a key chemical technique in the identification of different bacterial strains (Gutteridge and Norris 1979), although there are still certain practical problems in its universal adoption for this purpose. Likewise, difficulties in generic delimitation in bacteria have been overcome to a large extent by a combination of macromolecular analyses, of determining the nucleotide base ratios of bacterial DNA and of measuring the degree of DNA–DNA hybridization that occurs between different bacteria. An interesting recent application of these DNA data can be seen in the recognition of the Legionnaire's disease organism *Legionnella pneumophylla* as a distinct, new genus in the bacterial kingdom (Lattimer and Ormsbee 1981).

As with the bacteria, so with other lower plant phyla, chemical methods are being applied more and more frequently to problems of taxonomic delimitation. For example, among the lichens, pigment analysis by thin layer or gas chromatography is recognized as an essential feature of the taxonomic process. As Hawksworth (1976) has put it: "chemical investigations now form an integral part of all serious taxonomic studies in lichens and any taxonomic revision not considering chemical data is likely to be regarded as incomplete".

In the case of the higher plants, chemosystematics has been as widely applied as with other plant groups. However, the emergence of chemical techniques as alternative procedures for recognizing and distinguishing taxa has been less dramatic. This is largely because for most purposes conventional biological characters, derived from so-called "eyeball taxonomy", are sufficiently discriminating. The most obvious application of chemical methods among the angiosperms has been in those instances where morphological criteria fail to provide a key to separation. Thus chemical methods have won a place for cultivar discrimination, where they can be shown to be more rapid and reliable than other techniques. Similarly, chromatographic tests have been developed for the rapid identification of drug plants, in those common situations where only certain parts of the plant (e.g. roots) are available for analysis. There is no doubt that such procedures are competitive in terms of time, effort and equipment needed with the older anatomical procedures. In addition, chemistry has been important as an aid in the recognition of hybrid plants and it may occasionally provide the essential documentation of parental origin which is not provided by either morphological or cytological measurements.

The other major application of chemical data in practical taxonomy has been as an aid in revision where the available biological characters are ambiguous or conflicting in their distribution. Here, a drawback of chemical characters over most biological features is the fact that some of the necessary background information may be lacking, i.e. the distribution of a chemical in all relevant plants may not be known sufficiently to use that chemical diagnostically. However, as Cronquist (1980) puts it: "chemical characters are like other characters: they work when they work and they don't work when they don't work. Like all taxonomic characters, they attain their value through correlation with other characters. . .". Some examples will be given here where chemical characters work in solving problems of taxonomic arrangements at the species, generic and family levels.

Chemical characters were not extensively used in taxonomy until the development of chromatographic procedures during 1947–1954 provided the means for rapidly scoring organisms for the presence/absence of given chemicals or sets of chemicals. In recent times, more highly sophisticated analytical tools have been developed in phytochemistry, which have considerably extended and improved the chemical data available to taxonomists. Some of these newer techniques will be mentioned briefly in the following section. The major part of this review will be concerned with the higher plants and the use of chemical data for identification purposes and for solving simple problems of phenetic classification. The application of chemical data among plants in a phylogenetic context is treated separately in this volume by Kubitzki (see Chapter 14).

NEWER CHEMICAL TECHNIQUES

(1) Analysis of low molecular weight constituents

For the detection of low molecular weight constituents in plants, it is necessary to extract the tissue with an appropriate solvent and to apply some preliminary purification procedure, particularly chromatography, to the extract before determining the presence/absence or amount of a given substance. During such procedures, material may be lost, particularly if the compound is labile, or artifacts may be formed which can confuse the analytical procedure. There are clear advantages, therefore, in analysing crude plant extracts directly after extraction, if this can be done without interference by all the other substances that will be also present in the extract. A significant analytical advance has been made, therefore, by Pedersen (1978) in developing a procedure for the direct determination of quinols and quinones in crude plant extracts.

These two classes of natural substance are identified in alcoholic solution as their semiquinones and they can be specifically detected by electron spin resonance (ESR) spectroscopy. For some compounds, the ESR signal is entirely characteristic; thus for hydroquinone or for simple naphthoquinones such as juglone of the Juglandaceae ESR spectroscopy provides absolute proof of identity. For other compounds, the semiquinone nucleus acts as a label for a particular class of compounds, e.g. for the widely occurring caffeic acid esters (see Fig. 1). In favourable cases, mixtures of compounds can be separately detected and identified. The method is semi-quantitative in the sense that the magnitude of the ESR signal is directly related to the concentration of the quinol or quinone in the plant. The method has been already applied to the study of distribution patterns of several quinones and caffeic acid derivatives in

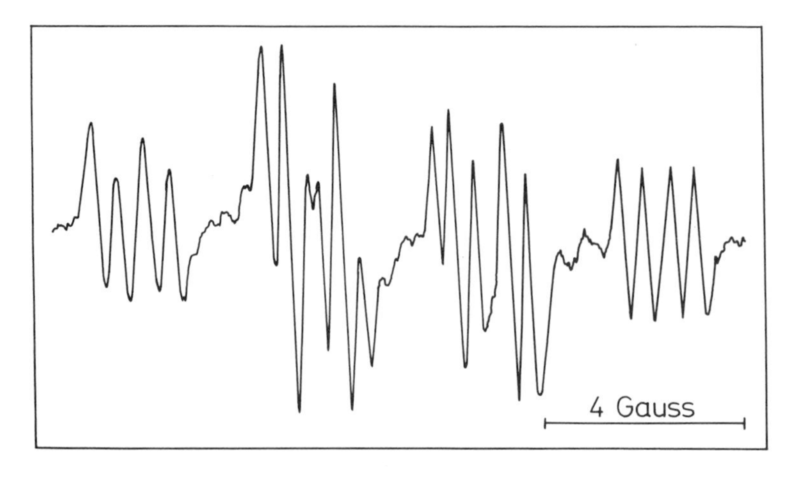

Fig. 1. Electron spin resonance spectrum of the semiquinone of dihydrocaffeic acid ester in a crude alcoholic leaf extract of a *Lycopodium* species.

higher plants (Pedersen 1978) and to the detection of hydroxyphenolic acids among species of the genus *Lycopodium* (Pedersen and Øllgaard 1982).

In analysing volatile plant constituents, routine procedures have been developed over the last decade for the direct injection of small samples of dried plant material into gas chromatographic (GC) apparatus. Direct injection is not only very convenient but it also avoids the possibility of artifact production that may occur during solvent extraction and subsequent extract concentration. The GC set up can be automated so that a series of samples can be analysed in turn without further attention of the operator. Using such procedures, it is not only possible to examine essential oil (terpene) variation among plant populations but also to determine terpene variations in different parts of the same plant.

An additional advantage of GC analysis of plant volatiles is that the procedure provides both qualitative and quantitative data on the chemicals present, in one and the same operation. Measurement of the peak area under each GC signal allows the calculation of the percentage concentration of a given constituent in the total volatile fraction. By employing an appropriate detector, the method can be made to be highly accurate. The extensive use of essential oil analysis in chemotaxonomic studies of gymnosperms (von Rudloff 1976) is at least partly due to great sensitivity of the GC procedure in recording these quantitative data.

No comparable procedure for quantifying the occurrence of non-volatile plant constituents has been available until very recently. It is true that quantitative determination of alkaloids or flavonoids has been possible by paper chromatography or TLC, but the procedures are elaborate (cf. Shellard 1968) and the accuracy and sensitivity are relatively low. A new technique of liquid chromatography developed during the last five years now provides an answer to this outstanding analytical problem. This is high performance liquid chromatography (HPLC), which serves for non-volatile substances that which GC has contributed to the analysis of plant volatiles.

An example of the HPLC of plant flavonoids is shown in Fig. 2, where two traces are shown of the separation of the methylated flavones present in the crude peel extract of two *Citrus* cultivars (Bianchini and Gaydou 1980). Although the same five flavones are present in the tangerine and the orange the concentrations

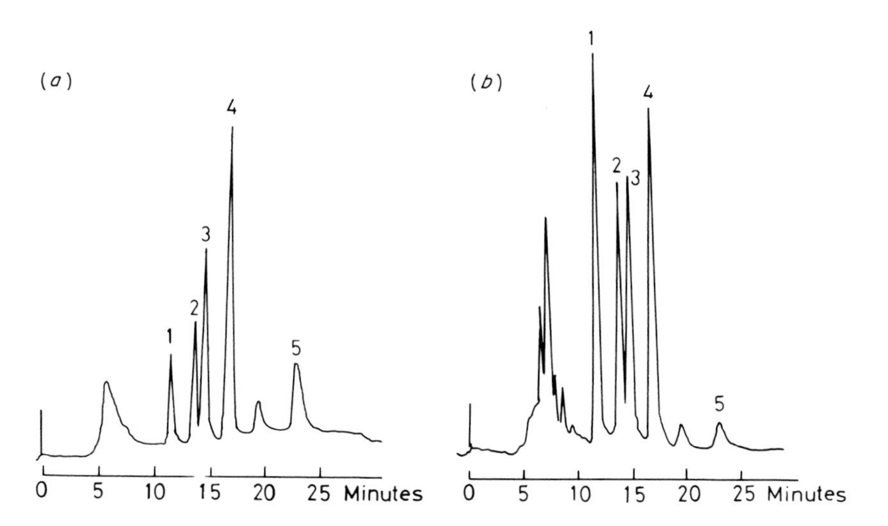

Fig. 2. HPLC traces of methylated flavones in tangerine (*a*) and orange (*b*) peel. Compounds 1–5 are tangeretin, tetramethylscutellarein, 3,5,6,7,8,3′,4′-heptamethoxyflavone, nobiletin and sinensetin.

are very different and clearly distinguish the two cultivars. This procedure has obvious applications in cultivar identification and in characterising hybrids in *Citrus* generally. The power of HPLC as a quantitative analytical tool is illustrated in the work of Ivie *et al.* (1982) who used this technique to show that furanocoumarins do not normally occur in carrot tissues; at least, they were able to establish that if present, the levels were well below 0.5 ppm. Other examples of HPLC analysis of plant constituents will be given in later sections.

A notable feature of the chemotaxonomic scene of the last decade has been the wider use of herbarium material in plant surveys for flavonoids, alkaloids and other low molecular weight constituents. This has been partly because of refinements in phytochemical analysis, which have made it possible to detect substances in very small samples of plant material. There has been a real "breakthrough", for example, in the alkaloid field, where a range of appropriate techniques has been developed for this purpose (Phillipson 1982). Also, it has been increasingly appreciated that surveys for particular chemicals can be made properly representative by analysing those plants not available in living collections from herbarium specimens. Additionally, of course, herbarium curators have been willing to supply the necessary samples to phytochemists when they appreciate there is a taxonomic purpose behind the request.

Nevertheless, phytochemical analysis is normally destructive of what may be relatively precious resources. Also, in cases of rare plants or very small herbs, it may never be possible to supply sufficient herbarium tissue for standard chemical extraction. The possibility of phytochemical analysis without damaging the tissue is clearly an important future goal. It is clear from recent developments in spectral techniques that such analyses may well become commonplace in the next decade. Already, mass spectrometry has been used to analyse alkaloid mixtures directly in dried leaves, without prior separation by chromatography. In addition, a sample of coca leaf 1 mm square has been analysed successfully for the presence of cocaine by tandem mass spectrometry *without* prior extraction of the alkaloid (Kondrat and Cooks 1978). In addition, ^{13}C nuclear magnetic resonance techniques are increasingly being developed for the analysis of solid samples (coal, soil, human tissues, etc.) and it is likely that these techniques will eventually become available for the analysis of a range of chemical constituents in herbarium specimens. These methods will not destroy the structural integrity of the samples, which can then be returned intact to the herbarium sheet.

(2) Analysis of proteins and nucleic acids

Methods of separating and characterizing the macromolecules of the plant cell have developed just as rapidly in recent times as those used with low molecular weight constituents. What has been termed "the protein explosion" (Bray 1977) occurred when the proven technique of gel electrophoresis was combined with new procedures of isoelectric focussing. This powerful combination provided for the first time a reproducible and reliable 2-dimensional system of protein analysis. For example, by using electrophoresis in the presence of urea and ampholines in the first direction and sodium dodecyl sulphate electrophoresis in the second, O'Farrell (1975) was able to locate over 1000 individual proteins in a bacterial lysate. Since then, electrophoretic separations with improved resolving power have continued to be applied in taxonomic studies with much success. Probably at least 50% of all papers published on plant chemosystematics are concerned with protein or isozyme separations. The range of electrophoretic and related techniques available for protein separations have recently been reviewed by Andrews (1982).

Remarkable developments have also occurred in the fields of amino acid sequencing of proteins (Walsh *et al.* 1981) and of base sequencing of nucleic acids. Ribosomal DNA and RNA sequences are now available for a range of organisms (Brimacombe 1981) and it is already possible to make taxonomic comparisons from the data on different bacteria (Kandler and Schleifer 1980). Further discussion of the taxonomic significance of sequence data is, however, outside the scope of the present review. The interested reader can consult various chapters on the subject in Bisby *et al.* (1980).

CHEMICAL IDENTIFICATION OF PLANT CULTIVARS

(1) Crop plants

One of the most severely practical tasks a higher plant taxonomist may face is the recognition and separation of different cultivars of the same given taxon. In most cases, the material available for identification is restricted for practical reasons to particular parts of the plants, e.g. the seed or other storage tissue. Morphological or anatomical characters for separation may be non-existent, because of the genetic uniformity of the material, and in such cases chemical methods may come to the fore for "typing" cultivars of a given crop plant.

By far the most important procedure here is gel electrophoresis of the plant proteins. Because the variety and number of different proteins that can be revealed by an appropriate separation technique is so considerable, it may be possible to uniquely define each known cultivar of a series by its gel electrophoretic profile. A second procedure, which has not been so widely applied, is isozyme analysis; this involves staining gel separations for different enzyme activities and is best applied to the proteins of a young seedling rather than to the proteins of the seed itself. Secondary compounds may also be analysed in cultivars of crop plants, since they vary at least quantitatively from cultivar to cultivar; such analyses are, however, unlikely on their own to provide a complete procedure of cultivar separation.

Probably the widest use of gel electrophoresis for cultivar identification is with the seeds of cereals (Ladizinsky and Hymowitz 1979; Jensen 1981). Commonly, there are no visible morphological differences between the grains of different varieties so that the tedious process of growing plants to maturity would be necessary before identification by morphological characters would be possible. By contrast, chemical analyses can be carried out on a single grain and give an answer within 24–48 hours.

The regular monitoring of wheat varieties supplied by the farmer to the miller is of utmost importance, since only certain varieties have the right quality for making into bread (Ellis 1979). It is not surprising therefore to find that much attention has been given to the starch gel electrophoresis of wheat proteins to provide an answer to this need. Analysis can be limited to the proteins in the gliadin fraction of the seed protein, i.e. that fraction soluble in dilute acetic acid or in solvents containing sodium dodecyl sulphate (SDS). Forty-three gliadins can be recognized by gel electrophoresis (Ellis and Beminster 1977) and any one variety has about 20 gliadin bands.

An analysis of 29 wheat cultivars grown in the UK provided unique electropherograms for each, apart from three which were closely related ancestrally. Some idea of the variety patterns can be seen in Fig. 3, while the key bands used for recognizing four major cultivars are listed in Table I. The patterns were found to be consistent in any one cultivar, irrespective of the normal variations that occur in crop husbandry. Furthermore, individual grains of a given cultivar gave identical bands in all but three cases. In three varieties, "biotypes" were encountered where one or two of the characteristic bands of that particular variety were missing. The presence of these biotypes, however, provided no serious impediment to correct cereal identification. Indeed, gel electrophoresis is now an accepted procedure for rapidly and efficiently monitoring UK wheat varieties (Ellis 1979). It has also been applied successfully to the identification of the 88 wheat cultivars most commonly grown in the USA (Jones *et al.* 1982).

An alternative approach to studying wheat storage proteins is to determine isozyme patterns in 15-day-old seedlings. A study of isozymes in hexaploid wheat cultivars by Salinas *et al.* (1982) offers an interesting comparison. In this work, 38 cultivars of *T. aestivum* and one of *T. spelta* were analysed using both

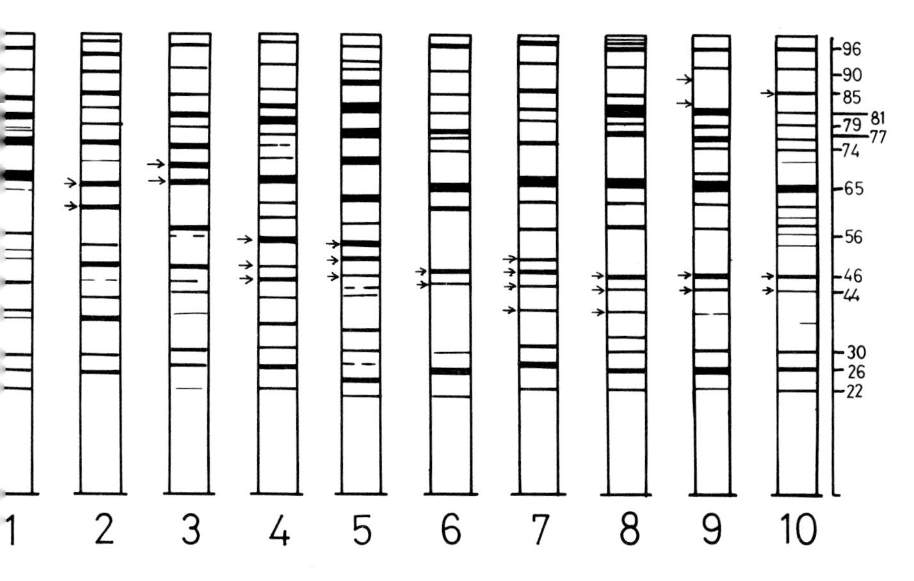

Fig. 3. Gel electrophoresis separations of gliadin proteins in seeds of selected wheat cultivars. 1 = "Maris Dove"; 2 = "Highbury"; 3 = "Bouquet"; 4 = "Clement"; 5 = "Maris Ranger"; 6 = "Maris Nimrod"; 7 = "Cappelle-Desprez"; 8 = "Maris Freeman"; 9 = "Aton"; 10 = "Maris Huntsman". (From Ellis 1979)

Table I. Taxonomic distinction between four wheat cultivars based on gliadin protein bands. (From Ellis 1979)

Cultivar	*Key to separation*	*Baking quality*
"Maris Huntsman"	strong band at 85	poor
"Clement"	strong pair of bands at 45 and 35	
"Highbury"	strong band at 60	excellent
"Bouquet"	strong band at 68	

British cultivars have about 20 gliadin bands each (see Fig. 3). Starch gel electrophoresis separates 26 of 29 UK cultivars.

starch and polyacrylamide gels. Some 15 enzyme activities were variously assayed and the patterns obtained were distinctive for 26 of the 39 cultivars. The remaining 13 cultivars fell into 5 pairs, with one group of three. Thus, the more complex procedure of isozyme separation provides a poorer key to cultivar identification than gliadin analysis. Nevertheless, it could be a valuable back-up procedure in cases where the gliadin system fails. On the other hand, the gliadin analysis of the wheat grain could be extended by 2-dimensional separation procedures, which would reveal many more proteins for cultivar identification. Such separations have been regularly employed in studies of the protein quality of wheat grains (Wall 1979).

Gel electrophoretic procedures have been applied to cultivar identification in other cereal crops, especially in barley and oats. With barley, Shewry *et al.* (1978) have used SDS-polyacrylamide gel electrophoresis at pH 8.9 to rapidly analyse the hordein fractions of 88 varieties. Twenty-nine patterns were obtained, with one to 25 cultivars in each group. Further subdivision of the largest group was achieved by a second electrophoretic run with SDS being replaced by urea and the pH changed to 4.6. In this case, the method does not give a complete separation of all varieties, but most can be identified by one or other procedure. Similar partial identifications of oat cultivars have been achieved by isozyme analyses of oat plants (Singh *et al.* 1973).

The use of electrophoretic methods for characterizing potato varieties is well known through the work of Stegemann (1979). By a combination of polyacrylamide gel electrophoretic patterns of total storage protein of the tuber and the patterns given by esterase isozymes, it has been possible to typify any given European potato variety. An index of 530 registered cultivars and their separation by electrophoretic patterns has been published (Stegemann and Loeschcke 1976). The cultivar "Maritta" is used as an internal standard for a given identification. The results of electrophoresis have been accepted in court cases involving disputes about cultivar designation or requiring the identification of a stolen potato crop. Other chemical procedures are also available for typification of potato varieties. The different patterns of anthocyanin pigments in the sprouts separate British cultivars into nine groups (Harborne and Swain 1979). In combination with other simple screening techniques, the anthocyanin data provide a useful key to variety testing among British potatoes (Brown and Moss 1976).

Yet other crop plants where chemistry has been applied to cultivar identification are those belonging to the genus *Brassica*. The earlier work of Vaughan and coworkers on the gel electrophoresis of albumin fractions of the seeds of cultivated and wild *Brassica* species is well known (cf. Vaughan and Denford 1968) and needs no reiteration here. More recent protein analyses in *Brassica* cultivars have been accomplished by using both the soluble seed proteins and certain isozymes, especially the esterases (Phelan and Vaughan 1976). The results showed *inter alia* that a taxon *B. alboglabra* of disputed specific rank is actually better treated as a member of the *B. oleracea* complex, at least from the chemotaxonomic viewpoint.

A second approach to cultivar identification in crucifer crops is through quantitative variations in their characteristic glucosinolates, or mustard oil glycosides. A study by Heaney and Fenwick (1980) of 18 Brussel sprout cultivars showed that this approach could be a useful adjunct to typing cultivars. Eleven glucosinolates were detected by GC and the concentrations varied significantly from cultivar to cultivar. Direct analysis of the glucosinolates appears to be preferred to the detection of the more volatile isothiocyantes formed on enzymic hydrolysis, because a variety of other by-products may be formed in this process. Seed glucosinolates have also been examined for taxonomic purposes in other crucifers besides *Brassica*, notably in *Cakile* (Rodman 1976), *Caulanthus* and *Streptanthus* (Rodman *et al.* 1981).

Cultivar identification of legume seeds by gel electrophoresis has been demonstrated for the runner bean *Phaseolus vulgaris* and the garden pea (*Pisum sativum*) (Boulter 1981), although this method does not appear to have been used much in practice. Serological studies of legume seed proteins have been widely applied to taxonomic problems at the subspecific level (Cristofolini 1981) and also for assigning species to their correct genera (Chrispeels and Baumgartner 1978). Recently, cultivars of *P. vulgaris* have been separated into several groups on the basis of isolectin differences, i.e. of variations in the structure of their recognition proteins (Felsted *et al.* 1981).

(2) Ornamental plants

Besides crop plants, cultivar identification is of considerable importance in ornamental species. A necessary protection on the patent rights of a new horticultural cultivar is the means of distinguishing it from other known cultivars and chemical methods may be crucial for such purposes. The identification of poinsettia cultivars *Euphorbia pulcherrima* is particularly difficult morphologically because of the very narrow genetic base. Attempts to separate them by electrophoresis of their proteins and the esterase isozymes were not successful. By contrast, the HPLC of the anthocyanins in the bracts (Fig. 4) has provided a valuable key to their typification. All the cultivars have the same five anthocyanins in the bracts, namely the 3-glucoside and 3-rutinoside of pelargonidin and the 3-glucoside, 3-galactoside and 3-rutinoside of cyanidin. Nevertheless, the concentrations of these five pigments and the total anthocyanin content (see Table II) vary sufficiently to allow separation of all 28 cultivars tested (Stewart *et al.* 1979). It is necessary to control the position of sampling, the maturity of the bract, the effect of light quality on growth and the effect of season. However, once this is done, the results are highly reproducible and can be used in practical

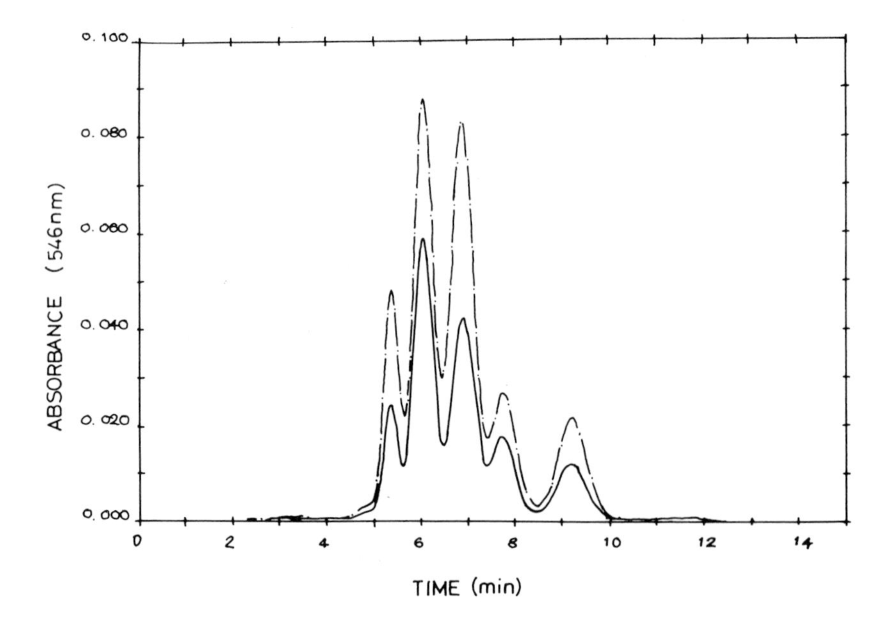

Fig. 4. HPLC traces of anthocyanins in two cultivars of poinsettia.

Table II. Anthocyanins in bracts of selected poinsettia cultivars as determined by HPLC. (Data from Stewart *et al.* 1979)

Cultivar	A1	A2	A3	A4	A5	*Total*
"Stoplight"	72	248	709	84	150	1263
"Red Baron"	38	160	364	194	189	945
"Rudolph"	63	162	331	158	187	901
"Ruff & Reddy"	45	167	418	115	148	893
"Prof. Laurie"	16	55	92	46	44	253
"Stoplight Pink"	17	68	94	42	23	244

Key: A1, cyanidin 3-galactoside; A2, cyanidin 3-glucoside; A3, cyanidin 3-rutinoside; A4, pelargonidin 3-glucoside; and A5, pelargonidin 3-rutinoside. Figures refer to ng pigment per mm^2 of bract.

identification. Flavonol glycosides are also present in the bracts and their concentrations can also be determined by HPLC (Stewart *et al.* 1980); however, the variations are not so clearcut and the data are not so readily applied to cultivar identification as those derived from the anthocyanins.

Finally, mention may be made of an instructive example of cultivar identification by chemical means in conifers. This is the work of Gough and Welch (1978) in solving the long standing horticultural problem of the true identity of a juvenile conifer, usually referred to as *Chamaecyparis obtusa* cv. Sanderi. A variety of other synonyms has been employed and the problem has been worked on extensively by taxonomists without their reaching any real agreement.

The problem was simply solved by GC of foliage extracts of this taxon and those of 47 species of Northern hemisphere Cupressaceae. The GC traces (Fig. 5) showed immediately that the plant cannot be *Chamaecyparis*, but is actually a fixed juvenile clone of *Thuja orientalis*, since it is closely identical in

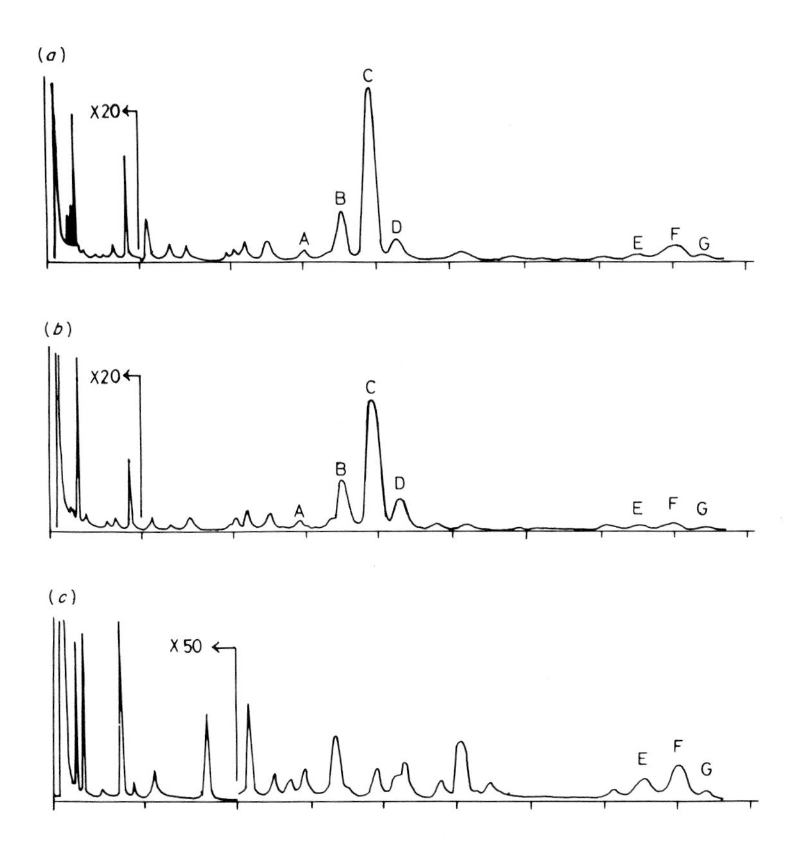

Fig. 5 GC leaf diterpenoid profiles of the cultivar "Sanderi" (*a*), *Thuja orientalis* var. *juniperoides* (*b*) and *Chamaecyparis obtusa* (*c*). Compounds A–D are diterpene acids; compounds E–G are diterpene alcohols.

terpenoid constituents to other juvenile forms of this species. The chemical profiles (Fig. 5) are of four diterpene acids: communic, sandaraco-pimaric, isopimaric and Δ^8-isopimaric acids. These acids are characteristic in *Thuja*, but are not found in *Chamaecyparis*. This chemical identification of cv. Sanderi as *Thuja orientalis* was shown by the authors to be completely consistent with all other available features of this problem taxon (Gough and Welch 1978).

CHEMISTRY OF PLANT HYBRIDS

Natural hybridisation between closely-related plant species is a well-known phenomenon (Stace 1975) and hybrids are usually identified by morphological characters, with occasional help from cytology. Problems of identification, however, often arise in the study of natural hybrid swarms, because of the possibilities of backcrossing occurring to one or other of the parents and of the difficulties of measuring the extent of introgression that may occur. Hybrids can be difficult to deal with taxonomically and in a number of well-known instances where hybrids have only been studied as herbarium specimens, they have not been recognized as such and have been given separate "species" names.

The value of chemistry in solving problems of hybrid identification lies in the likelihood that various putative parents have different chemicals profiles, so that it may be obvious from a chemical analysis that compounds characteristic of two particular parental species both appear on the chromatogram of the hybrid. Such a successful chemical analysis depends on the expectation that genes controlling the synthesis of the chemicals in question are additive in their effect in the hybrid. While this is most often the case, it has occasionally been found not to be so and some or all of the chemicals of one or other parent do not appear in the hybrid analysis.

While it is possible to analyse hybrids both for their low molecular weight constituents and for their protein and/or isozyme profiles, most attention has been given in plants to the flavonoids and related phenolics since these have been the most successful in general in revealing the presence of hybridization. Indeed, much of the early development of chemotaxonomy and of its acceptance by taxonomists was due to the discovery that hybridization among species in the legume genus *Baptisia* could only be fully documented from data derived from 2-dimensional chromatography of the flavonoids of either flower or leaf (Alston and Turner 1963). The diagnostic value of chemical studies of hybrids has also been demonstrated in a negative sense, i.e. in showing that hybrids recognized as such on morphological criteria were not, in fact, true hybrids. This refers to the studies of leaf terpene constituents in *Juniperus ashei* and *J.*

virginiana, where "hybrid populations" were found to be clinal variants of one of the two species, namely *J. virginiana* (Turner 1970).

Here, two examples from the more recent literature are chosen to indicate the continued value of chemical analysis with plant hybrids. The first example is taken from the work of Smith (1980) on silver and golden-backed ferns of the *Pityrogramma triangularis* complex. Here chemical analysis is particularly facile since these ferns exude flavonoids onto the surface of the fronds. The exudates can then be collected and analysed without any purification by TLC or preferably by HPLC. The pigments present include over 20 substances, based variously on chalcone, dihydrochalcone, flavanone or flavonol structures.

The type of separation of fern flavonoids that can be achieved on 1 cm lengths of pinnule in 15 to 20 min runs by HPLC is indicated in Fig. 6. This shows the quantitative analyses of the flavonoids of two *P. triangularis* var. *triangularis* chemotypes, of a var. *maxonii* chemotype and of a hybrid between the second var. *triangularis* chemotype and var. *maxonii*. Different chemotypes may also be recognized on cytological, morphological and distributional grounds. The

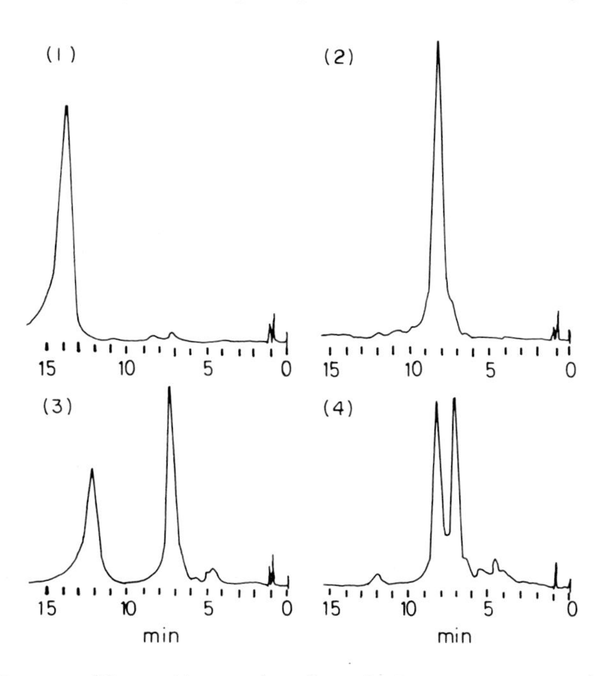

Fig. 6. HPLC traces of flavonoid separations from the *Pitryogramma triangularis* complex. *Key:* Trace (1) var. *triangularis* galangin 7ME type; Trace (2) var. *triangularis* Km 4′ ME, 74′ DME type; Trace (3) var. *maxonii* galangin type; Trace (4) var. *maxonii* × *triangularis* chemotype 2.

importance of the chemical analysis lies in the ease by which hybrids can be recognized through additive inheritance of the parental chemicals and the ease with which particular chemotypes can be discerned.

Taxonomically, the chemical analyses of frond surface flavonoids has aided species recognition. Thus *P. triangularis* var. *pallida* has been assigned specific rank, as *P. pallida*, from its highly distinctive flavonoid profile – it contains only flavanones – and the fact that it does not hybridize with any of the other forms of *P. triangularis*. Smith (1980) was able to conclude from these studies that "chemical analysis, especially the quantitative analysis of fully identified flavonoids by means of HPLC, has provided the key for understanding the complexities of this group of American ferns".

The use of storage protein bands or isozyme patterns in recognizing plant hybrids has recently been reviewed by Hurka (1980), who makes the point that complete additivity of the banding patterns of the parents does not always occur in the hybrid. In spite of this restriction, the method has much to commend it and these two procedures have been applied, separately or together, to a variety

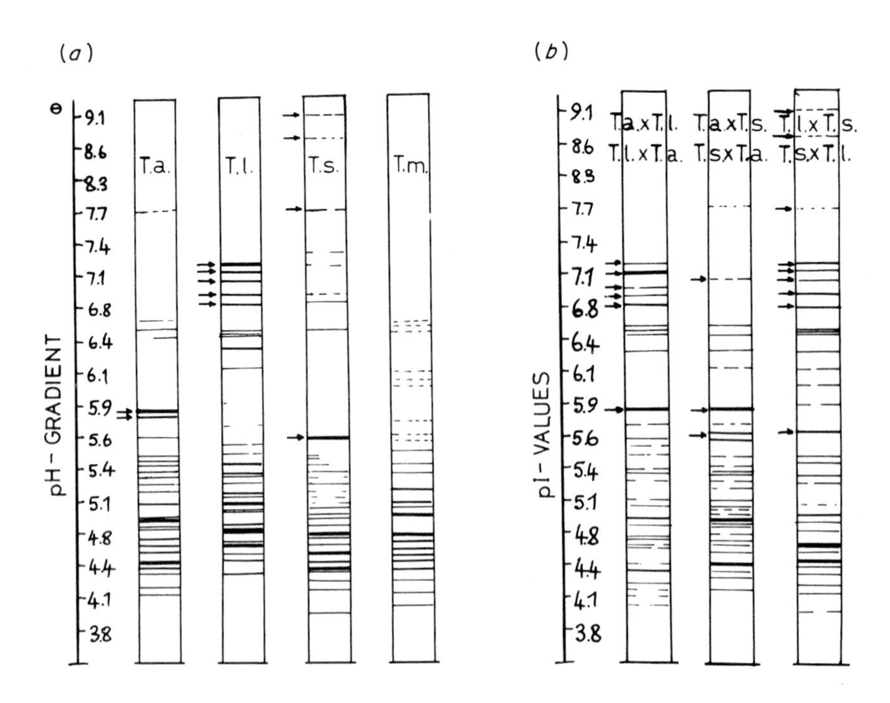

Fig. 7. Isoelectric focussing of *Typha* pollen proteins from four species (*a*) and various derived hybrids (*b*).

of hybrid taxa. The separation of different subunits of fraction 1 protein, the major leaf protein of green plants, by isoelectric focussing and the use of species pattern differences have been particularly successful in determining the origin of hybrids in a number of crop plants, e.g. in tobacco (Gray 1980). In a given hybrid, the small subunits of fraction 1 protein are inherited from both parents whereas the large subunits are inherited only from the maternal parent. This means that the fraction 1 protein profile yields additional information regarding the direction of the cross which gave rise to that particular hybrid plant.

Although most workers use seeds for storage protein analysis and young seedlings for isozyme patterns, the choice of tissue for extracting the proteins may be crucial to the success of hybrid identification. Root, tuber and seed kernel proteins and the proteins of phloem and latex have all been analysed. A recent analysis of proteins in four *Typha* species and artificial hybrids thereof was conducted on the fresh pollen (Krattinger *et al.* 1979). The results were quite successful in the sense that each species had a number of characteristic pollen proteins and these always appeared in the appropriate hybrid pollen (Fig. 7). The authors were able to conclude that whereas the morphological criteria for distinguishing intermediate forms in *Typha* were insufficient, the pollen protein profiles represented new, easily accessible characters by which F_1 hybrids could be unequivocally identified.

CHEMISTRY IN TAXONOMIC REVISION

While many papers are published regularly in the area of plant chemosystematics, most are not directly applicable to taxonomic revision. Often, the chemical data are incomplete in that compounds are recorded in some taxa but no information is provided on their presence/absence from other related plants. In spite of this, most of the data are broadly relevant since they extend our still fragmentary knowledge of the phytochemistry of the plant kingdom.

There are many cases in classification where chemistry might be usefully applied to problems of taxonomic revision, particularly when the biological information is ambiguous or conflicting. Few chemical studies have been directed at solving taxonomic problems in this way. This may be because, as with most other newer approaches to taxonomy, it is difficult to predict in a given case whether chemistry will help or not. The choice of chemical to study is very important and certain classes of compound lend themselves to such investigations more than others.

The advantages of combining chemical surveys of plants with active taxonomic studies of the same plants cannot be too greatly stressed. The taxonomist for his or her purpose may either freshly collect plant species from the wild or build up a living collection of the same plants. In either case, correctly verified plant material will be available for phytochemical analysis, in reasonable quantity. Secondly, the taxonomist can direct attention to borderline taxa which may deserve analysis, but which the phytochemist would otherwise ignore.

One example must suffice to indicate the benefits of such joint collaboration. A survey of flavonoids in leaves and inflorescences of Australian *Cyperus* species – a very difficult group of plants to identify or classify – was carried out at the same time as the group was being revised taxonomically. During this work, chemical and morphological data came together to indicate that the position of *Cyperus concinnus* in the section *Haspani* was unsatisfactory. Indeed, before the chemical data became available, one of the authors had already decided to reassign this species to section *Fusci* on morphological grounds. The available chemical data (Table III) (Harborne *et al.* 1982) point unambiguously to

Table III. Flavonoids of six Australian *Cyperus* species. (Data from Harborne *et al.* 1982)

Section/Species	*Inflorescence* Xanthones	Flavonoids	*Leaf* Xanthones	Flavonoids
Haspani				
C. haspan	Mn, Iso	—	Mn, Iso	Qu
C. prolifer	Mn, Iso	—	—	Qu
C. tenuispica	Mn, Iso	—	—	Qu
C. concinnus	—	Lu	—	Tr, Lu
Fusci				
C. difformis (a)	—	Lu	—	Lu
C. difformis (b)	—	Lu	—	Tr, Lu

Key: Mn, mangiferin; Iso, isomangiferin; Lu, luteolin; Tr, tricin; Qu, quercetin.

similarities in flavonoid pattern between *C. concinnus* and *C. difformis*, the only species in section *Fusci* which had been analysed. Conversely, the three species of section *Haspani* have the very distinctive occurrence of two glucoxanthones, mangiferin and isomangiferin, which do not appear anywhere else in *Cyperus* or in the Cyperaceae, as far as is at present known. It may be noted that the two xanthones occur in the inflorescence of all three species of *Haspani* section but appear in the leaf of only one of them. The importance of surveying other parts of the plants as well as the leaf is clearly indicated in this case.

While the species as the basic taxonomic unit remains relatively constant, the circumscription of the genus is less sure and is often the subject of revision. A commonplace in plant taxonomy is the decision to split some species from one genus and put them in a new genus or alternatively to abolish a genus by placing the species of two genera together. For example, Stearn (1977) decided to perform the latter operation among genera of the Oleaceae; he placed *Linociera* and *Chionanthus* together into *Chionanthus*, since the only clear cut difference separating the two known *Chionanthus* species was the deciduous habit of the leaves. This decision, however, contradicts newly available chemical data (Harborne and Green 1980) which indicate significant differences in leaf flavonoids between the two groups (Table IV). In particular, two of the three

Table IV. Generic delimitations and flavonoids of Oleaceae. (Data from Harborne and Green 1980)

Taxa	*Flavonols* 1	2	3	4	*Flavones* 1	2	3	4	*Flavanones*
Chionanthus									
C. retusus	−	−	−	−	+	+	−	+	eriodictyol
var. *serrulatus*	+	−	+	+	−	−	−	−	7-glucoside
C. virginicus	+	−	+	+	−	−	−	−	−
Linociera									
L. caribaea	+	−	−	−	+	+	−	−	−
L. foveolata	+	−	−	−	+	+	−	+	−
L. giordanoi	+	−	−	−	−	−	−	+	−
L. nilotica	+	−	−	−	+	−	+	−	−
L. quadristaminea	+	−	−	−	+	+	−	+	−
L. ramiflora	+	−	−	−	+	+	−	+	−

Chionanthus taxa accumulate flavanone derivatives which are highly distinctive within the Oleaceae and which do not occur anywhere in *Linociera sensu stricto*. While these chemical results are not conclusive, they do indicate that a further appraisal of the relationships between these two groups of plants may be necessary.

Another area of taxonomic dispute where chemistry may be able to make an unambiguous contribution is the placement of anomalous genera into their right families. One order of plants where there are numerous controversial genera is the Centrospermae (or Caryophyllales). There are particular difficulties in knowing whether certain plants belong to the Molluginaceae or should be better placed in the Aizoaceae or Phytolaccaceae (Ehrendorfer 1976). Here chemistry may be able to make a decisive contribution to placement, since there is a well-known dichotomy in pigment type between the Molluginaceae, which contains anthocyanins, and the other two families which are known to be betacyanin-producing (cf. Table V).

Table V. Plant families which constitute the order Centrospermae*

Core betalain families†	*Admitted because of betalain presence*
Aizoaceae	Cactaceae
Amaranthaceae	Didieraceae
Basellaceae	*Retained in spite of absence of betalains*
Chenopodiaceae	Caryophyllaceae
Nyctaginaceae	Molluginaceae
Portulacaceae	
Phytolaccaceae	

*Alternatively known as the order Caryophyllales.
†Betalain pigments include the purple betacyanins and the yellow betaxanthins.

Up to the present time, the two types of pigment are mutually exclusive in their occurrence. Thus, betacyanins have never been found in the two anthocyanin-containing families, the Caryophyllaceae and Molluginaceae, and likewise anthocyanins have never been authentically recorded in any betacyanin family. It is true that there have been several reports of anthocyanins in Amaranthaceae, Cactaceae and Phytolaccaceae (see e.g. Timberlake and Bridle 1975; Burret *et al.* 1981) but these are false. Thus, they conflict with the work of experts on betacyanin chemistry (e.g. Piattelli 1976) who report betacyanins in these same sources. Furthermore whenever these reports have been reinvestigated they have been shown to be wrong. For example, *Cactus opuntia* fruit was reported to be pigmented by anthocyanin by Duro and Condorelli (1971) but reinvestigation by this author of fruit tissue, supplied and authenticated by G. D. Rowley, showed that the usual betacyanins were the sole pigments.

The utility of the betacyanin/anthocyanin data for placing anomalous centrospermous genera is well illustrated in the cases of *Hectorella* and *Macarthuria*. *Hectorella caespitosa* is a monotypic New Zealand endemic which was originally placed in the Caryophyllaceae. Reinvestigation of its morphology and anatomy suggested that it more properly belonged in the Chenopodiaceae. This revision was nicely supported by chemical investigation, when the pigment induced in the root of the plant was identified as betacyanin-based (Mabry *et al.* 1978). Again, *Macarthuria* is a small genus of four Australian species which was originally placed in the Aizoaceae. However, *Macarthuria australis* has yielded an anthocyanin pigment, identified as cyanidin 3,5-diglucoside (T. J. Mabry and J. B. Harborne, unpublished result). It is therefore out of place in the betacyanin-producing Aizoaceae. Its affinities would seem to lie with the anthocyanin-containing Molluginaceae, assuming that its other characteristic features are not in disagreement with this.

Pigmentation is only one of several possible chemical approaches to these anomalous taxa and other chemical studies (e.g. of the flavonols and glycoflavones in the leaves, see Richardson 1981) may also yield data of value to taxonomists. It may be mentioned here that the presence of another type of chemical, the glucosinolate (or mustard oil glycoside) in the genus *Gyrostemon* once included in the Phytolaccaceae, served to prove that it was wrongly placed in the Centrospermae and more properly belonged with other glucosinolate-containing plants in the order Capparales. In this instance, its removal from the Centrospermae was supported by chromosomal data and by its lack of the centrospermous P-type plastid, a characteristic ultrastructural feature of this order (Goldblatt *et al.* 1976).

Besides operating at the generic level, chemistry may also have a significant impact on taxonomic decisions regarding family assignments. There are still a considerable number of usually small angiosperm families whose affinities are unclear and which are difficult to place with certainty into an ordered system of classification. One such family is the Julianaceae, a small group of two tree genera from Central and South America, the taxonomic relationships of which have been the subject of controversy for at least a century. From the wood anatomy and morphological features associated with wind pollination, this family has variously been placed in three different associations, near either the Anacardiaceae or the Burseraceae or the Juglandaceae.

A recent chemical analysis of the heartwood flavonoids provides clearcut data in support of the first of these associations. Thus the heartwood of *Amphipterygium adstrigens* (Julianaceae) yielded sixteen flavonoids, seven of which were 5-deoxyflavonoids. These rare substances occur quite characteristically in the Anacardiaceae but are not known in either the Burseraceae or the Juglandaceae. Indeed, all seven of these 5-deoxyflavonoids have been recorded widely in the tribe Rhoeae of the Anacardiaceae and one of them, the aurone rengasin, is only known otherwise from this family. By contrast, no evidence could be found in *Amphiptergium* of juglone, which is a characteristic quinone of Juglandaceae.

By using the new flavonoid data together with a reappraisal of the anatomical characters, Young (1976) decided to reduce the Julianaceae to subtribal rank and include it in the Anacardiaceae as the subtribe Julianiinae of the tribe Rhoeae. This decision has been accepted by Thorne (1981) in his latest revision of angiosperm classification and is also in line with recent serological investigations of these and related plants (Petersen and Fairbrothers 1979).

The above examples serve to indicate the range of practical taxonomic problems that chemical data may help to solve from the species to the family level. Although several ordinal arrangements of families based in part on chemical data are now generally accepted (cf. Cronquist 1980), the relationship of chemistry to plant classification above the family level is still a controversial one. For example, there is little acceptance among systematists of the strong chemical links that exist between the Umbelliferae (and the Apiales) and the Compositae (or Asterales). Again, Dahlgren's view (1981) that iridoid-containing families must all be closely related to each other and have to be placed in contiguous orders is not widely accepted by other taxonomists.

In many such instances, the chemical data are relatively incomplete and undoubtedly as more extensive and detailed chemical analyses are conducted on more and more plants, the impact of the chemical approach will be more considerable than at present. Since the majority of plant families have yet to be thoroughly surveyed for their chemical constituents, very much remains to be done.

REFERENCES

Alston, R. E. and Turner, B. L. (1963). "Biochemical Systematics". Prentice-Hall, Englewood Cliffs, N.J.

Andrews, A. T. (1982). "Electrophoresis: Theory, Techniques and Biochemical and Clinical Applications". Clarendon Press, Oxford.

Bianchini, J. P. and Gaydou, E. M. (1980). Separation of polymethoxylated flavones by straight-phase high performance liquid chromatography. *J. Chromatog.* **190,** 233–236.

Bisby, F. A., Vaughan, J. G. and Wright, C. A. (1980). "Chemosystematics: Principles and Practice". Academic Press, London, New York.

Boulter, D. (1981). Proteins in legumes. *In* "Advances in Legume Systematics" (R. M. Polhill and P. H. Raven, eds) pp. 501–512. Royal Botanic Gardens, Kew.

Bray, D. (1977). The protein explosion. *Nature* **267,** 481–482.

Brimacombe, R. (1981). Secondary structure and evolution of ribosomal RNA. *Nature* **294,** 209–210.

Brown, E. and Moss, J. P. (1976). The identification of potato varieties from tuber characters. *J. Nat. Inst. Agric. Bot.* **14,** 49–69.

Burret, F., Rabesa, Z., Zandonella, P. and Voirin, B. (1981). Contribution biochimique à la Systematique de l'ordre des Centrospermales. *Biochem. System Ecol.* **9,** 257–262.

Chrispeels, M. J. and Baumgartner, B. (1978). Serological evidence confirming the assignment of *Phaseolus aureus* and *P. mungo* to the genus *Vigna. Phytochemistry* **17,** 125–126.

Cristofolini, G. (1981). Serological systematics of the Leguminosae. *In* "Advances in Legume Systematics" (R. M. Polhill and P. H. Raven, eds) pp. 513–532. Royal Botanic Garden, Kew.

Cronquist, A. (1980). Chemistry in Plant Taxonomy. *In* "Chemosystematics: Principles and Practice" (F. A. Bisby, J. G. Vaughan and C. A. Wright, eds) pp. 1–27. Academic Press, London, New York.

Dahlgren, R. M. T. (1981). A revised classification of the angiosperms. *In* "Phytochemistry and Angiosperm Phylogeny" (D. A. Young and D. S. Seigler, eds) pp. 194–204. Praeger, New York.

Duro, F. and Condorelli, P. (1971). Anthocyanins in the fruit of *Cactus opuntia. Quad. Merceol.* **10,** 39–48.

Ehrendorfer, F. (1976). Systematics and evolution of Centrospermous families. *Pl. Syst. Evol.* **126,** 99–105.

Ellis, J. R. S. (1979). The application of starch gel electrophoresis of gliadin proteins to the identification of wheat varieties. *In* "Recent Advances in the Biochemistry of Cereals" (D. L. Laidman and R. G. Wyn Jones, eds) pp. 349–354. Academic Press, London, New York.

Ellis, J. R. S. and Beminster, C. H. (1977). Identification of British wheat varieties by means of starch gel electrophoresis. *J. Nat. Inst. Agric. Bot.* **14,** 221–231.

Felsted, R. L., Li, J. and Pokrywska, G. (1981). Comparison of *Phaseolus vulgaris* cultivars on the basis of isolectin differences. *Int. J. Biochem.* **13,** 549–558.

Ferguson, A. (1980). "Biochemical Systematics and Evolution". Blackie, Glasgow.

Goldblatt, P., Nowicke, J. W., Mabry, T. J. and Behnke, H. D. (1976). Gyrostemonaceae: status and affinity. *Bot. Notiser* **129,** 201–206.

Gough, L. J. and Welch, H. J. (1978). Nomenclatural transfer of *Chamaecyparis obtusa* 'Sanderi' to *Thuja orientalis* on the basis of phytochemical evidence. *Bot. J. Linn. Soc.* **77,** 217–221.

Gray, J. C. (1980). Fraction I Protein and plant phylogeny. *In* "Chemosystematics: Principles and Practice" (F. A. Bisby, J. G. Vaughan and C. A. Wright, eds) pp. 167–194. Academic Press, London, New York.

Gutteridge, C. S. and Norris, J. R. (1979). The application of pyrolysis techniques to the identification of micro-organisms. *J. Appl. Bacteriol.* **47,** 5–43.

Harborne, J. B. and Green, P. S. (1980). A chemotaxonomic survey of flavonoids in leaves of the Oleaceae. *Bot. J. Linn. Soc.* **81,** 155–167.

Harborne, J. B. and Swain, T. (1979) Flavonoids of the Solanaceae. *In* "The Biology and Taxonomy of the Solanaceae" (J. G. Hawkes, R. N. Lester and A. D. Skelding, eds) pp. 257–268. Academic Press, London, New York.

Harborne, J. B., Williams, C. A. and Wilson, K. L. (1982). Flavonoids in leaves and inflorescences of Australian *Cyperus* species. *Phytochemistry* **21,** 2491–2507.

Hawksworth, D. L. (1976). Lichen chemotaxonomy. *In* "Lichenology: Progress and Problems" (D. H. Brown, D. L. Hawksworth and R. H. Bailey, eds) pp. 139–184. Academic Press, London, New York.

Heaney, R. K. and Fenwick, G. R. (1980). Glucosinolate content of Brassica vegetables: a chemotaxonomic approach to cultivar identification. *J. Sci. Fd. Agr.* **31,** 794–801.

Hurka, H. (1980). Enzymes as a taxonomic tool: a botanist's approach. *In* "Chemosystematics: Principles and Practice" (F. A. Bisby, J. G. Vaughan and C. A. Wright, eds) pp. 103–122. Academic Press, London, New York.

Ivie, G. W., Beier, R. C. and Holt, D. L. (1982). Analysis of carrot for linear furocoumarins (psoralens) at the subparts per million level. *J. Ag. Fd. Chem.* **30,** 413–416.

Jensen, U. (1981). Proteins in plant evolution and systematics. *Prog. Bot.* **43,** 344–369.

Jones, B. L., Lookhart, G. L., Hall, S. B. and Finney, K. F. (1982). Identification of wheat cultivars by gliadin electrophoresis. *Cereal Chem.* **59,** 181–188.

Kandler, O. and Schleifer, K. H. (1980). Systematics of bacteria. *Prog. Bot.* **42,** 234–252.

Kondrat, R. W. and Cooks, R. G. (1978). Direct analysis of mixtures by mass spectrometry. *Analyt. Chem.* **50,** 81A–92A.

Krattinger, K., Rast, D. and Karesch, H. (1979). Analysis of pollen proteins of *Typha* species in relation to identification of hybrids. *Biochem. System. Ecol.* **7,** 125–128.

Ladizinsky, G. and Hymowitz, T. (1979). Seed protein electrophoresis in taxonomic and evolutionary studies. *Theor. appl. genet.* **54,** 145–151.

Lattimer, G. L. and Ormsbee, R. A. (1981). "Legionnaire's Disease". Marcel Dekker, New York.

Mabry, T. J., Neuman, P. and Philipson, W. R. (1978). *Hectorella*: a member of the betalain suborder Chenopodiineae of the order Centrospermae. *Pl. Syst. Evol.* **130,** 163–165.

O'Farrell, P. H. (1975). High resolution two-dimensional electrophoresis of proteins. *J. biol. Chem.* **250,** 4007–4021.

Pedersen, J. A. (1978). Naturally occurring quinols and quinones studied as semiquinones by electron spin resonance. *Phytochemistry* **17,** 775–778.

Pedersen, J. A. and Øllgaard, B. (1982). Phenolic acids in the genus *Lycopodium*. *Biochem. System. Ecol.* **10,** 3–10.

Petersen, F. and Fairbrothers, D. E. (1979). *Amphipterygium* – an amentiferous member of the Anacardiaceae. *Abst. Phytochem. Bull.* **12,** 28–29.

Phelan, J. R. and Vaughan, J. G. (1976). A chemotaxonomic study of *Brassica oleracea* with particular reference to its relationship to *Brassica alboglabra*. *Biochem. System. Ecol.* **4,** 173–178.

Phillipson, J. D. (1982). Chemical investigations of herbarium material for alkaloids. *Phytochemistry* **21,** 2441–2456.

Piattelli, M. (1976). Betalains. *In* "Chemistry and Biochemistry of Plant Pigments" (T. W. Goodwin, ed.) (2nd edn) pp. 560–596. Academic Press, London, New York.

Richardson, M. (1981). Flavonoids of some controversial members of the Caryophyllales. *Pl. Syst. Evol.* **138,** 227–233.

Rodman, J. E. (1976). Differentiation and migration of *Cakile*: seed glucosinolate evidence. *Syst. Bot.* **1,** 137–148.

Rodman, J. E., Kruckeberg, A. R. and Al-Shehbaz, I. A. (1981). Chemotaxonomic diversity and complexity in seed glucosinolates of *Caulanthus* and *Streptanthus*. *Syst. Bot.* **6,** 197–222.

Rudloff, E. von (1975). Volatile leaf oil analysis in chemosystematic studies of North American conifers. *Biochem. System. Ecol.* **2,** 131–167.

Salinas, J., Perez de la Vega, M. and Benito, C. (1982). Identification of hexaploid wheat cultivars on isozyme patterns. *J. Sci. Fd. Agr.* **33,** 221–226.

Shellard, E. J. (ed.) (1968). "Quantitative Paper and Thin-Layer Chromatography". Academic Press, London, New York.

Shewry, P. R., Pratt, H. M. and Miflin, B. J. (1978). Varietal identification of single seeds of barley by analysis of hordein polypeptides. *J. Sci. Fd. Agr.* **29,** 587–596.

Singh, R. S., Jain, S. K. and Qualset, C. O. (1973). Protein electrophoresis as an aid to oat variety identification. *Euphytica* **22,** 98–105.

Skinner, F. A. and Lovelock, D. W. (eds) (1979). "Identification Methods for Microbiologists" (2nd edn). Academic Press, London, New York.

Smith, D. M. (1980). Flavonoid analysis of the *Pityrogramma triangularis* complex. *Bull. Torrey. Bot. Club* **107,** 134–145.

Stace, C. A. (ed.) (1975). "Hybridisation and the Flora of the British Isles". Academic Press, London, New York.

Stace, C. A. (1980). "Plant Taxonomy and Biosystematics". Edward Arnold, London.

Stearn, W. T. (1977). Union of *Chionanthus* and *Linociera*. *Ann. Missouri Bot. Gard.* **63,** 355–357.

Stegemann, H. (1979). Characterisation of proteins from potatoes and the Index of European varieties. *In* "The Biology and Taxonomy of the Solanaceae" (J. G. Hawkes, R. N. Lester and A. D. Skelding, eds) pp. 279–284. Academic Press, London, New York.

Stegemann, H. and Loeschcke, V. (1976). Index of European potato varieties based on electrophoretic spectra. *Mitt. Biolog. Bundesanst. Land- u. Fortwirtsch.* **168,** 1–215.

Stewart, R. N., Asen, S., Massie, D. R. and Norris, K. H. (1979). The identification of Poinsettia cultivars by HPLC analysis of their anthocyanin content. *Biochem. System. Ecol.* **7,** 281–287.

Stewart, R. N., Asen, S., Massie, D. R. and Norris, K. H. (1980). The identification of Poinsettia cultivars by HPLC analysis of their flavonol content. *Biochem. System. Ecol.* **8,** 119–126.

Thorne, R. F. (1981). Phytochemistry and angiosperm phylogeny: a summary statement. *In* "Phytochemistry and Angiosperm Phylogeny" (D. A. Young and D. S. Seigler, eds) pp. 233–295. Praeger, New York.

Timberlake, C. F. and Bridle, P. (1975). Anthocyanins. *In* "The Flavonoids" (J. B. Harborne, T. J. Mabry and H. Mabry, eds) pp. 214–266. Chapman and Hall, London.

Turner, B. L. (1970). Molecular approaches to population problems at the infraspecific level. *In* "Phytochemical Phylogeny" (J. B. Harborne, ed.) pp. 187–206. Academic Press, London, New York.

Vaughan, J. G. and Denford, K. E. (1968). An acrylamide gel electrophoretic study of the seed proteins of *Brassica* and *Sinapis* species. *J. Exp. Bot.* **19,** 724–732.

Wall, J. S. (1979). The role of wheat proteins in determining baking quality. *In* "Recent Advances in the Biochemistry of Cereals" (D. L. Laidman and R. G. Wyn Jones, eds) pp. 275–312. Academic Press, London, New York.

Walsh, K. A., Ericsson, L.-H., Parmelee, D. C. and Titani, K. (1981). Amino acid sequencing of proteins. *Ann. Rev. Biochem.* **50,** 261–284.

Wright, C. A. (ed.) (1974). "Biochemical and Immunological Taxonomy of Animals". Academic Press, London, New York.

Young, D. A. (1976). Flavonoid chemistry and the phylogenetic relationships of the Julianaceae. *Syst. Bot.* **1,** 149–162.

14 Phytochemistry in Plant Systematics and Evolution

K. KUBITZKI

Institut für Allgemeine Botanik und Botanischer Garten, Universität Hamburg, Fedral Republic of Germany

Abstract: Various chemosystematic criteria that have been proposed are discussed, and examples given of their application. These include presence or absence of individual compounds, accumulation, and biosynthetic diversification. The classificatory and evolutionary importance of chemical data is discussed and special attention is paid to phytochemical diversification in the evolution and adaptation of plant populations. The presumed role of secondary metabolites in plant defence against herbivores and pathogens gives important clues as to their coevolutionary origin. Mechanisms which disturb the general correlation between chemical similarity and patristic relationship are described. It is concluded that phytochemical characteristics are by no means different from other adaptive traits of the plant populations, nor are the problems and difficulties of chemosystematics basically different from those of systematics in general.

INTRODUCTION

During the last two decades there has developed an enormous interest in phytochemistry as an aid in plant systematics and classification. This has been accompanied by a real avalanche of phytochemical information and it is most difficult to cope with the increasing knowledge of structural diversity of natural products and their distribution in plants. Although it has often been claimed that chemosystematics should consider the whole array of constituents present in a major taxon, in the study of a certain plant group hardly more than a limited set of data is available to form a basis for comparison with the different

Systematics Association Special Volume No. 25, "Current Concepts in Plant Taxonomy", edited by V. H. Heywood and D. M. Moore, 1984. Academic Press, London and Orlando.
ISBN 0 12 347060 9

classifications that have been proposed. In higher plants, for example, the classifications of Bentham and Hooker, Engler, Hutchinson, Sóo, Takhtajan, Cronquist, etc. may all be used in turn for comparison, and chemosystematics finds itself in the role of an impartial arbitrator. This procedure rests on the conviction that a classification, once correctly established, will be compatible with all subsequently added information. While this is mostly true and virtually forms the basis of the predictive value inherent to taxonomy, it has rarely been shown that reasonably broadly based classifications will remain really unaffected by subsequently added information as is claimed by the "non-specificity hypothesis" (Sokal and Sneath 1963), and that a classification based on one set of characters necessarily will be compatible with that based on another. On the contrary, there are examples in which the morphological and chemical traits of populations are not correlated (Payne 1976; Payne *et al.* 1973), which shows that different sets of characteristics may be subjected to different selective pressures.

CHEMOSYSTEMATIC CRITERIA

It has become trivial by now to state that identical, or similar, end products present in different plant taxa do not imply relationship if their biosynthetic origin is different; this is well illustrated, for instance, by the different pathways leading to naphtha- and anthraquinones in fungi and higher plants (Zenk and Leistner 1968). In practical plant taxonomy, however, judgements are made mainly on the basis of presence or absence of particular compounds, i.e. in a perplexingly static way. Among taxonomists the view is widely prevalent that isolated chemical properties are suited as taxonomic criteria only if they are supported by other characteristics (Dahlgren *et al.* 1981), i.e. the value of each chemical trait, just like that of each other character, has to be determined empirically (Cronquist 1980). In the light of the vast knowledge of secondary metabolism, in particular as to biosynthesis, regulation, etc. this view appears somewhat superficial, and indeed more precise definitions of what a chemical character is have been attempted. According to Hegnauer (1969a) the systematic significance of a certain compound would depend on the concentration in which it is accumulated in the plant. Microanalytical techniques have shown that many plant substances are widespread, or even ubiquitous, in low concentrations, but accumulate in large amounts only in rare cases. Hence accumulation is considered an important criterion. However, how then to evaluate the occurrence of more than 60 structurally related indole alkaloids in the different organs of *Catharanthus roseus* (Mothes *et al.* 1965), or of 30 slightly different glucosinolates in horse-radish roots (Grob and Matile 1980), most of which are present only in trace amounts? These cases show that the enzymatic equipment and, more fundamentally, the genetic information necessary for the synthesis of a wide array of the known representatives of a biogenetic group may be widespread within one lineage. Additional regulatory control will determine the pattern of accumulation of the set of constituents observed to characterize certain plants and populations. If sufficiently sensitive analytical procedures are used, allied species will prove to differ not in presence/absence of individual compounds, but rather in their quantitative relationship (Grob and Matile 1980), while major taxa will tend to emphasize different themes of secondary metabolism. Accordingly Gottlieb (1982), who is more concerned with the classification of the plant kingdom as a whole, considers biosynthetic diversification, and not accumulation, as the most important chemosystematic criterion.

CHEMISTRY IN CLASSIFICATION

The presence/absence criterion, inappropriate as it appears from the theoretical standpoint, can lead to useful results if taxa of low hierarchic level are studied. Their substances usually differ in such a way as to reflect the steps of biosynthetic diversification. Where chemosystematics complements plant population studies, deep insights into the nature and dynamics of the populations are possible that cannot be attained on the basis of morphological analysis alone. Alston and Turner's (1962) work on *Baptisia* is the classical example, and the application in plant population studies of 2-dimensional chromatography of flavonoids has since nearly become a matter of routine.

The presence/absence criterion is also applicable at the higher levels of the hierarchy and it is perhaps interesting to see to which degree major taxa are characterized by certain substances and to which degree chemical data have influenced the classification of major taxa of plants. If we leave apart the Algae in which pigmentation and photosynthetates have always been important criteria, the influence of chemical data on the classification of bryophytes and pteridophytes is negligible. It is true that we can perceive the emergence of some general trends, such as the predominance of flavone O-glycosides in Marchantiales vs. flavone C-glycosides in Jungermanniales (for a review see Frey 1980); different oligosaccharides in the major groups of pteridophytes (Kandler 1965); but the overall impression is that the strong systematic isolation of all these groups has long permitted their satisfactory taxonomic circumscription.

This is in sharp contrast to angiosperms which are a coherent group and the major taxa of which are continually changed in rank and rearranged by some authors (Cronquist 1968, 1981; Dahlgren 1975, 1980; Dahlgren *et al.* 1981; Takhtajan 1975, 1980; Thorne 1976, 1981). There is a prodigious suprafamilial superstructure being proposed by these authors considered to be of doubtful practical (Davis 1978) and even theoretical (Heywood 1977) value, and this is largely based on cryptic micromorphological and phytochemical evidence so that we have to ask how far-reaching the impact of chemical data in system making indeed is. The first to use chemical data consistently when dealing with the orders and families of angiosperms was Dahlgren (1975), although these data were used to characterize the families and orders rather than to rearrange them. During recent years the use of chemistry in classification has been accepted so widely that, for example, Cronquist (1981) has included pertinent information in his family descriptions in what he calls an "integrated system of classification of the angiosperms". On the whole, however, the impact of chemical data on the major classification of the angiosperms has been slight and has been restricted to some cases in which chemistry had the role of deciding between conflicting assignations. Apart from the case of the betalains, major realignments have been induced mainly by the distribution of two classes of substances, iridoids and glucosinolates. Both appear restricted to dicotyledons which makes a multiple origin for them unlikely. Consequently, Dahlgren in the various versions of his system has drastically changed the position of the families containing them while other authors have been more reluctant to rely on one character of doubtful value.

It was especially the order Cornales in Dahlgren's (1977, 1980) circumscription that has become a recipient for families between which, up to the present, little relationship, if any, has been noted. These families include such diverse elements as Symplocaceae, Icacinaceae, Escalloniaceae and several others and Dahlgren *et al.* (1981) stress their common possession of sympetalous perianths, unitegmic and tenuinucellate ovules and some embryological features together with iridoid compounds. I wonder, however, if some of the characters mentioned might not be grade characters rather than indicators of a patristic relationship. But even if these elements ultimately will prove to display phyletic coherence, will there not be closely allied taxa that instead of synthesizing iridoids will make different use of mevalonic acid? Within the Cornales, the Styracaceae, Aquifoliaceae and Diapensiaceae could be such examples (see also Huber 1963).

The erratic distribution of glucosinolates, which have a massive occurrence in the Cruciferae and Capparaceae but occur also in half a dozen or so other families, has puzzled systematists for many decades although it was accepted that the existence of these compounds in Tropaeolaceae and other families is almost certainly due to analogy. However, in 1977, in his broadly construed order Capparales Dahlgren included nearly all families producing these compounds. Only recently, he has removed from them Tropaeolaceae together with Limnanthaceae and Salvadoraceae to form two orders of unknown affinity. Again, while biochemical considerations would not militate against considering glucosinolate production as a homologous process in all taxa where it occurs (Rodman 1981), the morphological differences between several of these families make them a heterogeneous alliance.

While these are examples for the inappropriate application of the presence/absence criterion, I wish to point out that there are also cases in which very convincing chemical evidence that can resolve taxonomic controversies of long standing is largely ignored. This applies for example to the sharp chemical dichotomy between *Nelumbo* and Nymphaeaceae *sensu stricto* that necessitates their accommodation in separate orders (see Hegnauer 1969a); among the newer systems of angiosperm classification only that of Thorne (1981) has met this necessity. Another example is related to the position and natural relationship of the Bonnetiaceae. This family, in the circumscription of Hutchinson (1973), includes elements that previously had been accommodated partly in the Guttiferae, partly in the Theaceae. It has been shown that all genera of the Bonnetiaceae that have been studied (Rezende and Gottlieb 1973; Kubitzki *et al.* 1978) contain xanthones that are highly significant for the Guttiferae and are absent from Theacae; the fact that even the hydroxylation patterns fit completely in Bonnetiaceae and Guttiferae leaves two possibilities: either to include Bonnetiaceae into Guttiferae as a marginal subfamily, or to adjoin it in close association with Guttiferae. Again we see that among the modern systems only that of Dahlgren *et al.* (1981) matches this necessity, while in the treatments of Thorne (1981) and Cronquist (1981) Bonnetioideae and Kielmeyeroideae persist as widely separated subfamilies in the Theaceae and Guttiferae, respectively.

However, these are minor details and this criticism illustrates only that the whole array of systematic evidence has grown so vast to-day that it has become virtually impossible for one individual to critically evaluate systematic evidence from all potential sources instead of simply taking over the conclusions that have been drawn by others. Moreover, it is annoying that systematists tend to publish their systems without, or with insufficient documentation because it seems that for many of them the decisions that are made appear more important than the factual basis on which these rest. Because of this it is often hard, even for botanists, to see how uncertain is the basis for placing many angiosperm families, and that verv often they occupy their position for no reason other than tradition. It is therefore fortunate that our ignorance is sometimes frankly admitted and Dahlgren *et al.* (1981) mention that the position of several families is uncertain, while Thorne (1981) even attaches a score to each family in order to indicate the degree of confidence he places into its alignment or delineation.

TRACING EVOLUTION

Much more problematic than the use of chemical characteristics in a purely classificatory context are attempts to add an evolutionary interpretation to them, because the distinction between primitive and advanced character states is often extremely difficult. However, the systematic distribution of primitive character states, in particular their correlation with other allegedly primitive character states, and biogenetic reasoning have been used to this end. Even in the absence of experimental biosynthetic evidence, a careful comparison of all biogenetically related compounds in a major taxon can lead to some insight. The classical example is the work of Holger Erdtman (e.g. 1956) dealing with an array of biogenetically interrelated substances like stilbenes, flavonoids, lignans from the Coniferae that has produced a highly meaningful, dynamic picture of this plant group. More recently, the importance of such comparative phytochemical studies has been exemplified by Kisakürek and Hesse (1980) who have carefully collated and arranged in a systematic fashion all records of indole alkaloids derived from secologanin which are centred in the Loganiaceae, Apocynaceae and Rubiaceae. Increasing structural complexity is the main criterion for arranging the different structural types and the taxa that produce them and this again leads to a picture which is informative as to the details of the evolutionary relationships between these families.

Due to the wide distribution of flavonoids among green land plants attempts have been made to determine their general value as evolutionary markers and again distribution, correlation and biosynthesis have been the major criteria that have been used to this end. On this basis, evolutionary schemes for the flavonoids have been presented by Harborne (e.g. 1967, 1977) and Swain (e.g. 1975) but, as Gornall and Bohm (1978) have stated it is not clear whether these schemes are meant to apply to the entire plant kingdom, just to angiosperms, or to a certain group of angiosperms. It is this ambiguity that has led to discrepancies in the evaluation of certain substances, e.g. 3-deoxyanthocyanidins or C-glycosyl flavones, which sometimes were considered primitive sometimes advanced. These contradictions can be resolved if one determines the range to which the concepts primitive and advanced apply: they are relative concepts and have a meaning only within the context of the taxon under study, the primitive condition being the character state of the – mostly fictitious – most recent common ancestor of the group under study.

Unfortunately, flavonoids can vary drastically even at low hierarchic levels, i.e. between species, or even within species (Abd-Alla *et al.* 1980). Therefore, it is not surprising that there emerge disappointingly few generalities when the distribution of flavonoids in the angiosperms is carefully collated, as has been done by Gornall *et al.* (1979). On the other hand, fine results are obtainable if statements are restricted to the taxon under study, especially if cases of chemical convergence and parallelism are recognized as such by reference to morphology. For example, within the Dilleniaceae (Gurni (1979) found a trend towards methylation of the flavonoid nucleus in positions 4′, 7 and others in strong correlation with a trend towards the formation of sulphates and glucuronides.

An approach which to my mind has proved to be highly successful is that of O. R. Gottlieb (for a summary see Gottlieb 1982) who started from a consistently biogenetic classification of all compounds of a given biogenetic group. This work aims at bringing the items of information widely scattered in the chemical literature into an intelligible context. In this classification, an increasing number of biosynthetic steps runs parallel with constitutional, configurational and substitutional specialization of the compounds concerned and at the same time with the diminishing frequencies of their occurrence in nature. The introduction of a codification of biochemical specialization allows one to overcome, what is for many biologists, the cryptic nature of chemical formulae, and to make comparisons of the different degrees and directions of chemical specialization. Additional parameters are introduced in order to characterize the degree of advancement of substances that Gottlieb believes to constitute a systematic criterion completely independent from morphology which is thus suited for testing different classifications of the same taxon. In this approach neither absence/presence, nor quantity is used as the systematic criterion, but diversification of a given biogenetic group within a taxon.

However, the importance of this work goes well beyond the narrow matter of deciding between conflicting classificatory schemes of particular plant taxa: it is highly interesting to see that at the beginning of each major evolutionary line there are primary precursors from which the chemical diversity within each line originates while the precursors themselves originate via blocking of reaction steps, i.e. by an increasing abridgement of the biosynthetic pathway. Both simplification of primary metabolic pathways and diversification of secondary metabolites are operative, their occurrence generally taking place at different hierarchic levels.

I have already mentioned that taxonomic procedures are often criticized for their lack of clarity and it is understandable that there has been a movement towards repeatable, objective methodology. Much time and care has been spent on the elaboration of cladistic techniques, but a glance into journals like *Systematic Botany* and *Systematic Zoology* shows that the elaboration of this methodology is an issue of unending controversy. An especially moot point seems to be the distinction of primitive and derived character-states, and it appears that only one of the formalized approaches – out-group analysis (Stevens 1980, 1981; Wheeler 1981) – seems to be a reliable way of assessing it. Since, however, to this end the cladistic relationships of at least three groups has to be known: that of the taxon being examined, its sister group, and of at least one taxon that is cladistically more primitive, there is little chance of applying this method successfully in angiosperms for which this knowledge is usually absent (see Davies 1978). Humphries and Richardson (1980) have exemplified cladistic methodology for chemical systematics and one cannot escape the suspicion that with regard to chemical characters parallelism is still more frequent and uniquely derived characters, synapotopies, seem to be much rarer than hitherto recognized. Moreover, the necessity of pressing chemical information into polarized character states from the very beginning introduces such an element of uncertainty into the analysis that it is small wonder that the results of such enterprises are far from convincing.

CAUSES AND MECHANISMS OF PHYTOCHEMICAL CHANGE

So far I have dealt with phytochemical patterns and their use in systematics and evolution. There remains the question for the causes and mechanisms of phytochemical changes that have led to the present-day phytochemical diversity. For a long time, the view of secondary compounds as chemical waste products had been prevalent. This has now been superseded by the recognition of their ecological significance although some peculiarities of these substances such as their turn-over rates appear rather enigmatic in this context (Seigler 1977). Nowadays, secondary metabolites are looked upon as having originated as defence substances against herbivores but often it is not acknowledged that many flavonoids probably owe their omnipresence to their spectral properties which enable them to act efficiently as UV-screens in order to prevent photodestruction of nucleic acids and/or proteins (Swain 1977; Gottlieb 1982). This major difference implies two different mechanisms of chemical variation: flavonoids, xanthones and other substances with systems of conjugated double bonds are free to vary structurally as long as their filter function is maintained while substances that protect against herbivores and pathogens are subject to continuous selection by these agents. In the latter case, a coevolutionary race which includes the overcoming of a toxic, or otherwise repellent, barrier of the plant by the animal or microbe and the acquisition of a novel protective agent by the plant has proven to be a reasonable postulate (Rosenthal and Janzen 1979). It may be mentioned that there has been proof of differential herbivory in experiments using material of different terpene composition (Rice *et al.* 1978; Langenheim *et al.* 1978, 1980) which appears to be a plausible mechanism for fixing quantitative variation of the chemical composition in plant populations.

Much of the immense diversity of secondary metabolites thus appears as the result of reciprocal evolution between plant populations and their herbivores or pathogens. Different plant taxa clearly emphasize different themes of secondary compounds, and the herbivore plus pathogen pressure selects for increasing numbers of compounds. However, as Levin (1976) has emphasized, the shift from one defence system to another which will become necessary under this selective pressure will be the more effective the less the new chemical weapon previously has been employed by another member of the community. Therefore there is much pressure for each member of the community to be chemically different from all others. Species must partition the realm of chemical diversity in order to coexist (Levin 1976). Because of this, the pathways of phytochemical change within each lineage do not depend only on the coevolutionary relationship that is involved but also on the nature of defences already deployed by other members of the community. Since closely related species tend to deploy the same biogenetic group of metabolites as key chemical barriers, such species, if sympatric, can be expected to be more different from each other than allopatric ones. In fact, within six species of western African members of Rutaceae/Toddalioideae, Waterman *et al.* (1978) found that the species with similar alkaloids and triterpene chemistry have different ecological ranges while the species coexisting with each other are chemically diverse.

Further examples of allelochemical diversification have been found in the alkaloid chemistry of sympatric species of desert legumes (Cates and Rhoades 1977), in the glucosinolate composition of eight sympatric species of Crucifers (Rodman and Chew 1980), and in the neolignan and pyrone composition of 18 closely related species of the Lauraceae (Gottlieb and Kubitzki 1981b). In the latter case, chemically similar species tend to be allopatric while overlap in distribution is paralleled by chemical diversification.

However, not only chemical differences between species but also the chemical polymorphism within species might be of an adaptive nature. There are emerging patterns in the distribution of cyanogenic vs. acyanogenic forms of *Lotus corniculatus*, *Trifolium repens*, *Achillea millefolium*, *Juncus* spp., *Pteridium aquilinum*, and other species (Jones 1972; Cooper-Driver *et al.* 1977; Zandee 1976; Fikenscher and Hegnauer 1977) that follow certain environmental gradients and perhaps it can be said that selection decides whether the defence system has to be maintained through the synthesis of secondary compounds, or whether the energy available is better invested in growth rate, propagation, etc. The metabolic cost of the invariable presence of the chemical defence system might be too high (Jones 1973; Raven 1979).

Most chemical changes within major plant taxa are gradual and there is no question that by and large there is a neat correlation between chemical similarity and patristic relationships. However, this picture is sometimes blurred by chemical switch-overs that lead to the sudden appearance of a novel type of substances within a lineage. This is exemplified by the mutually exclusive occurrence of neolignans vs. pyrones within a single genus of the Lauraceae (Gottlieb and Kubitzki 1981a), and of diterpene-based alkaloids of the taxine type in Taxaceae vs. diterpenoid resins in most other conifers (Hegnauer 1962). Such situations have led to the statement that it is the homology of biosynthetic pathways, not the substances produced, which is a plausible indication of affinity (Birch 1973; Gottlieb 1980). Such switchovers may be much more frequent than hitherto recognized, but it must be stressed that they can only be recognized if there is an independent criterion for assessing the relationship between the chemically divergent forms which will be morphology in most cases.

Also the origin of novel compounds by the merging of otherwise separate biosynthetic pathways must be envisaged as forming a possible mechanism for rapid phytochemical changes (McKey 1980). The indole alkaloids, for example, are composed of an indole component derived from tryptophane, and a monoterpenoid based on an iridoid compound. Iridoids are not only ubiquitous within the families producing complex indole alkaloids but are also present in related families that lack these alkaloids. It is conceivable that this coupling reaction, resulting in a fundamentally new type of highly toxic alkaloids, could have been due to the appearance of a single new enzymatic activity, perhaps arising through relaxed substrate specifity of existing enzymes which catalyze similar condensations in other pathways (McKey 1980).

CONCLUSIONS

From the above it becomes obvious that integrated taxonomic and chemical studies hold the greatest potential in chemosystematics because morphological evidence often permits one to decide between several plausible chemical hypotheses and *vice versa*. If chemosystematics expands its outlook into the realm of comparative phytochemistry and tries to incorporate aspects like biosynthesis and coevolution, its scientific potential will grow enormously (Alston 1966). Perhaps we no longer need a more or less superficial incorporation of some rough chemical information into plant systematics, and certainly the less so, the more chemical evidence is consonant with that from other sources. Instead, integrated taxonomic and chemical studies of carefully selected plant groups appear much more promising in order to provide insight into variation, ecological significance and evolutionary origin of chemical traits. Raven (1979) has made a similar claim for the progress in plant population biology.

Although the methods of chemical characterization and structural elucidation are very different from the methods used by the plant taxonomists, chemical traits of the plant appear to be not so different from the plant's other variable traits: adaptation of the population to its environment can be achieved, for instance, through changes of pollinating and fruit dispersing agents, through the timing of reproductive events, through changes of the defence chemistry, etc. A problem is that sometimes we succeed in unveiling the adaptive significance of the traits used as a taxonomic criterion, but mostly this remains hidden to us. Since there is evidence that the size of the effective breeding population in which gene-flow occurs is very limited (Raven 1979), and that species owe their phenetic homogeneity perhaps rather to standardizing selection than to the possession of a common gene-pool (Ehrlich and Raven 1969), one must conclude that chemical patterns even at low hierarchic levels can be affected by parallelism and convergence. Again, this is not even restricted to a certain category of adaptive traits, but is true of all adaptations. Therefore, the problems and difficulties of chemosystematics are in no way different from those of plant systematics and population biology in general.

ACKNOWLEDGEMENTS

The author is most grateful to Professor O. R. Gottlieb (University of São Paulo) and Professor H. Huber (University of Kaiserslautern) for reading and commenting on this paper.

REFERENCES

Abd-Alla, M. F., El-Negoumy, S. J., El-Lakany, M. H. and Saleh, N. A. M. (1980). Flavonoid glycosides and the chemosystematics of *Eucalyptus camaldulensis*. *Phytochemistry* **19**, 2629–2632.

Alston, R. E. (1966). Chemotaxonomy or biochemical systematics? *In* "Comparative Phytochemistry" (T. Swain, ed.) pp. 33–56. Academic Press, London, New York.

Alston, R. E. and Turner, B. L. (1962). New techniques in the analysis of complex natural hybridization. *Proc. Nat. Acad. Sci. (Wash.)* **48,** 130–137.

Birch, A. J. (1973). Biosynthetic pathways in chemical phylogeny. *In* "Chemistry in Botanical Classification", Nobel Symp. Vol. 25 (C. Bendz and J. Santesson, eds) pp. 261–270. Academic Press, London, New York.

Cates, R. G. and Rhoades, D. F. (1977). Patterns in the production of antiherbivore chemical defenses in plant communities. *Biochem. Syst. Ecol.* **5,** 185–193.

Cooper-Driver, G., Finch, S., Swain, T. and Bernays, E. (1977). Seasonal variation in secondary plant compounds in relation to the palatibility of *Pteridium aquilinum*. *Biochem. Syst. Ecol.* **5,** 177–183.

Cronquist, A. (1968). "The Evolution and Classification of Flowering Plants", 396 pp. Nelson, London.

Cronquist, A. (1980). Chemistry in plant taxonomy: an assessment of where we stand. *In* "Chemosystematics: Principles and Practice" (F. A. Bisby, J. G. Vaughan and C. A. Wright, eds) pp. 1–27. Academic Press, London, New York.

Cronquist, A. (1981). "An Integrated System of Classification of Flowering Plants", 1262 pp. Columbia University Press, New York.

Dahlgren, R. (1975). A system of classification of the angiosperms to be used to demonstrate the distribution of characters. *Bot. Notiser* **128,** 119–147.

Dahlgren, R. (1977). A note on the taxonomy of the "Sympetalae" and related groups. *Publ. Cairo Herbar.* **6** & **7,** 83–102.

Dahlgren, R. (1980). A revised system of classification of the angiosperms. *Bot. J. Linn. Soc.* **80,** 91–124.

Dahlgren, R., Rosendal-Jensen, S. and Nielsen, B. J. (1981). A revised classification of the angiosperms with comments on correlation between chemical and other characters. *In* "Phytochemistry and Angiosperm Phylogeny" (D. A. Young and D. S. Seigler, eds) pp. 149–199. Praeger, New York.

Davies, P. H. (1978). The moving staircase: a discussion of taxonomic rank and affinity. *Notes Roy. Bot. Gard. Edinb.* **36,** 325–340.

Ehrlich, P. R. and Raven, P. H. (1969). Differentiation of populations. *Science* **165,** 1228–1232.

Erdtman, H. (1956). Organic chemistry and conifer taxonomy. *In* "Perspectives in Organic Chemistry" (Sir Alexander Todd, ed.) pp. 453–494. Interscience, New York.

Fikenscher, L. H. and Hegnauer, R. (1977). Die Verbreitung der Blausäure bei den Cormophyten. 11. Mitteilung: Über die cyanogenen Verbindungen bei einigen Compositae, bei den Oliniaceae und in der Rutaceen-Gattung *Zieria*. *Pharm. Weekblad* **112,** 11–20.

Frey, W. (1980). Systematik der Bryophyten. *Progress in Botany* **42,** 306–322.

Gornall, R. J. and Bohm, B. A. (1978). Angiosperm flavonoid evolution: a reappraisal. *Syst. Bot.* **3,** 353–368.

Gornall, R. J., Bohm, B. A. and Dahlgren, R. (1979). The distribution of flavonoids in angiosperms. *Bot. Notiser* **123,** 1–30.

Gottlieb, O. R. (1980). Micromolecular systematics: principles and practice. *In* "Chemosystematics: Principles and Practice" (F. A. Bisby, J. G. Vaughan and C. A. Wright, eds) pp. 329–352. Academic Press, London, New York.

Gottlieb, O. R. (1982). "Micromolecular Evolution, Systematics and Ecology: An Essay into a Novel Botanical Discipline", 165 pp. Springer Verlag, Berlin.

Gottlieb, O. R. and Kubitzki, K. (1981a). Chemosystematics of *Aniba*. *Biochem. Syst. Ecol.* **9**, 5–12.

Gottlieb, O. R. and Kubitzki, K. (1981b). Chemogeography of *Aniba* (Lauraceae). *Pl. Syst. Evol.* **137**, 281–289.

Grob, K. and Matile, P. (1980). Capillary GC of glucosinolate-derived horseradish constituents. *Phytochemistry* **19**, 1789–1794.

Gurni, A. A. (1979). "Vergleichend-phytochemische Untersuchungen an den Flavonoiden der Dilleniaceen", 82 pp. Doctoral thesis, University of Hamburg.

Harborne, J. B. (1967). "Comparative Biochemistry of the Flavonoids", 383 pp. Academic Press, London, New York.

Harborne, J. B. (1977). Flavonoids and the evolution of the angiosperms. *Biochem. Syst. Ecol.* **5**, 7–22.

Hegnauer, R. (1962). "Chemotaxonomie der Pflanzen", Vol. 1, 517 pp. Birkhäuser Basel.

Hegnauer, R. (1969a). Chemical evidence for the classification of some plant taxa. *In* "Perspectives in Phytochemistry" (J. B. Harborne and T. Swain, eds) pp. 121–138. Academic Press, London, New York.

Hegnauer, R. (1969b). "Chemotaxonomie der Pflanzen", Vol. 5. 506 pp. Birkhäuser, Basel.

Heywood, V. H. (1977). Principles and concepts in the classification of higher taxa. *In* "Flowering Plants. Evolution and Classification of High Categories" (K. Kubitzki, ed.) pp. 1–12. *Plant Syst. Evol.*, Suppl. 1. Springer Verlag, Vienna.

Huber, H. (1963). Die Verwandtschaftsverhältnisse der Rosifloren. *Mitt. Bot. Staatssamml. München* **5**, 1–48.

Humphries, C. J. and Richardson, P. M. (1980). Hennig's method and phytochemistry. *In* "Chemosystematics: Principles and Practice" (F. A. Bisby, J. G. Vaughan and C. A. Wright, eds) pp. 353–378. Academic Press, London, New York.

Hutchinson, J. (1973). "The Families of Flowering Plants", 968 pp. Clarendon Press, Oxford.

Jones, D. A. (1972). Cyanogenetic glycosides and their function. *In* "Phytochemical Ecology" (J. B. Harborne, ed.) pp. 103–124. Academic Press, London, New York.

Jones, D. A. (1973). Co-evolution and cyanogeneis. *In* "Taxonomy and Ecology" (V. H. Heywood, ed.) pp. 213–242. Academic Press, London, New York.

Kandler, O. (1965). Möglichkeiten zur Verwendung von C^{14} für chemotaxonomische Untersuchungen. *Ber. dt. Bot. Ges.* **77**, (62)–(73).

Kisakürek, M. V. and Hesse, M. (1980). Chemotaxonomic studies of the Apocynaceae, Loganiaceae and Rubiaceae, with reference to indole alkaloids. *In* "Indole and Biogenetically Related Alkaloids" (J. D. Phillipson and M. H. Zenk, eds) pp. 11–26. Academic Press, London, New York.

Kubitzki, K., Mesquita, A. L. and Gottlieb, O. R. (1978). Chemosystematic implications of xanthones in *Bonnetia* and *Archytaea*. *Biochem. Syst. Ecol.* **6**, 185–187.

Langenheim, J. H., Stubblebine, W. H., Lincoln, D. E. and Foster, C. E. (1978). Implications of variation in resin composition among organs, tissues and populations in the tropical legume *Hymenaea*. *Biochem. Syst. Ecol.* **6**, 299–313.

Langenheim, J. H., Foster, C. E. and McGinley, R. B. (1980). Inhibitory effects of different quantitative compositions of *Hymenaea* leaf resins on a generalist herbivore *Spodoptera exigua*. *Biochem. Syst. Ecol.* **8,** 385–396.

Levin, D. A. (1976). The chemical defences of plants to pathogens and herbivores. *Ann. Rev. Ecol. Syst.* **7,** 121–159.

McKey, D. (1980). Origins of novel alkaloid types: A mechanism for rapid phenotypical evolution of plant secondary compounds. *Amer. Naturalist* **115,** 745–754.

Mothes, K., Richter, J., Stolle, K. and Gröger, D. (1965). Physiologische Bedingungen der Alkaloid-Synthese bei *Catharanthus roseus* G. Don. *Naturwissenschaften* **52,** 431.

Payne, W. W. (1976). Biochemistry and species problems in *Ambrosia* (Asteraceae-Ambrosieae). *Pl. Syst. Evol.* **125,** 169–178.

Payne, W. W., Geissman, T. A., Lucas, A. J. and Saitoh, T. (1973). Chemosystematics and taxonomy of *Ambrosia chamissonis*. *Biochem. Syst.* **1,** 21–33.

Raven, P. H. (1979). Future directions in plant population biology. *In* "Topics in Plant Population Biology" (O. T. Solbrig, S. Jain, G. B. Johnson and P. H. Raven, eds) pp. 461–481. Macmillan, London.

Rezende, C. M. A. da M. and Gottlieb, O. R. (1973). Xanthones as systematic markers. *Biochem. Syst.* **1,** 111–118.

Rice, R. L., Lincoln, D. E. and Langenheim, J. H. (1978). Palatability of monoterpenoid compositional types of *Satureja douglasii* to a generalist molluscan herbivore, *Ariolimax dolichophallus*. *Biochem. Syst. Ecol.* **6,** 45–53.

Rodman, J. E. (1981). Divergence, convergence and parallelism in phytochemical characters: the glucosinolate-myrosinase system. *In* "Phytochemistry and Angiosperm Phylogeny" (D. A. Young and D. S. Seigler, eds) pp. 43–79. Praeger, New York.

Rodman, J. E. and Chew, F. S. (1980). Phytochemical correlates of herbivory in a community of native and naturalized Cruciferae. *Biochem. Syst. Ecol.* **8,** 43–50.

Rosenthal, G. and Janzen, D. H. (eds) (1979). "Herbivores. Their Interaction with Secondary Plant Metabolites", 717 pp. Academic Press, London, New York.

Seigler, D. (1977). Primary roles for secondary compounds. *Biochem. Syst. Ecol.* **5,** 195–199.

Sokal, P. R. and Sneath, P. H. A. (1963). "Principles of Numerical Taxonomy", 359 pp. W. H. Freeman, San Francisco.

Stevens, P. F. (1980). Evolutionary polarity of character states. *Ann. Rev. Ecol. Syst.* **11,** 333–358.

Stevens, P. F. (1981). On ends and means, or how polarity can be assessed. *Syst. Bot.* **6,** 186–188.

Swain, T. (1975). Evolution of flavonoid compounds. *In* "The Flavonoids" (J. B. Harborne, T. J. Mabry and H. Mabry, eds) pp. 1096–1126. Chapman and Hall, London.

Swain, T. (1977). Secondary compounds as protective agents. *Ann. Rev. Plant Physiol.* **28,** 479–501.

Takhtajan, A. (1975). "Ausbreitung und Evolution der Blütenpflanzen", 189 pp. G. Fischer Verlag, Stuttgart.

Takhtajan, A. (1980). Outline of the classification of flowering plants (Magnoliophyta) *Bot. Rev.* **46,** 225–359.

Thorne, R. F. (1976). A phylogenetic classification of the angiospermae. *Evolutionary Biology* **9,** 33–106.

Thorne, R. F. (1981). Phytochemistry and angiosperm phylogeny. *In* "Phytochemistry and Angiosperm Phylogeny" (D. A. Young and D. S. Seigler, eds) pp. 233–276. Praeger, New York.

Waterman, P. G., Meshal, I. A., Hall, J. B. and Swaine, M. D. (1978). Biochemical systematics and ecology of the Toddalioideae in the central part of the West African forest zone. *Biochem. Syst. Ecol.* **6,** 239–245.

Wheeler, Q. D. (1981). The ins and outs of character analysis: A response to Crisci and Stuessy. *Syst. Bot.* **6,** 297–306.

Zandee, M. (1976). Beobachtungen über Cyanogenese in der Gattung *Juncus. Proc. Kon. Ned. Akad. Wetensch. Amsterdam C,* **79,** 529–543.

Zenk, M. H. and Leistner, E. (1968). Biosynthesis of quinones. *Lloydia* **31,** 275–292.

Data Processing and Taxonomy

15 | Taximetrics To-day

J. McNEILL

Department of Biology, University of Ottawa, Ottawa, Canada

Abstract: Taximetrics is defined as the application of multivariate numerical techniques in systematics. The growth and development of taximetrics are described and its contribution to methodological precision noted. Taximetric procedures are explicable and repeatable, even although their diversity can produce an almost overwhelming variety of possible classifications. The nature and purpose of the investigation must determine the taximetric methodology.

Much of systematics involves pattern recognition and examples are given of taximetric applications seeking pattern from data accumulated at the level of individual specimens, populations or taxa. Applications to other taxonomic situations, such as in the identificatory phase, are discussed. The nature of biological variation, whether at the level of the individual, the population or that of higher taxa, is such that numerical assessment is essential. The future of systematics rests on more effective pattern recognition through the use and development of taximetric techniques, including those of numerical cladistics.

INTRODUCTION

Fifteen years ago, a very similar symposium to this one was held in Liverpool. It was entitled "Modern Methods in Plant Taxonomy" (Heywood 1968a) and included four chapters dealing with aspects of numerical taxonomy (Cullen 1968; Dale 1968; Johnson and Holm 1968; Stearn 1968). My task in this volume

Systematics Association Special Volume No. 25, "Current Concepts in Plant Taxonomy", edited by V. H. Heywood and D. M. Moore, 1984. Academic Press, London and Orlando.
ISBN 0 12 347060 9

on current concepts in plant taxonomy is to look at the field in terms of the developments that have taken place in these last 15 years.

We should probably first remind ourselves of the state of numerical taxonomy in 1967. Sokal and Sneath's (1963) "Principles of Numerical Taxonomy", the first bible of "NT", had been read and was being assimilated and applied. Many of its followers looked on NT as the New Taxonomy, and indeed for some it seemed to be the New Testament. Sokal and Sneath's (1963) frontispiece, with its flow-chart going from the specimens to the classification, epitomized the aims and purposes of numerical taxonomy in those days. The role of NT was to take the subjectivity, the mumbo-jumbo or mystique some might say, out of taxonomy and replace it with an objectivity that would lay to rest once and for all the unscientific appeal to authority in matters of taxonomic judgement.

It seemed that one needed only to observe one's plants or animals, select a sufficiently large number of attributes, score or measure them, put these data "into the computer", and out would come an objective classification. One obvious weakness of this scenario is the nature of that "black box" in the middle, rather mysteriously termed in those days "the computer". What exactly was it that the computer was being asked to do and what indeed were the criteria that should determine how it should be programmed?

Sokal and Sneath (1963) did, of course, put their methodology into context and suggested procedures that seemed at the time most appropriate to different types of taxonomic problem. But all too many practitioners were much less sophisticated and their choice of method was commonly dictated by the computer programs that happened to be available to them. If several programming options were tried and these produced somewhat different results, the solution all too often was to select that which best matched the traditional taxonomic treatment. With such a restrictive criterion, could any worthwhile contribution be expected from these numerical methods, particularly when they entailed the extra effort of consistent character scoring over the entire study group? This is certainly the explanation for there being very many more bad papers on numerical taxonomy published in the sixties and early seventies than there were good ones and, in turn, for some of the opprobium that was fixed on numerical taxonomy by many more traditional taxonomists. Nevertheless, as Heywood noted many years ago (Heywood 1964, 1968b), numerical studies did force taxonomists to look more critically at the characters they used and did provide the stimulus for the appraisals of taxonomic methodology that have so characterized the past 15 years and which are still very much with us.

THE NATURE OF TAXIMETRICS

(1) Phenetic classification and numerical taxonomy

The most striking feature of numerical taxonomy in 1967, however, was the fact that it was almost entirely predicated on a phenetic approach to classification. This is not the place to try to characterize phenetic classification. This is well described by Jardine and Sibson (1971: 136–138) and is discussed in relation to alternative approaches by McNeill (1980). Briefly, a phenetic classification (also known as a natural or general-purpose classification) is one in which a large number of general statements can be made about the classes recognized. It is one in which "the constituent groups describe the distribution among organisms of as many features as possible" (Farris 1977), including features that were not used in the original classification, i.e. it seeks to incorporate a predictive component (McNeill 1980). Phenetic classifications have generally been constructed using overall similarity (Sneath and Sokal 1973) but it has been argued (Farris 1980) that better phenetic classifications can be developed in other ways. This is certainly possible, but by no means proven (cf. McNeill 1982).

Although there has been this strong association between the growth of numerical taxonomy and a phenetic approach to classification, it is important to emphasize that the two are not synonymous. Phenetic classification, obviously, need not be numerically based. Most traditional botanical classification, at least since the latter part of the 18th century, has utilized a phenetic approach. Indeed, this is equally true of most plant taxonomy to-day. Likewise, numerical taxonomy, although it developed in the context of phenetic philosophy, is not by any means restricted to the construction of phenetic classifications.

(2) Objectivity: reality or illusion?

What, then, has happened in these last 15 years? One of the most obvious features is that, with the growth of the field, there has been a mushrooming of the number and diversity of methods available for numerical classification. Sneath and Sokal (1973) provide a "taxonomy of clustering methods" which demonstrates the potential for 256 (2^8) different clustering methods all starting from a matrix of pair-wise similarities or dissimilarities. Although not all of these methods are appropriate in biological taxonomy, very many are, and when these are combined with the wide choice of measures of similarity assessment, the number of phenograms that could be produced from the same data set becomes very large (cf. McNeill 1979a). Indeed, papers have been published with an almost embarassing wealth of analysis. For example, Baum (1974) reports that he obtained 80 different phenograms using eight different clustering methods (one with 9 variants) on five subsets of the same data on 27 species of *Avena*. Although he presents a plausible case for selecting 16 of these as "admissable", not everyone would necessarily accept his criteria. Similarly the differences between the phenograms presented by McNeill (1975a) for species of the Portulacaceae tribe Montieae (*Claytonia*, *Montia*, etc.) are considerable and, like Baum, McNeill had to invoke external criteria for choice. This is a far cry from the early idea that *the* one objective classification would result from a numerical study. Numerical taxonomy may lack the degree of objectivity that was at one time claimed for it, but it is certainly not subjective in the way that traditional methods are, with their unexplained neural assessment of similarities and subsequent grouping. The procedures of numerical taxonomy are clearly explicable and hence repeatable. This is, in itself, a very great advance on uncritical appeal to authority. What the past 15 years have established is that "the computer" is not a replacement authority, and that is as it should be.

(3) The scope of taximetrics

My title is "Taximetrics to-day" and, so far, I have considered only the historical development of numerical taxonomy. For practical purposes the two terms are synonymous, but I use "taximetrics" to emphasize that I am considering all aspects of quantitative taxonomy, or at least those that use multivariate methods, and am not confining myself to numerical phenetics. A field that involves all applications of quantitative methods in systematics cannot be covered comprehensively in the space I have available in this volume. "Numerical Taxonomy" (Sneath and Sokal 1973), the new and expanded version of the 1963 "bible", has nearly 600 pages. There has been a steady output of books and symposium volumes on various aspects of the topic. Of these, Cole (1969), Lockhart and Liston (1970), Jardine and Sibson (1971), Estabrook (1974), Everitt (1974), Pankhurst (1975), and Clifford and Stephenson (1975) may be mentioned. This spate of publications has by no means ended and recent books that provide useful introductions to current taximetric techniques are those by Neff and Marcus (1980), Gordon (1981), Dunn and Everitt (1982) and Felsenstein (1983). The burgeoning field of numerical taxonomy is manifest, not only in these textbooks and symposia, but also in many reviews, of which that by Duncan and Baum (1981) has a particularly extensive bibliography.

So, instead of trying to provide an encyclopaedic view of taximetrics, I will concentrate instead on a few aspects of taxonomy in which I believe numerical methods can and are making important contributions. I will describe these briefly and point out some current trends.

TAXIMETRIC METHODS AND TAXONOMIC PURPOSE

(1) Taximetric applications

In considering the current role of numerical methods in taxonomy, I would place the greatest emphasis on *purpose*. It always was naïve to think in terms of putting one's data into "the computer" and expecting the "right" answer to come out. With the plethora of methods available to us to-day, and with almost infinite scope for devising new ones, it is now manifestly absurd to imagine that one can embark on a taximetric investigation without a framework in which the numerical techniques are to be applied. What is the purpose of our systematic research? Is there a model to explore or a hypothesis that can be tested? There may well not be and this should not, of itself, be a concern, so long as the structure of the problem can be defined. Hypothesis-testing, whatever some may claim, is not the *sine qua non* of science. Much of taxonomy involves, I believe, pattern recognition. We are trying to discern pattern and trying then to summarize it in a manner that will be useful to other students of biology. Taxonomy is, in part, a service industry, that tries to provide the best and most cost-efficient mapping of plant and animal diversity. If our purpose is to use numerical methods to discern meaningful pattern, it is essential that we chose appropriate ones. We must consider the type of pattern that we can reasonably expect from the nature of our sampling, from its ecological and geographical parameters, from evolutionary theory, etc.

If our sample is all the herbarium specimens of a particular well-marked genus, in which species limits have always been problematical, we want a method that *can* produce well-defined groups, but that will not inevitably *force* the specimens into clusters that may be artefacts. On the other hand, if we want to group distinct species or cultivars into useful classes for communication of information, for recognition, or for summarizing of relationship, we *may* want to find a limited number of polythetic groups even at the cost of pushing a few genuine outliers into that larger group with which they have the closest affinity.

In other cases there is no reason to expect or to want discrete clusters. There may be no evidence of barriers to interbreeding and no suggestion of any discontinuities in the variation. The most that might be expected would be dense regions in what is otherwise a continuum. In such a situation most clustering techniques would be inappropriate, and, instead, one of the many ordination techniques, such as principal components analysis (PCA) or principal coordinates analysis (PCO), should be considered. Because they take account of character covariance, these techniques are much to be preferred to the less than satisfactory efforts to extend univariate analysis by polygonal graphs and the like, methods rightly criticized by Snaydon (this volume, Chapter 11).

Not all of taxonomy involves pattern recognition, however. The pattern may have already been established by intensive research, and we may now be seeking ways to express it more conveniently, or to communicate it to others. Populations, clearly delimited on chromosome number or configuration, may be distinguishable morphologically by techniques such as discriminant analysis (DA), and, moreover, the discriminant axes then used to assign, say, herbarium specimens to the appropriate cytodeme, even although chromosome data cannot be obtained. Methods of automatic identification, including computer-generated keys are also in this category.

I shall try to exemplify some of these different purposes to which taximetric methods can be applied and also refer to some of the techniques that are particularly appropriate to different purposes. I do this largely from work with which I have been associated, not because of any lack of wider choice, but because it is that with which I am most familiar.

(2) Grouping of individuals

(*a*) Spiranthes *in northern Ontario*. In 1976, R. C. Simpson, an observant naturalist, noticed unusual plants of the genus *Spiranthes* (lady's tresses) growing in Killbear Provincial Park in the Parry Sound District of northern Ontario. This prompted Dr P. M. Catling, who was then studying the genus as part of his Ph.D thesis, to sample the *Spiranthes* in the Park and to make detailed measurements on 34 specimens, representing the four species that are found in that area, *S. casei*, *S. cernua*, *S. lacera* and *S. romanzoffiana*. These are well distinguished by flower colour and time of flowering (cf. Catling and Cruise 1974; Simpson and Catling 1978), and so it was possible to assign the specimens to species prior to making the measurements. The only exception was a sample of the unusual plants originally found by Simpson which were thought to be hybrids between *S. lacera* and *S. romanzoffiana*. The measurements were made to allow numerical studies, designed to explore this hypothesis, to be carried out (Simpson and Catling 1978), but the data are particularly instructive for a different reason. The qualitative characters of flower colour are usually lost in herbarium material and so, in studies on *Spiranthes*, numerical clustering techniques are often necessary for satisfactory taxonomic treatment. In this case we can look at the performance of such techniques with the hindsight of knowing the "right" answer, but deliberately excluding that information from the numerical study.

Cluster analysis, using the group average method (UPGMA), of the 34 specimens (9 from *S. lacera*, 8 from each of the other three species, and one, the putative hybrid) reflected the known taxonomy reasonably well. Apart from the isolation of one specimen of *S. romanzoffiana* and the possible association of the hybrid with *S. lacera*, the members of each species clustered together (Fig. 1). The groups were by no means unambiguous, however, as is readily exemplified by the three vertical lines in Fig. 1, which reflect optimal splitting levels using Lefkovitch's loss function minimization strategy (Lefkovitch 1976a, b). There is the choice of 4, 12 or 30 groups, and even if one rejects the 30 as a

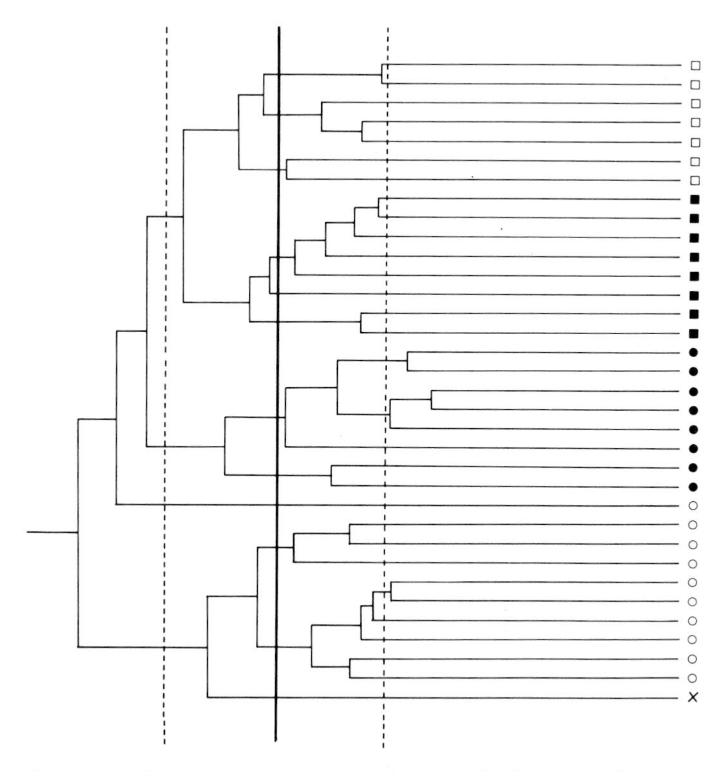

Fig. 1. Phenogram, based on group average (UPGMA) clustering of 34 specimens of *Spiranthes* sampled from Killbear Provincial Park, Parry Sound District, Ontario, Canada. The symbols represent the species and putative hybrid distinguishable on floral characters not used in the analysis (□, *S. romanzoffiana*; ■, *S. cernua*; ●, *S. casei*; ○, *S. lacera*). The vertical lines represent optimal splitting levels (see text).

near trivial solution, neither the 4 nor the 12 groups accurately reflect the classification known to be correct.

This is a not uncommon situation in numerical studies at the level of grouping individual organisms. The clustering is generally plausible, but few, if any, clear cut, unambiguous groupings are present. Why should this happen? Not, I believe, because of any inadequacies in the methodology, but simply from the nature of the data being used.

(*b*) *Pattern recognition not goodness of fit*. These data are derived in a relatively unbiased manner from the whole organism. For any one individual they are a function of its environment and of its particular genetic individuality, as well as of the genomic structure of its population and of other populations that might be considered conspecific. It is only these populationally based components that biological classification generally seeks to reflect. In producing a classification, we are not seeking a best fit to the original data so much as trying to discern the most clear cut pattern that the data suggest.

This is one of the reasons why space-dilating methods, such as Lance and Williams's (1967) flexible sorting, that generate well-defined groups, have tended to meet with more favour among taxonomists than space-contracting ones, like nearest neighbour (single linkage), that tend to chain the OTUs (cf. Williams *et al.* 1971a, b; McNeill 1979a). But space-dilating methods have their drawbacks. In this *Spiranthes* example, they tend to place the "odd" specimen of *S. romanzoffiana* with *S. cernua* and to imbed the hybrid so closely with *S. lacera* that its distinctiveness is overlooked.

(*c*) *Iterative weighting*. What alternatives are there? One is an iterative weighting procedure such as that developed by Hogeweg (1976) or that suggested by McNeill (1979b). In this, an initial clustering by a space-conserving method such as group average is used to generate "incipient groups", which generate character weights. In the *Spiranthes* study this was done using Lefkovitch's optimal splitting strategy referred to above, choosing the 12 groups option as being non-trivial, and yet without too much risk of "forcing" potentially unnatural groupings from the start. The characters were then weighted on the basis of the between to within group variance. Because the F ratio itself is unbounded for characters that are invariant within groups, it is not a very satisfactory measure, and, instead, weights were derived from the added variance component between groups. (For details see Barkworth *et al.* 1979, where the same weighting procedure is used in a different context). Clustering is then carried out using the weighted characters and the process continued in an

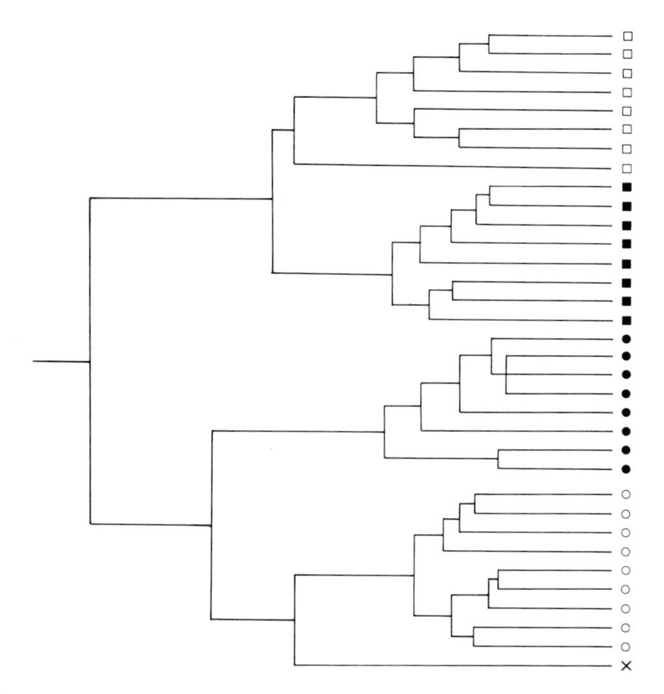

Fig. 2. Phenogram, as in Fig. 1, but after two iterations of the iterative weighting procedure described in the text.

iterative fashion until (and if) there is no change in the clusters produced. Figure 2 shows one of the intermediate results, in which the correct grouping is seen to be taking shape.

Although iterative weighting is a potentially useful approach, in that it selects that portion of the data that reflects pattern, it does share with space-dilating methods the possibility of creating artificial clusters if the incipient groups are themselves heterogeneous. In general, the only check on this, like other clustering techniques, is whether the resultant groups "make sense" according to some external criteria, such as ecology, geography, etc. This is what Williams (1967) calls the hypothesis-generating role of numerical taxonomy.

(*d*) *Rotation of component axes.* Other ways of selecting a pattern-revealing subset of the data that are more readily explicable biologically, may be utilizable. For example, in the *Spiranthes* case, projection onto the first two principal components of a PCA, reveals three, or possibly four, clusters with a parallel trend in each (Fig. 3). This turns out to be in part a size component, that may

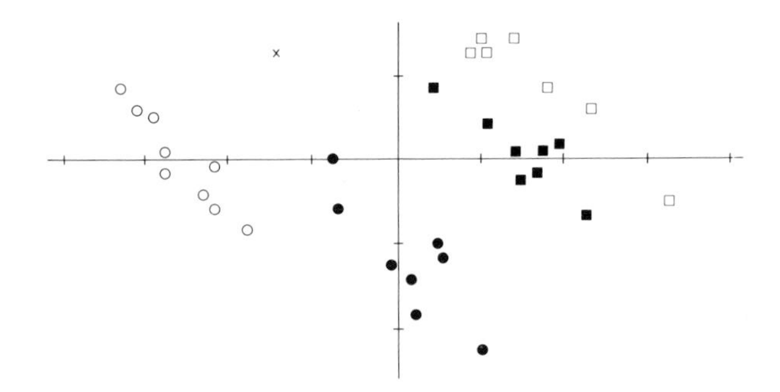

Fig. 3. Plot of the projection of 34 *Spiranthes* specimens onto the first two principal axes of a PCA (source and symbols as in Fig. 1).

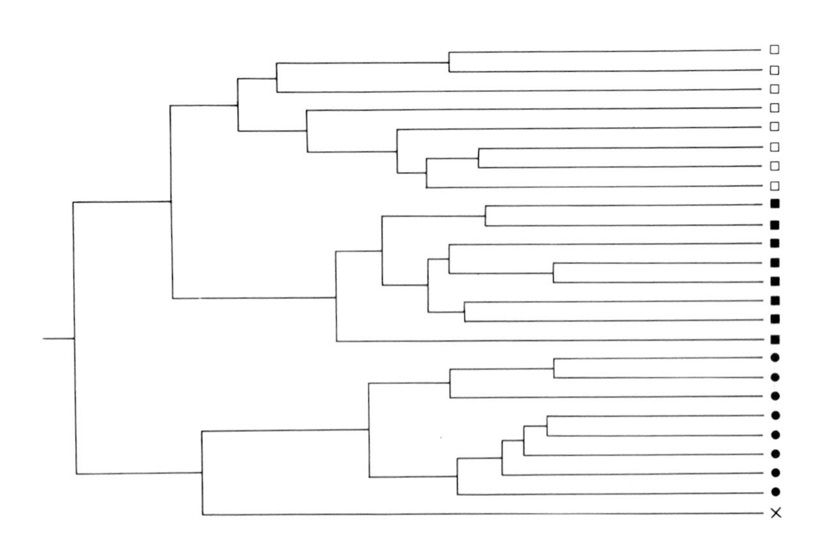

Fig. 4. Phenogram, as in Fig. 1, but based on the PCA scores, with those on the rotated first principal axis excluded (see text). This axis not only reflected within cluster variation but also separated *S. lacera*, and so, in consequence, the 9 specimens of that species are excluded.

reflect the degree of moisture and nutrient in the immediate environment of the plants concerned. To this extent, it is not, therefore, a component of classificatory interest. The same dimension does, however, also separate *S. lacera* from the other species. By rotating principal axes 1 and 2 by –0.7 radians (approximately –40°), this size trend becomes parallel to the new first axis. For further analyses of the specimens, other than those of *S. lacera*, this axis can be discounted and only the rotated second axis along with the subsequent axes used. Figure 4 shows the result of group average clustering on these data after removal of the rotated first axis. The results are obviously much more in line with the known correct solution, than those that used all of the original data on this group of OTUs.

(3) Hierarchical ordering of taxa

(*a*) *Portulacaceae tribe Montieae*. The tribe Montieae of the family Portulacaceae is a group of some 35–40 predominantly North American species, but with two in the southern hemisphere and one (*Montia fontana*, water blinks) very widely distributed throughout the world (Jage 1979, fig. 641). Taxonomic treatments have ranged from recognizing one genus (Boivin 1967) to 10 or 11 (Nilsson 1967; Yurtsev in Yurtsev and Tsvelev 1972). Most often two (*Montia* and *Claytonia*) are accepted, but with varying delimitation as is seen in the vacillation in the British and European literature over the names of the two commonly introduced species. These appear as *Calytonia perfoliata* and *C. alsinoides* (a segregate of *C. sibirica*) in the first edition of Clapham, Tutin and Warburg's *Flora of the British Isles* (Clapham *et al.* 1952), as *Montia perfoliata* and *M. sibirica* in the second edition (Clapham *et al.* 1962) and in *Flora Europaea* (Walters 1964), and now back to *Claytonia* in the most recent edition of "Hegi" (Jage 1979). These changes reflect similar vicissitudes in the North American literature, probably now resolved through the work of Swanson (1966) and myself (McNeill 1975a). The latter exemplifies some of the potential and some of the problems in applying numerical techniques in the grouping of species.

(*b*) *How many genera?* Because the numerical studies have already been described in detail (McNeill 1975a), only the outline will be presented here. Essentially, the situation is that summarized in Fig. 5. There is a distinct group of 24 OTUs (species or species complexess that are well-distinguished regardless of the clustering method used. This corresponds to the genus *Claytonia* as recognized by Swanson (1966), and includes the widely introduced ruderals, *C. sibirica sensu lato* and *C. perfoliata*. The genus is well characterized by the

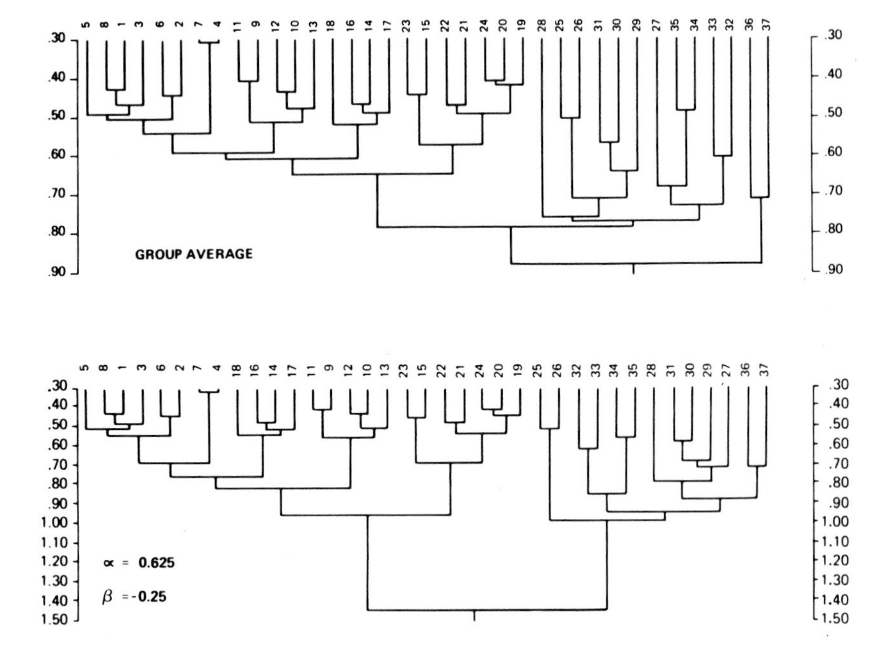

Fig. 5. Phenograms of 37 species or species groups of the Portulacaceae tribe Montieae. The upper diagram is based on group average (UPGMA) clustering and the lower on flexible sorting with $\beta = -0.25$. Numbers 1–24 represent the species of *Claytonia*; for further details see McNeill (1975a).

presence, on an otherwise leafless scape, of a pair of leaves or bracts (sometimes connate) below the inflorescence.

The grouping of the remaining 13 OTUs, however, differs widely according to the clustering method used. Those, such as nearest neighbour or group average, that are not space-dilating, suggest a very heterogeneous assemblage with several taxa of equal rank to *Calytonia*. This approximates to Nilsson's (1967) and Yurtsev and Tsvelev's (1972) positions. Space-dilating methods, such as flexible sorting and incremental sum of squares, suggest a single group parallelling *Claytonia*. This is the group that I recognize as *Montia* (McNeill 1975a).

The conceptual basis for these alternative interpretations is analysed in McNeill (1979a). The Nilsson position, with 9 or 10 genera for 13 species, is undoubtedly a better fit to the OTU by character data analysed in McNeill (1975a). Equally, *Montia*, as recognized by McNeill, appears to be a "natural" taxon in the sense that its members are more similar to each other than any is to, say, species of *Claytonia*. In cladistic terms, *Montia* would seem at least paraphyletic, if not necessarily holophyletic (cf. Ashlock 1971; Wiley 1981: 82–92).

McNeill (1979a) argues, that in grouping species into higher taxa, taxonomy has a simplifying role as well as one of reflecting measured phenetic relationships. Assessment of the optimality of classifications in terms of summarizing character state distributions with the creation of a minimal number of new (higher) taxa, seems to be a promising area of investigation (McNeill 1982).

(*c*) *What of cladistics?* Having raised the possibility of *Montia sensu* McNeill being paraphyletic and perhaps not holophyletic, it is only reasonable to consider whether cladistic techniques might not provide a more definitive answer.

The group has not been analysed cladistically, largely because the majority of the data represent continuous measurements rather than the binary character states, presumed in most cladistic writings. It would certainly be instructive to do so, but, because of the relative paucity of qualitative data, it is evident that the cladograms that emerged, would be based, every bit as much as the space-dilating phenograms, on a subset of the total character-state data that are currently available. Moreover, in this case, the subset would be arbitrary, through the omission of all quantitative characters.

Cladistics is the subject of a separate chapter in this volume (Humphries and Funk, Chapter 17) and for that reason, and because I have considered elsewhere the role of cladistics in classification (McNeill 1980, 1983), I will restrict my comments here to issues that relate to studies such as that on the Montieae.

That a cladistic analysis uses only a subset of all the data, whether because of its current inability to handle continuous variables, or because of its restriction to character states considered to be apomorphic, is not of itself a limitation. The latter restriction may even be a strength, as cladists argue, in that pattern may be more easily discernible thereby. This seems to be the crux of the matter. Can a pattern be discerned that is expressable in terms of a hierarchical classification that will provide a useful summary of the overall variation? A cladogram that implies an identifiably separate origin for every one of its n species, and hence one that requires a minimum of $2n$–4 higher taxa, if it is to dictate the classification, is no better at describing pattern than a chained nearest neighbour phenogram. Despite the theoretically plausible pattern-seeking cladistic model of nested synapomorphies, there is, in practice, no necessary reason for the character subset selected for the cladistic analysis to be any more reliable or representative of the genome than that used for the phenetic one. This is because of limitations in character selection, and problems of determining homology and character-state polarity.

In both cases, one hopes that the data are the best that can be obtained, but in both cases, they can only be a fuzzy representation of the information which is sought. The test of either is the clarity and stability (with new data) of the patterns it suggests.

(4) Continuous variation

The fact that there is no reason to expect the presence of the discontinuities that provide the basis for taxonomic recognition is no cause for rejecting taximetric methods. Indeed, in such situations, multivariate analyses are often very informative. In the genus *Capsicum*, for example, current morphological and cytogenetic research suggest that the very diverse assemblage of cultivated chili and bell peppers are referable to four or perhaps five species (Heiser and Pickersgill 1969). Of these, *C. pubescens* and *C. baccata* are relatively distinct, but cultigens or near cultigens identified with the remaining three species are hard to disentangle morphologically. A study of morphological features of wild and cultivated members of this group (and of *C. baccata*) revealed an interesting pattern of variation (Pickersgill *et al.* 1979).

Both clustering and ordination techniques were employed. A PCO analysis revealed that there did indeed appear to be character intergradiation, but that there is not a continuum of variation. Instead, the wild representatives are relatively distinct and seem to fall into three clusters corresponding to *C. annuum*, *C. frutescens* and *C. chinense* (Pickersgill *et al.* 1979, figs. 54.4 and 54.7). The PCO reveals that it is only in the cultivated representatives that the confusion exists. Moreover, a clear wild–cultivated component can be discerned running at an angle to the first and second principal coordinates. As with the *Spiranthes* study, it would evidently be informative to look at the components of variation that remain after the wild–cultivated dimension has been removed.

(5) Other taxonomic situations

Numerical techniques are applicable in almost every taxonomic situation and it is clearly not possible to give examples of them all. Three situations are, however, sufficiently common as to justify mention, particularly because of the apparent similarity of at least two of them to the ordination methods already described.

The *Spiranthes* specimens, discussed above, were considered as individuals whose best classification was being sought. If, instead, they are looked on as four sets of samples from known populations, canonical variates analysis (see, e.g. Gordon 1981) would be the appropriate method to use. Projection of the specimens onto the first two canonical axes results in four clusters that are so "tight", that the individual specimens cannot be distinguished in a page-size plot. Likewise, a group average cluster analysis on the Mahanolobis distances provides four very strongly clustered groups with the hybrid separate but relatively closely associated with *S. casei*.

Discriminant analysis is closely related, but here the contribution is to the identificatory rather than the classificatory phase of taxonomy. Otherwise the situations are rather similar as, for example, where unambiguous classificatory characters exist, but are not to be found in much of the material that must be identified. As well as the polyploid situation referred to above, this could be applicable to immature animals or plants in a vegetative state.

Other applications of taximetrics to the identificatory phase of taxonomy are key-generating algorithms and on-line identification programs. The former have wide utility, particularly if a data matrix is being generated for other purposes. The latter has a potential role where rapid identification of a relatively small number of taxa has high economic value (McNeill 1975b). This would be true of poisonous fungi, especially when fragmentary or even partially digested, but I am not aware of any practical scheme yet operating.

CONCLUSIONS

In reviewing the role of numerical taxonomy in systematics to-day, I have deliberately emphasized two things and I have consciously excluded one other.

The emphases that I have placed have been on real data recorded from real specimens, living or dead, and on the fact that these data are not all equally suited for taxonomic representation. In saying this, one of the contrasts I am making is with the sets of entirely two-state characters that phylogeneticists would have us believe characterize all taxa from species upwards. Nature, in my experience, is not as tidy as that, and in any classificatory endeavour it behoves us to pay close attention to variability, whether within an individual, a population, or a taxon. Numerical techniques give us the opportunity to do so.

The other contrast is with what I will call the "goodness of fit" school. Logically, even the most meticulously acquired data cannot be expected to be reflected faithfully in a classification. Biologically, they are structured to include an "error term", be it environmental or mutational. The role of classification is to express the clearest pattern that the data suggest, and to discount the "error term".

And that brings me to my major exclusion and also, perhaps, to the main growth area of taximetrics to-day, namely numerical cladistics. The almost bewitching appeal of cladistic analysis lies in its simple explanation of pattern. For this reason it rightly deserves the attention it is currently receiving. My references to real data and to within-taxon variation make it clear that I do not think that pen and pencil cladistics on tiny suites of poorly understood (and largely underscored) characters will lead to another revolution in systematics. But what very well may, is a development of numerical techniques, that will seek, by whatever method is appropriate to the available data, an analysis of pattern, pattern that is, as we all agree, the product of evolution.

ACKNOWLEDGEMENTS

I am grateful to Dr P. M. Catling, now of the Biosystematics Research Institute, Agriculture Canada, Ottawa, for permission to use his *Spiranthes* data, and to Ms K. Pryer and Mr G. Ben-Tchavtchavadze for assistance in the preparation of the illustrations. I am indebted to Mr L. P. Lefkovitch and the Statistics and Engineering Research Institute of Agriculture Canada for some of the computer programs used and to Dr F. J. Rohlf, State University of New York, Stony Brook, for others. Some of the research described was carried out while I was a research scientist with Agriculture Canada and their support and, more recently, that of the Natural Sciences and Engineering Research Council (NSERC) of Canada are acknowledged.

REFERENCES

Ashlock, P. H. (1971). Monophyly and associated terms. *Syst. Zool.* **20,** 63–69.

Barkworth, M. E., McNeill, J. and Maze, J. (1979). A taxonomic study of *Stipa nelsonii* (Poaceae) with a key distinguishing it from related taxa in western North America. *Canad. J. Bot.* **57,** 2539–2553.

Baum, B. R. (1974). Classification of the oat species (*Avena*, Poaceae) using various taximetric methods and an information-theoretic model. *Canad. J. Bot.* **52,** 2241–2262.

Boivin, B. (1967). Enumération des plantes du Canada. III. Herbidées, 1er partie: Digitae: Dimerae, Liberae. *Naturaliste Can.* **93,** 583–646.

Catling, P. M. and Cruise, J. E. (1974). *Spiranthes casei*, a new species from northeastern North America. *Rhodora* **76,** 526–535.

Clapham, A. R., Tutin, T. G. and Warburg, E. F. (1952). "Flora of the British Isles". Cambridge University Press, Cambridge.

Clapham, A. R., Tutin, T. G. and Warburg, E. F. (1962). "Flora of the British Isles" (2nd edn). Cambridge University Press, Cambridge.

Clifford, H. T. and Stephenson, W. (1975). "An Introduction to Numerical Classification". Academic Press, London, New York.

Cole, A. J. (ed.) (1969). "Numerical Taxonomy". Academic Press, London, New York.

Cullen, J. (1968). Botanical problems in numerical taxonomy. *In* "Modern Methods in Plant Taxonomy" (V. H. Heywood, ed.) pp. 175–183. Academic Press, London, New York.

Dale, M. B. (1968). On property structure, numerical taxonomy and data handling. *In* "Modern Methods in Plant Taxonomy" (V. H. Heywood, ed.) pp. 185–197. Academic Press, London, New York.

Duncan, T. and Baum, B. R. (1981). Numerical phenetics: its uses in botanical systematics. *Ann. Rev. Ecol. Syst.* **12,** 387–404.

Dunn, G. and Everitt, B. S. (1982). "An Introduction to Mathematical Taxonomy". Cambridge University Press, Cambridge.

Estabrook, G. F. (ed.) (1974). "Proceedings of the Eighth International Conference on Numerical Taxonomy". W. H. Freeman, San Francisco.

Everitt, B. (1974). "Cluster Analysis". Heinemann, London.

Farris, J. S. (1977). On the phenetic approach to vertebrate classification. *In* "Major patterns in vertebrate classification" (M. K. Hecht, B. M. Hecht and P. C. Goody, eds) pp. 823–850. Plenum Press, New York, London.

Farris, J. S. (1980). The information content of the phylogenetic system. *Syst. Zool.* **28,** 483–519.

Felsenstein, J. (ed.) (1983). "Numerical Taxonomy: Proceedings of a NATO Advanced Studies Institute". NATO Advanced Study Institute Series G (Ecological Sciences), No. 1. Springer Verlag, Berlin, Heidelburg, New York.

Gordon, A. D. (1981). "Classification". Chapman and Hall, London.

Heiser, C. B. and Pickersgill, B. (1969). Names for the cultivated *Capsicum* species (Solanaceae). *Taxon* **18,** 277–283.

Heywood, V. H. (1964). I. General principles: Introduction. *In* "Phenetic and Phylogenetic Classification" (V. H. Heywood and J. McNeill, eds) pp. 1–4. Systematics Association, London.

Heywood, V. H. (1968a). "Modern Methods in Plant Taxonomy". Academic Press, London, New York.

Heywood, V. H. (1968b). Plant taxonomy today. *In* "Modern Methods in Plant Taxonomy" (V. H. Heywood, ed.) pp. 3–12. Academic Press, London, New York.

Hogeweg, W. (1976). Iterative character weighting in numerical taxonomy. *Comput. Biol. Med.* **6,** 199–211.

Jage, H. (1979). Familie Portulacaceae. *In* "Hegi Illustrierte Flora von Mitteleuropa" (K. H. Rechinger, ed.) Band 3, Teil 2, pp. 1183–1221. Paul Parey, Berlin, Hamburg.

Jardine, N. and Sibson, R. (1971). "Mathematical Taxonomy". John Wiley, London.

Johnson, M. P. and Holm, R. W. (1968). Numerical taxonomic studies in the genus *Sarcostemma* R.Br. (Asclepiadaceae). *In* "Modern Methods in Plant Taxonomy" (V. H. Heywood, ed.) pp. 199–217. Academic Press, London, New York.

Lance, G. N. and Williams, W. T. (1967). A general theory of classificatory sorting strategies. 1. Hierarchical systems. *Computer J.* **9,** 373–380.

Lefkovitch, L. P. (1976a). A loss function minimization strategy for grouping from dendrograms. *Syst. Zool.* **25,** 41–48.

Lefkovitch, L. P. (1976b). Hierarchical clustering from principal coordinates – an efficient method for small to very large numbers of objects. *Math. Biosc.* **31,** 157–174.

Lockhart, W. R. and Liston, J. (eds) (1970). "Methods for Numerical Taxonomy". Amer. Soc. Microbiol., Bethesda, Maryland.

McNeill, J. (1975a). A generic revision of Portulacaceae tribe Montieae using techniques of numerical taxonomy. *Canad. J. Bot.* **53,** 789–809.

McNeill, J. (1975b). A botanist's view of automatic identification. *In* "Biological Identification with Computers" (R. J. Pankhurst, ed.) pp. 283–289. Academic Press, London, New York.

McNeill, J. (1979a). Structural value: a concept used in the construction of taxonomic classifications. *Taxon* **29,** 481–504.

McNeill, J. (1979b). The application of iterative character-weighting in numerical taxonomy. *Bot. Soc. Amer. Misc. Ser. Publ.* **157,** 63.

McNeill, J. (1980). Purposeful phenetics. *Syst. Zool.* **28,** 465–482.

McNeill, J. (1982). Phylogenetic reconstruction and phenetic taxonomy. *Zool. J. Linn. Soc.* **74,** 337–344.

McNeill, J. (1983). The future of numerical methods in plant systematics: a personal prospect. *In* "Numerical Taxonomy: Proceedings of a NATO Advanced Studies Institute". NATO Advanced Study Institute Series G. (Ecological Sciences), No. 1, pp. 47–52. (J. Felsenstein, ed.). Springer Verlag, Berlin, Heidelberg, New York.

Neff, N. A. and Marcus, L. F. (1980). "A Survey of Multivariate Methods for Systematics". Privately published, New York (for American Society of Mammalogists).

Nilsson, O. (1967). Studies in *Montia* L. and *Claytonia* L. and allied genera. III. Pollen morphology. *Grana Palynol.* **7,** 279–363.

Pankhurst, R. J. (ed.) (1975). "Biological Identification with Computers". Academic Press, London, New York.

Pickersgill, B., Heiser, C. B. and McNeill, J. (1979). Numerical taxonomic studies on variation and domestication in some species of *Capsicum*. *In* "The Biology and Taxonomy of the Solanaceae" (J. G. Hawkes, R. N. Lester and A. D. Skelding, eds) pp. 679–700. Academic Press, London, New York.

Simpson, R. C. and Catling, P. M. (1978). *Spiranthes lacera* var. *lacera* × *S. romanzoffiana*, a new natural hybrid orchid from Ontario. *Can. Field-Nat.* **92,** 350–358.

Sneath, P. H. A. and Sokal, R. R. (1973). "Numerical Taxonomy". W. H. Freeman, San Francisco.

Sokal, R. R. and Sneath, P. H. A. (1963). "Principles of Numerical Taxonomy". W. H. Freeman, San Francisco.

Stearn, W. T. (1968). Observations on a computer-aided survey of the Jamaican species of *Columnea* and *Alloplectus*. *In* "Modern Methods in Plant Taxonomy" (V. H. Heywood, ed.) pp. 219–224. Academic Press, London, New York.

Swanson, J. R. (1966). A synopsis of relationships in Montioideae (Portulacaceae). *Brittonia* **18,** 229–241.

Walters, S. M. (ed.) (1964). Portulacaceae. *In* "Flora Europaea" (T. G. Tutin, V. H. Heywood, N. A. Burges, D. H. Valentine, S. M. Walters and D. A. Webb, eds) pp. 114–115. Cambridge University Press, Cambridge.

Wiley, E. O. (1981). "Phylogenetics: the Theory and Practice of Phylogenetic Systematics". John Wiley, New York.
Williams, W. T. (1967). Numbers, taxonomy, and judgment. *Bot. Rev.* **33,** 379–386.
Williams, W. T., Clifford, H. T. and Lance, G. N. (1971a). Group-size dependence: a rationale for choice between numerical classifications. *Computer J.* **14,** 157–162.
Williams, W. T., Lance, G. N., Dale, M. B. and Clifford, H. T. (1971b). Controversy concerning the criteria for taxonometric strategies. *Computer J.* **14,** 162–165.
Yurtsev, B. A. and Tsvelev, N. N. (1972). Novȳe taksonȳ iz severovostochnoí. *Bot. Zh.* **57,** 644–647.

16 | Automated Taxonomic Information Systems

F. A. BISBY

*Biology Department, University of Southampton,
Southampton, England*

Abstract: The dream of producing an automated taxonomic information system has been with us since the early 1970s. It is to store a central computerized taxonomic data-base that can be connected to programs which carry out a range of taxonomic activities such as data-retrieval, description writing, identification, phenetic classification and cladistic analysis. Several types of taxonomic data might be included but of prime interest are data describing organisms or taxa. The data-base would be revised frequently and the retrieval facilities would allow the other taxonomic activities to be carried out on selected subsets of data.

In reality a number of experimental floristic and monographic projects have operated with varied levels of success and with implementation that has proved time-consuming. Examples are given from "Flora de Veracruz", "Flora North America", the "European Documentation System", Watson's Grass and Caesalpinioideae Databases and the "Vicieae Database Project".

Attention is drawn to problems of data structure and data-base structure which must be overcome if significant progress is to be made. In what data structure can we codify descriptions so as to accommodate both the difficulties of descriptions themselves and the structural demands of the various taxonomic activities planned? In what data-base structure can we interconnect files containing different taxonomic information so as to permit automated operations involving several files together?

THE DREAM

How far back can we trace the dream of an automated taxonomic information system? An information system does not appear to have been one of the objectives of Sokal and Sneath when their book introduced the use of computers in

Systematics Association Special Volume No. 25, "Current Concepts in Plant Taxonomy", edited by V. H. Heywood and D. M. Moore, 1984. Academic Press, London and Orlando.
ISBN 0 12 347060 9

taxonomy in 1963. The use of computers for numerical phenetics had started by 1957 (Sneath 1957) but was followed only after a gap by cladistics (Edwards and Cavalli-Sforza 1964), key-making (Morse 1968), data retrieval (Estabrook and Brill 1969), and description writing (Porter *et al.* 1973). Separate from these innovations were the introduction of computers for drawing dot-maps from phytogeographical records by Perring and Walters (1962), and the start of museum curatorial data-bases by Creighton and Crockett (1971) and Cutbill and Williams (1971). All of these activities have in common that they require the entry of a multivariate data table on which the operations are performed.

The dream is to connect all of these facilities together to make an automated taxonomic information system that will allow us to hold, disseminate and analyse taxonomic data, and in which these data may be revised frequently. The first steps towards its realization were taken by Shetler (1971) who described the "Flora North America" on which he was working as an information system, and Morse (1974) who wrote a series of connecting programs for operations on what he called taxonomic data matrices. At its simplest a taxonomic information system would hold a central data file and pass on all or selected parts of this to other taxonomic operations. The aims would be both the efficient completion of normal taxonomic tasks, such as answering retrieval queries or printing descriptions, and the provision of new facilities, for instance the regular revision of descriptions or the manufacture of keys purpose-made for certain regions. The high input effort would be amortized over the large number of users and uses, an advantage over piecemeal activities.

(1) Information content

There is a wide range of taxonomic information that can be formed into a data-base and incorporated into an information system. I have made a rough classification of data-bases according to data content into the five classes shown in Fig. 1. I am concerned in the remainder of this paper primarily with just one of these, *descriptive data-bases* or information systems that contain a descriptive data-base. Such data-bases contain descriptive features of the organisms themselves, their morphology, cytology or chemical composition. Many taxonomic projects span several of the classes but it must be the handling of variation amongst organisms that is central to a truly taxonomic information system. By making this restriction I shall have to pass over a number of projects of great interest such as, for example, the "Med-Checklist" project (Greuter *et al.* 1981) where nomenclatural and phytogeographical but not descriptive records are collated in a computer-generated catalogue.

CURATORIAL

Accession data for items in collections as found in Museums, Herbaria, Botanic Gardens, Germplasm collections and Zoos.

PHYTO-GEOGRAPHIC

Distribution data for taxa as found in Atlases, and in Inventory, Distribution and Conservation projects.

NOMENCLATURAL

Names, sometimes with details of authority, publication and typification and often containing valid names cross-indexed to synonyms.

BIBLIOGRAPHIC

Bibliographic details of publications relating to taxonomy.

DESCRIPTIVE

Descriptions of plants or taxa such as details of morphology, anatomy, cytology or phytochemistry.

Fig. 1. Classification of taxonomic data-bases by content.

If we plan an information system that will provide a comprehensive service, whether this be for taxonomists within the profession, or to scientists and others outside, we may well wish eventually to connect together data-bases in my five classes as suggested in Fig. 2. How to connect these is the second of two problems which I discuss later.

(2) Information meaning

There is an important distinction to be made between textual and codified descriptive data-bases. It relates to the detailed content and subsequent plans for usage. At one level, that of a *textual data-base*, we simply enter taxon-descriptions in machine-readable form as a text, and the main use is to print or display this on request. At a more complicated level, a *codified data-base*, we codify taxon-descriptions in such a way that taxonomic analyses can be carried out in addition to printing and display. In a textual data-base the entry might for instance be a word, phrase or page of text, say the morphological description of a species. It may be stored, handled and retrieved by machine, but any interpretation or comparison of its taxonomic meaning can only be made by

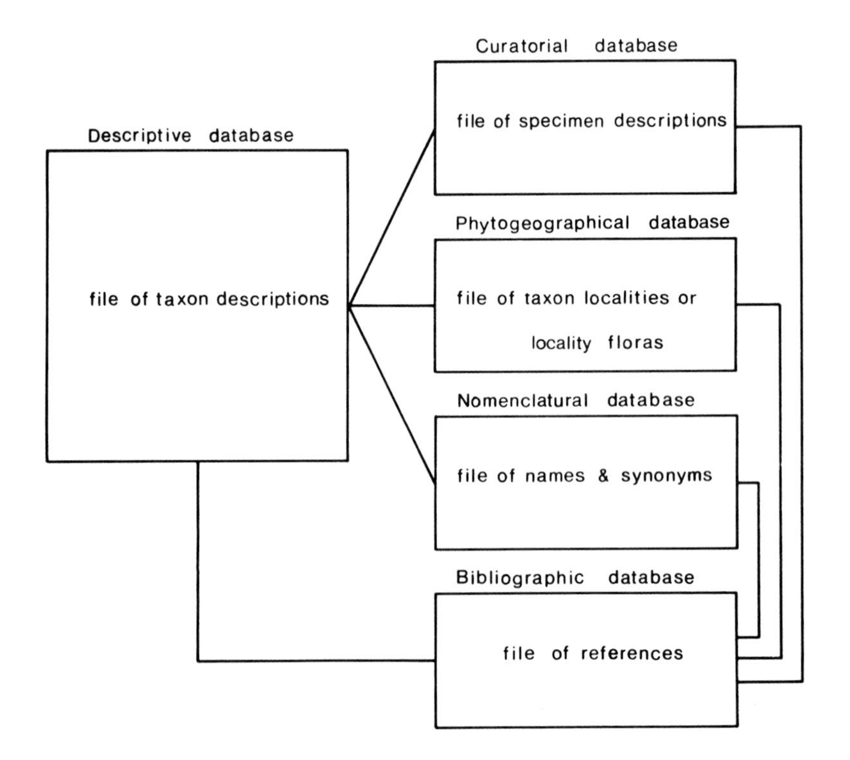

Fig. 2. Possible connections amongst taxonomic data-bases.

its being presented to a human for reading. A codified data-base contains words, numbers, or codes as entries, but in contrast, these are structured in such a way that at least some of its taxonomic meaning is available for machine operations. For instance a quantitative character – *standard petal length*, and a qualitative character – *anther attachment*, would be handled differently. The first would be stored as a number and handled as such, by subtraction or examination of ranges in classificatory or key-making activities, and as a value with units in description printing. The second might be stored as one of a known number of codes (say *basifixed* = 1, *medifixed* = 2, *dorsifixed* = 3), compared by matching in classificatory or key-making activities, and de-coded and associated with the anther in description printing.

A machine holding a textual data-base is providing an alternative to a book. If we call the entries pages, then we may flick through, search or sort the pages in a variety of ways and when we have the appropriate page, read it. The page may be updated frequently by the owner, and as a user we may reach that page instantaneously from a terminal or viewdata set far away. A codified data-base potentially offers all of these facilities, plus a great deal more in the way of taxonomic manipulation. The additional possibilities are detailed descriptor-based retrieval, printing of selected parts of descriptions and selected lists or catalogues, the use of comparative data for numerical taxonomy, and the use of comparative data to construct keys or aid directly in identification. The difference then is essentially between disseminating a finished taxonomic product in a textual data-base, and not only disseminating but actually forming taxonomic products with a codified data-base.

Data retrieval services would enable a user to extract data subsets (e.g. a list of seed characteristics) for just those taxa with selected properties (e.g. with spines and pinnate leaves). The resulting data subset might be used directly as

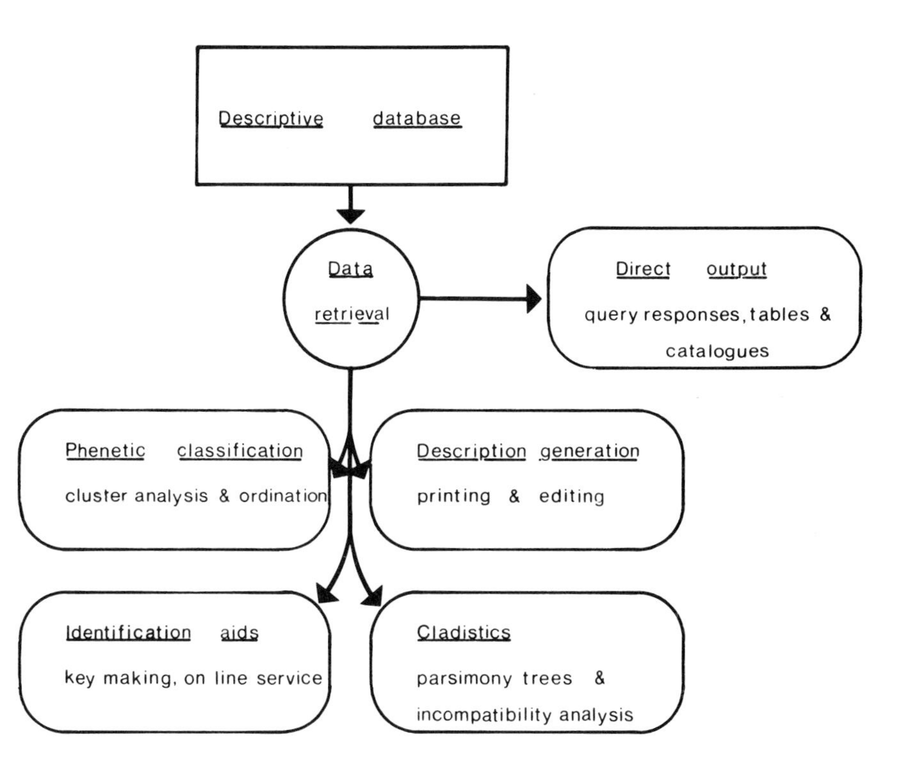

Fig. 3. Idealized flow diagram for descriptive data in a taxonomic information system.

a retrieval service (in our example, to print a table of seed characters for the selected taxa which the enquirer could take away) or it might be kept in the machine and submitted to further taxonomic analysis or manipulation. In a floristic context these could be the manufacture of identification aids (such as key-making or on-line identification services) or description generation. In a monographic context these might be phenetic classification (such as cluster analysis or ordination) or cladistic studies. An idealized flow diagram linking these activities is given in Fig. 3.

(3) Floras, monographs and interconnections

In practice taxonomists have thought of taxonomic information systems in terms of Floristic and Monographic projects. Taxonomists are used to working in one or other of these modes. However, a fast developing area of computer technology is the ability to connect different varieties of data-base held in different computers at different sites. We should not close our minds either to the idea that an investigator might gather data from many data-bases, as he currently gathers it from many publications, or that data-bases might be interconnected so that a species description might be both part of a Flora and of a Monograph. How many botanic gardens would like to connect their curatorial data-bases to a system that automatically alerted them to name changes?

(4) Information services

I believe that an important part of improving taxonomic information provision lies in clarifying which products are intended for use within the profession (as is clearly the case with "Index Kewensis") and which are intended for use by the public (as is the case with Clapham *et al.* 1952). The same distinctions will be evident in descriptive data-bases: some will provide a service to the public, others will be for use by taxonomists, yet others by both. The service for the public might not be by public access to a central public data-base: these require an exceptionally foolproof and simple set of commands and are effectively limited in the services that can be offered. An alternative is for taxonomists to operate the data-bases themselves at an information centre, making a variety of publications and querying services available to the public through correspondence or by telephoned messages. Here a small team of people, supported by a computer information system, would make themselves experts on the taxa in question and then both work on the taxonomy and provide an information service to the public.

THE REALITY

The goals described above were essentially recognized by the mid-1970s but implementations have proved time-consuming and troublesome. I am aware of five major projects so far in higher plant taxonomy and these illustrate a pleasing diversity of approach, although hardly realization of the dream.

(1) Floristic projects

The Flora de Veracruz project developed in about 1967 to make a Flora of the poorly known State of Veracruz in Mexico. It involved the daunting task of collating information from materials distributed widely in the world's herbaria, of organizing botanical exploration and collection in the state itself, and this with a shortage of skilled taxonomists. At an early stage it was decided to use the latest information handling techniques (Scheinvar and Gómez-Pompa 1969; Gómez-Pompa and Nevling 1973), leading first to the printing of herbarium labels, then a herbarium curatorial data-base (Gómez-Pompa *et al.* 1975), and subsequently to the production of dot distribution maps (Fonseca and Gómez-Pompa 1973). Despite this I think it fair to say that progress was hampered by the lack of a program to handle taxonomic descriptions and it was with some reluctance that the project embarked on the publication of a conventional Flora made up of fascicles starting with the Hamamelidaceae (Sosa 1978). The descriptions are prepared by hand and just the specimen citations and distribution maps by computer.

The difficulty that "Flora de Veracruz" encountered in handling taxonomic descriptions, one that I shall return to, is that taxonomic descriptions involve a complexity of data structure beyond that allowed for in standard retrieval programs. Taxon variability, dependent characters, inconsistent terminology and the way these change as one moves to different levels in the taxonomic hierarchy are at the root of the problem (Allkin 1980a).

At a later stage "Flora de Veracruz" made use of the Pankhurst (1975–1979) suite of programs in which a common data format, which allows for variability in qualitative characters, is used for programs concerned with identification and description generation. The programs were used for a morphological data-base on the angiosperm families of Veracruz to produce descriptions and identification aids primarily as an aid to botanists and trainees working on the project (Allkin 1980b). Thus far the project loosely approaches my information centre concept. Despite the limited scope of the descriptive data-base and the limited services to the public, an information system is being built up to support the activities of a team that is both working on the taxonomy and providing an information centre to the public.

The "Flora North America" project (FNA) started as a very large-scale collaborative project (Shetler 1966) to write a continental Flora that would match achievements in Eurasia, the newly completed Flora of the USSR (Komarov *et al.* 1934–64) and "Flora Europaea" which was than in progress (Tutin *et al.* 1964–80). By 1969 the idea, not only of using computers to aid in the production of a Flora in book form, but also of starting a continuing FNA information system had taken hold (Shetler 1971). A descriptive data-base was to be connected with phytogeographical, curatorial, nomenclatural, bibliographical and biographical data-bases. The project was ahead of its time and, whatever the critics said, the period 1969–1973 was a highly productive one in which many problems were faced for the first time and novel experiments carried out.

FNA produced a data structure for a codified descriptive data-base designed primarily to meet the needs of printed descriptions but also to allow retrieval and comparison (Shetler 1975). The enormity of the task in listing characters for optional use in a complete description of taxa in a large Flora is shown by the fact that their list contained 28 000 characters. A peculiarity of the structure was that character-states, mostly descriptive adjectives, were available for use with any characters (Porter *et al.* 1973). Whilst it might seem important to use, for instance, pubescence adjectives uniformly, the fact that any given character does not have a fixed set of character-states creates problems. The structure did allow for variability and for adjectival modifiers such as *mostly* and *rarely*. FNA also specified an arrangement of connections between files, essentially a hierarchy using the Generalized Information System (GIS) then available from IBM (Krauss 1973). For instance, nested beneath each species of the taxon file would be curatorial records for herbarium sheets and type collections. What I have not been able to find is any reference to whether these major innovations were used satisfactorily before the project closed through lack of funds in 1973 (Irwin 1973; Walsh 1973). An interesting paper by Shetler (1974) recounts some of the difficulties and lessons learned.

Lastly we come to the European Floristic Taxonomic and Biosystematic Documentation System (ESFEDS) which started in Reading under Heywood and Moore's direction early in 1982. Clearly we are only seeing the first moves of what should be a major development. Already they have taken a rather different stance choosing to start with established nomenclatural and geographical records and to place these in what will become a public data-base. Although it will reside in a central computer the data-base can be accessed via telephone lines by viewdata-type TV terminals all over the country and even world-wide (Heywood *pers. comm.*). A feature of the viewdata-type system is that screens (or "pages") of text are presented to remote users with a choice of pointers. If the user selects one of these pointers he moves on automatically to another page and so on. Thus the present EDS data-base will allow the user to retrieve accepted names, synonyms and doubtful names in a hierarchial arrangement from a remote viewdata terminal.

The plans for the development of the EDS project list eleven aims, including what I think of as seven types of data-base (nomenclatural, phytogeographical, ecological and phytosociological, bibliographical, conservation status, check-list and morphological). The initial emphasis will be on completing the nomenclatural and phytogeographic data-bases drawn from "Flora Europaea" and then moving to the bibliography, and in particular the documentation of literature containing taxonomic, distributional and nomenclatural changes for species found in Europe (Heywood and Moore *pers. comm.*).

We should note that whilst all three of these floristic projects are milestones, we are far from reaching the goals that I sketched earlier. "Flora de Veracruz" and FNA point to the intractible difficulties of handling taxonomic descriptions and in the EDS project it is as yet too early to see whether an extensive codified descriptive data-base will be included, and if so, how the difficulties are to be met.

(2) Monographic projects

The first monographic data-bases for higher plants have been produced by Watson and collaborators in Canberra. These consist of codified descriptive data-bases in DELTA format (Dallwitz 1980) which can be used by the CONFOR program to generate printed descriptions and to supply data to a key-making program. The DELTA format is intended as a format for communication of taxonomic data to a variety of programs, and incorporates a number of important facilities such as the ability to record variability in both qualitative and quantitative characters and the ability to carry extra text comments attached to characters or character-states. The CONFOR program does not, however, allow data retrieval, either for direct use as enquiry responses, or for the extraction of subsets with a selected character-state for entry to other programs.

One of the DELTA format data-bases is of morphological and anatomical characters for the genera of grasses (Watson and Dallwitz 1981), the contents of which have been made available as public data-bases by publishing in English in a book (Watson and Dallwitz 1980), by printing in English on microfiche (Watson and Dallwitz 1982), and by offering DELTA format coded magnetic tapes (Macfarlane and Watson 1982). The descriptions generated by CONFOR are clumsily verbose in that the character name and character-state are printed for every character, but the tedium of reading them is easily compensated by the pleasure of seeing truly comparable descriptions, something not seen in conventional publications (Watson 1971; Pankhurst 1975). The lack of retrieval facility however means that a visual search of the book or the microfiche is needed which may not be much less time-wasting then searching a conventional monograph. The use of microfiches is however an excellent innovation. They are sufficiently cheap to permit replacement after the data-base is revised and they are transmitted cheaply through the post.

A second monographic data-base has been compiled by Watson for the genera of the Leguminosae subfamily Casesalpinioideae and distributed as a microfiche containing descriptions and keys attached to the paper published in "Advances in Legume Systematics" (Watson 1981). It contains comparative information on 71 morphological characters and 45 additional characters such as leaf and wood anatomy, chemistry and pollen ultrastructure. He offers one of the services that I had envisaged for an information centre, to make keys tailor-made for particular uses by selecting the taxa and/or the characters to be included.

Lastly, the "Vicieae Database Project" (VP), lead by Dr White and myself at Southampton, aims to produce a monographic descriptive data-base and to experiment with using this to help produce a variety of products and services for the public as if in a miniature information centre (Bisby 1981). The descriptions refer to the species of the tribe Vicieae, many of which are of commercial or applied biological use. At an early stage we decided to use several programs that were already available, so that we could build the data-base from the start in 1979, and avoid the temptation to become involved in a major programming exercise. The result is that we use three at present unconnected programs. The EXIR program holds three data-bases: morphological records arranged as 99 characters (see Fig. 4), phytochemical records arranged as 88 characters (see Fig. 5) and geographical records. These may be used for direct retrieval of lists and responses to queries, or for indirect retrieval where the response lists are made available as computer files for other programs as shown in Fig. 6. Dr White has written a program EXIRPOST which reorganizes these retrieval response files, either for tabulation and printing (as illustrated in Figs 4 and 5) or for passing on to other programs. Two other programs used are Dallwitz's CONFOR for producing descriptions, such as that illustrated in

QUERY put to EXIR program

PRINT: (Number, Binomial, Life span, Inflorescence) FOR species WITH Genus, Lathyrus*

RESPONSE after tabulation with EXIRPOST program

Number	Binomial	Life span	Inflorescence
104	Lathyrus americanus	perennial	multiflowered
121	Lathyrus cabrerianus	unknown	multiflowered
149	Lathyrus hasslerianus	unknown	multiflowered
154	Lathyrus hookeri	unknown	multiflowered
183	Lathyrus linearifolius	perennial	multiflowered
188	Lathyrus macropus	unknown	multiflowered
189	Lathyrus macrostachys	perennial	multiflowered
190	Lathyrus magellanicus	perennial	multiflowered
193	Lathyrus multiceps	perennial	variable
222	Lathyrus paranensis	annual	variable
237	Lathyrus pusillus	annual	variable

Fig. 4. An example of a query put to the morphological file of the "Vicieae Database" and the response produced.

QUERY

Print the names, flower colours and appropriate references for Vicia species with Malvidin and Delphinidin present in the flowers.

RESPONSE

Genus	Species	Flower colour	Ref Mv/ANCN	Ref Dp/ANCN
Lathyrus	angulatus	Purple	Pe60	Pe60
Lathyrus	clymenum	Red	Pe60	Pe60
Lathyrus	hirsutus	Purple	Pe60	Pe60
Lathyrus	linifolius	Red	Pe60	Pe60
Lathyrus	odoratus	Purple	Pe60	Pe60
Lathyrus	tingitanus	Purple	Pe60	Pe60
Lathyrus	vernus	Purple	Pe60	Pe60
Pisum	sativum	Variable	Mu+61&Ha+71	Mu+61&St+72
Vicia	biennis	White	Ha+71	Ha+71
Vicia	cracca	Blue	Ha+71	Ha+71
Vicia	disperma	Purple	Ha+71	Ha+71
Vicia	dumetorum	Purple	Ha+71	Ha+71
Vicia	pseudo-orobus	Blue	Is+78	Is+78
Vicia	unijuga	Blue	Is+78	Is+78
Vicia	villosa	Blue	Ha+71	Ha+71

Pe60 Pecket,R.C.,1960,New Phytol. 59,138

Mu+61 Mumford,F.E.,Smith,D.H.,& Castle,J.E.,1971,Pl.Physiol.Lancaster 36,752

Ha+71 Harborne,J.B.,Boulter,D.,& Turner B.L.,1971,"Chemotaxonomy of the Leguminosae".Academic Press:London & New York.

St+72 Statham,C.M.,Crowden,R.K., & Harborne,J.B.,1972,Phytochemistry 11 1083-1088.

Is+78 Ishikura,N., & Shibata,M.,1978,Bot.Mag.Tokyo 86,1-4.

Mv/ANCN Malvidin
Dp/ANCN Delphinidin

Fig. 5. An example of a query put to the chemotaxonomic file of the Vicieae "Database" and the response produced.

Fig. 7, and SYNONYMS for organizing a synonymized nomenclatural data-base.

The first phase of the VP project has been achieved (Bisby *et al.* 1983). The data-bases are operational with morphological data from our own observations, phytochemical data from collections whose identity has been rigorously checked and geographical data from the literature, for a gradually increasing fraction of the 330 taxa. Of the difficulties with descriptive data, two have been avoided (by describing only closely related species we can standardize terminology easily and do without dependent characters) and the third, taxon variability, overcome by an awkward and verbose method of codifying the data before entry to EXIR. The result is that we can, for instance, retrieve details of species with a prostrate habit either including or excluding those for which prostrate habit is a variable occurrence.

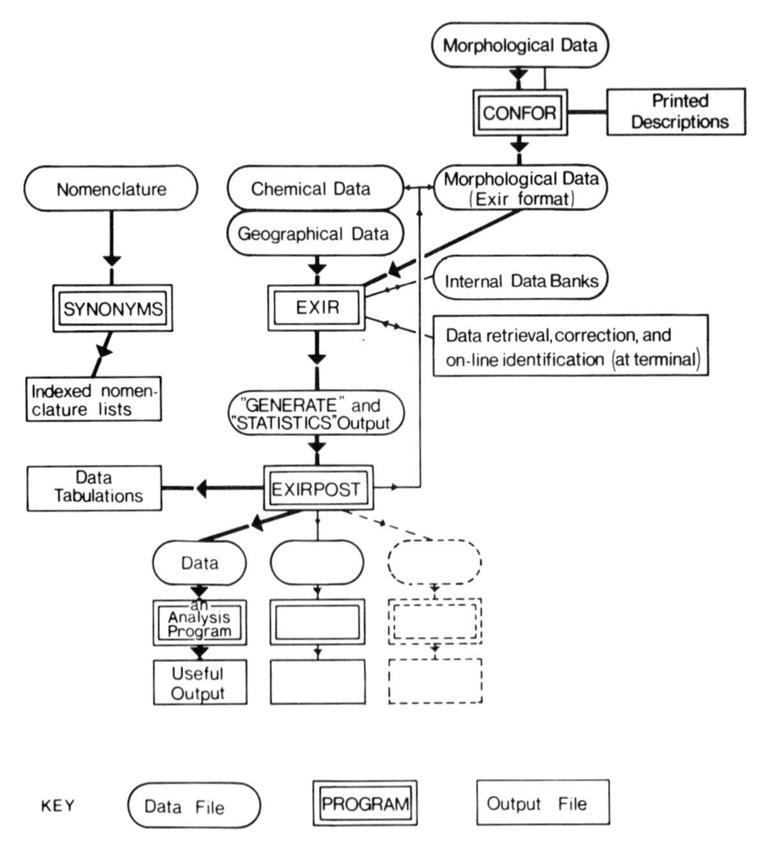

Fig. 6. Flow diagram for data handled by computer in the "Vicieae Database Project".

9. 745 Vicia pseudo-orobus.

(1) perennial . (3) ascending or erect ; OR climbing or scrambling . (4) height or maximum length of shoot 105 cm. (5) stem (lower or middle parts) terete, ridged or many-sided. (6) leaflet abaxial surface glabrous TO densely hairy. (7) leaflet adaxial surface glabrous; OR sparsely hairy. (8) leaf 48 mm long TO 98 mm long . (9) petiole length 2 mm TO 8 mm. (10) leaflet 29 mm long TO 75 mm long. (11) leaflet maximum width 10 mm TO 36 mm. (12) leaflet length-maximum width ratio 1.9 TO 3. (13) leaflets ovate <sometimes narrowly>; OR elliptical. (14) members of each stipule pair similar in shape. (15) stipule shape dentate . (16) stipule abaxial nectary absent. (17) petiole channelled. (18) middle and upper mature leaves with leaflets. (19) tip of mature leaf tendrillous. (20) colour of mature leaf after

Fig. 7. Part of a species description generated in the "Vicieae Database Project" by the CONFOR program.

We are now involved in the second phase of taking data from EXIR into programs for other taxonomic operations, and using the data-base and its products, our seed and living collections, and our human expertise to offer an experimental information service to a study group of people with interests in the Vicieae. The taxonomic products planned include descriptions, catalogues or lists, keys and phenetic analyses for retrieved subsets of the data.

It seems clear from these monographic examples that the path towards an automated information system is easier in monographic than in floristic work. It is not just that the scale of operation is smaller with hundreds rather than thousands of taxa. It is also because the complexity of making truly comparable descriptions is a whole order less: it is possible to find a character set with homologous comparisons that can be made for most if not all of the taxa. Dependent characters tend to involve only one or a few levels of dependency and can therefore be dealt with by cross-multiplying the characters. Also by choosing a single taxonomic rank, the many difficulties related to the taxonomic hierarchy are avoided. It is also possible for all of the observations of any one character to be made by one person or in one team, thus making the achievement of data consistency a less demanding task.

DATA STRUCTURE

One of the major difficulties encountered in the moves towards automated taxonomic information systems is the structure to be given to plant descriptions.

How should one structure the elements of a codified description so as to accommodate the full information of a taxonomic description, and how can this be done to permit satisfactory manipulations of the data in subsequent analyses?

(1) Character structure

The standard taxonomic data matrix or multivariate table used in numerical taxonomy is a rectangular table with taxa as rows and taxonomic characters as columns. If each character is thought of either as a quantitative character with values, e.g. *Number of stamens* or *Leaf length in cm*, or as a qualitative character with character-states, e.g. *Life form: annual*, *biennial*, or *perennial*, then we may record each character for each taxon and enter the appropriate character-values or character-states in the table. This is the simple data structure assumed for many taxonomic programs, including classificatory programs, some cladistic analyses, some early identification programs and the TAXIR/EXIR series of retrieval programs and many commercial programs. However, such a structure omits two major complexities of taxon description, within-taxon variation and dependent characters.

Within-taxon variation must necessarily be included in full descriptions and is the cause of many problems in identification. The simplest way of including this variation is to allow the entry of multiple values or states in the table, or the entry of ranges. So instead of entering a single leaf length and life-form say for *Vicia johannis*, we enter *leaf length: 3–6 cms* and *life-form: annual* or *biennial*. These multiple entries are allowed in DELTA format and were anticipated in the FNA data-bases. They do however destroy the strictly rectangular arrangement of entries and so rule out the file-inversion used by TAXIR/EXIR to make an efficient retrieval mechanism. The Pankhurst programs, which are limited to qualitative characters, incorporate a device for coding these multiple entries to give a single code to be stored. The character-states are subcoded as powers of two (1, 2, 4, 8, 16, etc.) and the subcodes for all character-states found in the taxon added together to yield the code actually stored. Thus if the character-states of *Life form* are subcoded *annual* = *1*, *biennial* = *2*, *perennial* = *4*, we would store the code *3* for *Vicia johannis*, *6* for a taxon that was *biennial* or *perennial*, and *7* for *annual* to *perennial*. On the VP morphological data-base we have used a different device. Quantitative characters are split into two descriptors such as *Leaf length min* and *Leaf length max* to store the minimum and maximum measurements. Qualitative characters are split into one descriptor for each character-state, thus:

Life form 1: annual, not annual.
Life form 2: biennial, not biennial.
Life form 3: perennial, not perennial.

Our *Vicia johannis* would therefore be scored as follows:

Life form 1: annual, Life form 2: biennial, Life form 3: not perennial.

Despite being awkward and verbose this enables us to distinguish species that are always annual (by requesting those that are *annual* AND *not biennial* AND *not perennial*) from those that are only sometimes annual (by requesting those that are *annual* AND (*biennial* OR *perennial*)).

A dependent character is a character whose applicability depends on another character. Thus a character *Stipule indumentum: glabrous* or *pilose* would be dependent if there was another character *Stipules: present* or *absent*, in the sense that the stipule indumentum could only be applicable in cases where stipules were present. Workers using standard taxonomic data matrices have either entered a special *not-applicable* value for inapplicable cases or they have cross-multiplied the characters in question to obtain a single character of more states. In our example this would yield *Stipules: present and glabrous, present and pilose* or *absent*, which could be abbreviated to *Stipules: glabrous, pilose* or *absent*. Both solutions cause some difficulties in phenetic analyses where it is intended to "give all characters equal weight". The *not-applicable* solution contributes two characters to comparisons of plants with stipules, but only one to comparisons of exstipulate plants, thus giving variable weights in different comparisons. Conversely cross-multiplication downweights two characters into one. The cross-multiplication solution is awkward to use in retrievals as the simple request for plants with stipules present or absent has to be modified. Cross-multiplication also becomes very unwieldly if there are several levels of dependence. The quantity of dependent characters and the number of levels of dependence are likely to be greater the wider the diversity of organisms to be described consistently; this is a major difficulty in floristic data-bases, but only a minor one in monographic data-bases.

Pankhurst (unpublished) points out that many of the dependencies naturally form a hierarchy and in a pilot data-base for a sample of British species he uses up to nine levels. A certain character-state scored for a character high in the hierarchy, such as say the absence of leaves, determines that many other characters are inapplicable. Pankhurst makes use of this to avoid wasting storage on the characters lower down the hierarchy known to be inapplicable in such a case. As the overall fraction of characters that are inapplicable is likely to be very large in floristic descriptions this may prove a useful structure.

(2) Character formulation

Several authors distinguish raw *observations* (Bisby 1970), and *features* (Allkin 1979) from the structured comparison known, at least in taximetrics, as a *character* with *character-states* or *character-values*. These characters take several forms, such as *quantitative characters* or a variety of *qualitative character* types (Watson *et al.* 1967). It seems accepted now that there are usually several ways in which given observations can be formulated into characters, and that the choice of the best of these is an important skill developed by taxonomists (Watson 1970; Bisby and Nicholls 1977; Allkin 1979). The different formulations produced are the result of variations in human perception as well as intentional attempts at meeting varied objectives.

Further, there are many steps between the first observations and the final use of characters in classifications or leads in keys. There is a continuing process of character formulation; incorporating fine-tuning as the taxonomist learns more of the variation pattern, and alterations as he formulates the same observations into possibly quite different *sister characters* (Allkin and White 1982): *retrieval characters*, *classificatory characters*, *identification leads*, *descriptive characters*, *synapomorphies* and so on. Some sister characters may differ substantially in meaning and some just in wording to meet the different styles of say printed descriptions and retrieved lists. This formulation and reformulation process requires the full skills of a taxonomist even if, once defined, the changes are made by machine. Shetler (1975), for example, describes the differences between description characters and identification leads, and Hennig (1966:93) discusses the search for transformation conditions and apomorphies.

So the question arises as to which stage or stages of the character formulation process should provide the descriptors for the stored data-base? One strategy is to aim for a compromise character set that satisfies all five purposes outlined in Fig. 3: direct retrieval, description generation, phenetic classification, identification aids and cladistics. This ranges from being satisfactory for some clear binary discontinuities to impossible for complex, variable characteristics. Another strategy is to enter data drawn from certain observations in several forms designed for the different purposes. For example we have entered morphological data in two forms in our "Vicieae Project" although automatic conversion from one to the other is being programmed. A third method is to give one application priority over others, as in the 'Flora North America" plans. And last is a strategy not yet available on the appropriate programs, to enter data for a *master character* containing the full detail of foreseeable utility (as in Bisby and Nicholls 1977), and then to generate within the machine new versions of this data for sister character definitions. Merged character-states, changed phraseology of character or character-state names, and multi-state characters generated from ranges on quantitative characters are some of the possibilities that would not require further observations. Thus a flower colour character might require many states to be merged for use in identification with a wide range of taxa (*white* contrasted with *pink to purple to blue*) but be of more specific use (as *white*, *pink*, or *blue*) in a restricted subset in which species of variable flower colour were absent. However splitting character-states or combining pairs of characters would require the supply of further data as for instance the additional observation of wing-spot colour needed to expand a character *Wing-spot*: *present* or *absent* to *Wing-spot*: *purple*, *brown* or *absent*. Whilst adding considerably to the complexity of the programs, the ability to formulate sister characters would allow not only alteration of characters for particular tasks, but also alteration of characters within particular data subsets.

(3) Character selection

It is important to remember when designing a descriptive data-base for what FNA called "generalised" use that not all descriptors will be of value in all of the taxonomic uses. To some extent we should not worry as to how highly variable characters will be utilized in phenetic analysis, or obscure variables turned into identification characters – they should be rejected for given uses.

Difficulties in structuring and formulating characters should be seen in the context of different selections being used for different purposes. For example we may feel that every description should include plant height and flower colour. If so these characters must be recorded in the data-base even though we may know that one is too variable within taxa and the other too lacking in discontinuities for either to be of value in classification or identification in the group under study. Methods of selecting characters and character-sets of identificatory value are discussed by Allkin (1979), and of classificatory value by Bisby (1970) and Bisby and Nicholls (1977). I list the difficulties and variables above to emphasize that as yet there is, as far as I know, no program which has a suitable data structure and is able to pass the data to the full range of manipulations.

DATA-BASE STRUCTURE

Most taxonomic data-bases have the simple structure of a rectangular table, such as a taxonomic data matrix, stored in the computer as what is called a *flat file*. Descriptor data is entered for each of a set of items, such as taxa or specimens in taxonomic examples. However in trying to make optimal use of computers, many taxonomic projects find that in practice they use several different flat files. Gómez-Pompa *et al*. (1975), for instance, compiled both a nomenclatural file and a specimen file in the early stages of the "Flora Veracruz" project. On our "Vicieae Database Project" we have a nomenclatural file, several descriptive taxon files and a garden-material file. These are presently stored and operated quite independently. So, for instance, if we wish to make a list of our living and seed collections of the *Lathyrus* species known from Portugal, we require two computer retrievals – first we list which *Lathyrus* are known from Portugal from a taxon file, and then we list which accessions we have for that list of species from the garden material file.

For some years now software experts have been producing data-base management systems (dbms) designed to allow not only the handling of several files, but also operations such as retrievals and editing involving simultaneous use of several files (Fry and Sibley 1976). So far these dbms's have had rather little use in taxonomy, possible because they have been available only through commercial sources and were mainly designed for the large scale operations of large companies. An exception is the GIS system, an early example of a dbms, used for "Flora North America" (Krauss 1973). That early dbms's were not used has probably been fortunate; there has been a fantastic growth industry and concomitant improvements in dbms development in the last few years. For holding codified data-bases one can now choose between dbms's with hierarchical, network or relational structures, and some of these are becoming available even for microcomputers.

The strictly nested structure of a *hierarchical dbms* will be familiar to taxonomists (Tsichritzis and Lochovsky 1976). Despite the superficially attractive resemblance that a hierarchy of files has to the taxonomic hierarchy, it does limit efficient retrievals to those that use the hierarchy. In our Vicieae example a species file would contain species that lead to a garden-material file. But a request for details of specimens of Portuguese species would be answered very inefficiently if they belonged to many species in several genera. The search would involve searching in and out of files at three levels many times.

The obvious suggestion that one create a hierarchy of descriptive files for the categories of the taxonomic hierarchy illustrates nicely how the question of data-base structure is connected to the previous one of data structure. There are elements of taxon-variation which would be included as within-taxon variation in a single rank data-base, but which might be inserted without within-taxon variation in a hierarchy. For instance, most members of the genus

Cytisus have plain yellow petals but there is one species with white petals, and one with bicoloured yellow petals. In a generic data-base *Cytisus* would have to be recorded as variable for petal colour, but in a hierarchy this could be improved by omitting flower colour for the genus but specifying it for each of the species:

Cytisus – petal colour: variable *Cytisus scoparius – petal colour: yellow*
Cytisus ingramii – petal colour: bicoloured yellow
Cytisus multiflorus – petal colour: white

A *network dbms* allows a network of interconnections to be specified when the files are created (Taylor and Frank 1976). The principal difficulty in adopting such a structure is to predict at the outset what kinds of retrievals will be needed. Preliminary enquiries from members of the users panel for our "Vicieae Database Project" have impressed us with the variety and unpredictability of angles from which applied biologists have interests in our plants, and this would seem to mitigate against a network dbms.

Lastly, reliable *relational dbms's* have just recently become available. In theory they are ideal: they allow interconnections between any entries in one flat file and any entries in others (Chamberlin 1976). In practice early versions have proved troublesome or not given the flexibility expected, but newer versions are being used with confidence. In our Portuguese *Lathyrus* example a single request would lead to the list being provided. The dbms would respond to the request by obtaining a list of those specimens in the specimen file whose species name corresponded with the species names of taxa in the taxon file for which occurrence was listed for Portugal.

It is too early to say which types of dbms will be used or indeed how successful they will prove for the variety of taxonomic uses: I have heard that the "Flora de Veracruz" have just switched their curatorial data-base to a relational dbms (Allkin *pers. comm.*); we are looking seriously at two relational dbms's for developments from our present "Vicieae Database" (Allkin and White 1982) and I understand that Brill and Estabrook (*pers. comm.*) are working on a near-relational version of TAXIR.

CONCLUDING REMARKS

I believe the dream and the reality of automated taxonomic information systems are worth the attention of all plant taxonomists for two reasons. First, when automated systems start to make a real contribution to taxonomy it will be to an underexplored and undervalued area: information services to the public. Most of us are paid from public funds to provide such a service and yet the means of doing this has hardly entered the twentieth century. Second, it is widely acknowledged in industry that a dbms works for an organization only as well as it succeeds in modelling that organization's flow of information. So for those interested in describing or modelling the information handling role of the taxonomic profession this is a matter of great interest.

ACKNOWLEDGEMENTS

I am grateful to Dr M. E. Adey, Dr R. J. White, and Mr M. Freeston for commenting on this paper and particularly to Dr R. Allkin for many stimulating discussions which have contributed extensively to my views.

REFERENCES

Allkin, R. (1979). "The Evaluation and Selection of Plant Characteristics for use in Computer-aided Identification". Ph.D. thesis, Polytechnic of Central London.

Allkin, R. (1980a). "Some Difficulties of Computer-stored Taxonomic Descriptions". INIREB, Mexico (Flora de Veracruz Internal Document No. 8030036), 26 pp.

Allkin, R. (1980b). "The Computer-assisted Production of Aids to the Identification of the Angiosperm Families of Veracruz: An Introduction and brief Outline". INIREB, Mexico (Flora de Veracruz Internal Document No. 8130023).

Allkin, R. and White, R. J. (1982). "Design Criteria for a Computer Program to Facilitate the Acquisition, Storage, Retrieval and Reformatting of Biological Descriptions". Southampton University (Biology Dept. Internal Report).

Bisby, F. A. (1970). The evaluation and selection of characters in Angiosperm taxonomy: an example from *Crotalaria*. *New Phytol.* **69,** 1149–1160.

Bisby, F. A. (1981). Vicieae Database Project. *Taxon* **30,** 732.

Bisby, F. A. and Nicholls, K. W. (1977). Effects of varying character definitions on classification of Genisteae (Leguminosae). *Bot. J. Linn. Soc.* **74,** 97–122.

Bisby, F. A., White, R. J., Macfarlane, T. D. and Babac, M. T. (1983). The Vicieae Database Project: experimental uses of a monographic taxonomic database for species of vetch and pea. *In* "Numerical Taxonomy" (J. Felsenstein, ed.) pp. 625–629. Springer Verlag, Berlin.

Chamberlin, D. D. (1976). Relational Database Management Systems. *Computing Surveys* **8,** 43–66.

Clapham, A. R., Tutin, T. G. and Warburg, E. F. (1952). "Flora of the British Isles". Cambridge University Press, Cambridge.

Creighton, R. and Crockett, J. J. (1971). SELGEM: a system for collection management. Smithsonian Institution, Washington (*Information Systems Innovations*, **2**).

Cutbill, J. L. and Williams, D. B. (1971). *In* "Data Processing in Biology and Geology" (J. L. Cutbill, ed.) pp. 105–113. Systematics Association Special Volume 3, Academic Press, London, New York.

Dallwitz, M. J. (1980). A general system for coding taxonomic descriptions. *Taxon* **29,** 41–46.

Edwards, A. W. F. and Cavalli-Sforza, L. L. (1964). Reconstruction of evolutionary trees. *In* "Phenetic and Phylogenetic Classification" (V. H. Heywood and J. McNeill, eds) pp. 67–76. Systematics Association (Publication No. 6), London.

Estabrook, E. G. and Brill, R. C. (1969). The theory of the TAXIR Accessioner. *Math. Biosciences* **5,** 327–340.

Fonseca, S. O. and Gómez-Pompa, A. (1973). Ensayo de procesamiento de datos para la Flora de Veracruz. *An. Inst. Biol. Univ. Nat. Autón. México* **44,** 9–28.

Fry, J. P. and Sibley, E. H. (1976). Evolution of Database Management Systems. *Computing Surveys* **8,** 7–42.

Gómez-Pompa, A. and Nevling, L. I. (1973). The use of electronic data processing methods in the Flora of Veracruz Program. *Contr. Gray Herb.* **203,** 49–64.

Gómez-Pompa, A., Toledo, J. A. and Soto, M. (1975). Electronic data processing of herbarium specimens data for the Flora of Veracruz Program. *In* "Computers in Botanical Collections" (J. P. M. Brenan, R. Ross and J. T. Williams, eds) pp. 35–52. Plenum Press, London, New York.

Greuter, W., Burdet, H. M. and Long, G. (eds) (1981). "Med-Checklist: I, Pteridophyta". O.P.T.I.M.A., Geneva, Berlin.

Hennig, W. (1966). "Phylogenetic Systematics". University of Illinois Press, Urbana, Ill.

Irwin, H. S. (1973). Flora North America: austerity casualty? (editorial). *Bioscience* **23,** 215.

Komarov, V. L. *et al.* (eds) (1934–1964). "Flora of the USSR", Vols. 1–30. Academy of Sciences of the USSR, Moscow.

Krauss, H. M. (1973). The Information System Design for the Flora North America Program. *Brittonia* **25,** 119–134.

Macfarlane, T. D. and Watson, L. (1982). The classification of Poaceae subfamily Pooideae. *Taxon* **31,** 178–203.

Morse, L. E. (1968). Construction of identification keys by computer (Abstract). *Amer. J. Bot.* **55,** 737.

Morse, L. E. (1974). Computer programs for specimen-identification, key construction and description-printing using taxonomic data matrices. *Publs. Mus. Michigan State Univ., Biol. Ser.* **5,** 1–128.

Pankhurst, R. J. (1975). Identification methods and the quality of taxonomic descriptions. *In* "Biological Identification with Computers" (R. J. Pankhurst, ed.) pp. 237–250. Systematics Association Special Volume 7, Academic Press, London, New York.

Pankhurst, R. J. (1975–1979). Internal documents 1–7 from R. J. Pankhurst. British Museum (Nat. Hist.), London.

Perring, F. H. and Walters, S. M. (1962). "Atlas of the British Flora". Nelson, London.

Porter, D. M., Kiger, R. W. and Monahan, J. E. (1973). "A Guide for Contributors to Flora North America Part II". Smithsonian Institution, Washington, D.C. (FNA Report 66).

Scheinvar, L. and Gómez-Pompa, A. (1969). Algunos metodos automaticos para la elaboracion de etiquetas de herbario. *Bol. Soc. Bot. México* **30,** 73–93.

Shetler, S. G. (1966). Meeting of Flora of North America committee. *Taxon* **15,** 255–257.

Shetler, S. G. (1971). Flora North America as an information system. *BioScience* **21,** 524–532.

Shetler, S. G. (1974). Demythologizing biological data banking. *Taxon* **23,** 71–100.

Shetler, S. G. (1975). A generalized descriptive data bank as a basis for computer-assisted identification. *In* "Biological Identification with Computers" (R. J. Pankhurst, ed.) pp. 197–236. Systematics Association Special Volume 7, Academic Press, London, New York.

Sneath, P. H. A. (1957). The application of computers to taxonomy. *J. Gen. Microbiol.* **17,** 201–226.

Sokal, R. R. and Sneath, P. H. A. (1963). "Principles of Numerical Taxonomy". W. H. Freeman, London.

Sosa, V. (1978). "Flora de Veracruz: I, Hamamelidaceae". INIREB, Xalapa, Mexico.

Taylor, R. W. and Frank, R. L. (1976). CODASYL Database Management Systems. *Computing Surveys* **8,** 67–103.

Tsichritzis, D. C. and Lochovsky, F. H. (1976). Hierarchical Data-base Management: A Survey. *Computing Surveys* **8,** 105–123.

Tutin, T. G., Heywood, V. H., Burges, N. A., Moore, D. M., Valentine, D. H., Walters, S. M. and Webb, D. A. (eds) (1964–1980). "Flora Europaea", Vols. 1–5. Cambridge University Press, Cambridge.

Walsh, J. (1973). Flora North America Project nipped in the bud. *Science* **179,** 778.

Watson, L. (1970). Representation of Taxonomic Data. *In* "Data Representation" (R. S. Anderson and M. R. Osborne, eds) pp. 108–112. University of Queensland Press, Brisbane.

Watson, L. (1971). Basic taxonomic data: the need for organisation over presentation and accumulation. *Taxon* **20,** 131–136.

Watson, L. (1981). An automated system of generic descriptions for Caesalpinioideae, and its application to classification and key-making. *In* "Advances in Legume Systematics" (R. M. Polhill and P. H. Raven, eds) pp. 65–80 (in Part 1) plus microfiche (back cover of Part 2). Royal Botanic Gardens, Kew.

Watson, L. and Dallwitz, M. J. (1980). "Australian Grass Genera – Anatomy, Morphology and Keys". The Australian National University Research School of Biological Sciences, Canberra.

Watson, L. and Dallwitz, M. J. (1981). An automated data bank for grass genera. *Taxon* **30,** 424–429.

Watson, L. and Dallwitz, M. J. (1982). "Grass Genera: Descriptions" (3rd edn, microfiche). Research School of Biological Sciences, Australian National University, Canberra.

Watson, L., Williams, W. T. and Lance, G. N. (1967). A mixed data approach to Angiosperm taxonomy: the classification of Ericales. *Proc. Linn. Soc. Lond.* **178,** 25–35.

17 | Cladistic Methodology

C. J. HUMPHRIES

Botany Department, British Museum (Natural History), London, England

and

VICKI A. FUNK

National Museum of Natural History, Smithsonian Institution, Washington, USA

Abstract: During the last 18 years, since the publication of Hennig's book – "Phylogenetic Systematics" (1966) – there has been a major revolution in systematic methodology. Hennig's work is significant because in his search for a general theory of systematics he developed a method which resolved the meaning of relationship in comparative biology. The method, which has been labelled cladistics, defines monophyletic groups based on internested sets of synapomorphies. It is an empirical method, for, in applying the parsimony criterion, it chooses among alternative hypotheses of relationship (cladograms) on nothing other than their explanatory power. Because of these factors it has developed into a general approach to natural classification. The main achievements include (1) the concepts of synapomorphy and monophyletic groups, (2) the recognition of the distinction between systematic method and evolutionary theory, (3) the establishment of parsimony as the basis for cladogram construction, and (4) the realization that synapomorphy can be defined in terms of homology and that such homologies can be used to characterize monophyletic groups. Details of these developments, using botanical examples, will be given. Riedl's (1979) three tests for homology are examined: conjunction (homologues may not exist in one organism), similarity (checks of topography, ontogeny and composition) and congruence (with other homologies). Homologies defined by these three tests become group defining homologies or synapomorphies. Characters (apparent homologies) that fail the congruence and similarity tests are the result of parallel or convergent evolution and are *ad hoc* statements. Determining whether

Systematics Association Special Volume No. 25, "Current Concepts in Plant Taxonomy", edited by V. H. Heywood and D. M. Moore, 1984. Academic Press, London and Orlando.
ISBN 0 12 347060 9

or not a character is an apomorphy can be resolved by ontogeny or, in the absence of ontogenetic data, by the sampling of more taxa – out-group analysis.

The application of cladistics to botanical problems is becoming increasingly more popular but it is still in its infancy. Not only is the method demanding, but also, for many botanists, it is quite unacceptable because of its apparent failure to accommodate reticulate evolution or hybridization. We hope that by discussing the basic concepts of cladistics and by examining a range of botanical examples, some of the resistance of the botanical community can be overcome.

INTRODUCTION

Several chapters in this volume are concerned with data gathering in systematics. This aspect is the basis of all research but because how we use the data has some influence on how the data are gathered, the methods and philosophy of systematics must be considered. Other than basic taxon delimitation, we should be most interested in how data can be utilized to express interrelationships among different organisms. Basically there are two ways – intuitive and methodological. Methods can be further subdivided into two groups: cladistics and phenetics.

Cladistics covers a wide range of methods of phylogenetic inference which for someone with a really broad imagination could start with the work of Haeckel (1866). However, the initiation of the field, as we understand it today, really began in 1950 when Hennig published his "Grundzüge einer Theorie der Phylogenetischen Systematik". This played no influence whatsoever on the English-speaking systematic community until the ideas were first discussed by Kiriakoff (1959, 1962) and the revised English translation of Hennig's book "Phylogenetic Systematics" was made available by Rainer and Zangerl in 1966. Wider discussion of Hennig's methods have since been published, particularly in Europe, through the influence of Brundin's (1966) work on chironimid midges. Early adherents of Hennig's methods include Koponen (1968, 1973) on mosses, and Nelson (1969) on fishes. Recent applications of Hennig's methods for plants include Bremer (1976a, b, 1978a, b), Bremer and Wanntorp (1978, 1979a, 1981), Humphries (1979, 1981a, b), Parenti (1980), Humphries and Richardson (1980), Wiley (1980), Bolick (1981), Churchill (1981), Funk (1981), Sanders (1981) and Hill and Crane (1982). Textbooks discussing Hennig's methods include Eldredge and Cracraft (1980), Nelson and Platnick (1981) and Wiley (1981).

It is worth mentioning briefly other major developments in cladistics, which for reasons of space will not be discussed in detail here. Somewhat independently from Hennig, Wagner (1961, see also 1980) developed a method of phylogenetic inference used for ferns, the groundplan-divergence method, which gave rise to developments by Farris (1970) and Kluge and Farris (1969) in producing algorithms for computing Wagner trees and networks (see Jensen, 1981 for review) as used in botanical studies by Baum (1975, 1977) and Nelson and Van Horn (1975). Other computer algorithms which go under the general heading of parsimony methods (e.g. Camin and Sokal, 1965) take inspiration from early solutions to "travelling salesman" problems of Kruskal (1956) and Prim (1957) as discussed in Sneath and Sokal (1973). Here it is significant to mention an early application of parsimony methods by Boulter (1973) and Boulter *et al.* (1972) for computing phylogenies from amino acid sequences of proteins in higher plants.

Another branch of cladistics with a totally different philosophy from Hennig's method is the compatibility method (Estabrook *et al.* 1976a, b; Estabrook and Anderson, 1978; Gardner and La Duke, 1978; Meacham, 1980). This has been reviewed in two separate symposia (see Funk and Stuessy 1978; Farris and Kluge 1979; Duncan 1980; Duncan *et al.* 1980) so will not be considered by us.

The elements of Hennigian and transformed cladistics will be discussed here. The basis of this field, to quote Patterson's (1981) catchphrase is to "determine the distribution of homologies parsimoniously", or the grouping of taxa into monophyletic sets and subsets by shared derived evolutionary novelties or synapomorphies. We consider that the concept of monophyly is central to the understanding of a natural classification and that groups, based on shared primitive characters (and which are paraphyletic) and convergent groups (which are polyphyletic) have no place in systematics.

Whilst many would agree with the aims of cladistics, doubts are raised as to its efficacy as a method for determining relationships. The best way to assess a method is to apply it to a variety of data. Generally there are three steps in a cladistic study: (1) the choice of taxa and sampling of characters, (2) the determination of synapomorphies (which can be divided into the grouping and polarizing of characters), and (3) a search for the most robust cladogram incorporating the least number of assumptions (parsimony). All three steps have had various problems ascribed to them, such as how to determine homologies, problems of hybridization, problems of parallel and convergent evolution, and cladogram construction, arising mostly as a result of a lack of understanding the method or as a result of not distinguishing the aims of systematics (the perception of pattern) from the aims of evolutionary interpretation (the understanding of process).

The idea that the analysis of homologies and the determination of relationships are part of systematics and do not require evolutionary interpretations has been dubbed "transformed cladistics" (see Platnick 1979; Nelson and Platnick 1981). This means, as Platnick (1979) points out, if contemporary cladistics has to be summarized, classifications are based on synapomorphies and cladistics has nothing to do with speciation or with geological age. Platnick (1979) has three principles justifying cladistics: (*i*) nature is ordered in a single specifiable pattern represented by a branching hierarchical diagram, (*ii*) the pattern is estimated by sampling characters and finding replicated nested sets of synapomorphies, (*iii*) our knowledge of evolutionary history, like written classifications are derived from the pattern. This chapter is entirely about pattern analysis, and will be considered under two major headings, the relationship between synapomorphy and homology and the criteria for assessing apomorphies.

SYNAPOMORPHIES AND HOMOLOGY

One can redefine homology in a cladistic sense. As Simpson (1959) remarked homology is "the first and greatest generalization in anatomy", a point reiterated by Bock (1974) when he said that homology is "without question the most important principle in comparative biology". Homology is central to any discussion of comparative biology because it is the basis upon which the hypotheses of relationship are formulated. The term has had a tortuous history of changing concepts that has resulted in a range of classical, evolutionary, phenetic, cladistic and utilitarian definitions (see review in Patterson, 1982a). One can redefine homology and use it to clarify the concept of synapomorphy, after which it can be used as the central definition of cladistics: homology is the relation, or property, which characterizes monophyletic groups (and is also called a synapomorphy). There are, however, two uses of homology and they must first be distinguished in the light of our approach to systematics.

(1) Approaches to systematics

Eldredge (1979) distinguishes between two approaches to systematics which he called taxic and transformational. The first is concerned with the patterns of diversity, the second with the process of change. Cladistics is a taxic approach which specifies hierarchies of taxa based on character distributions, while transformational homologies imply character-change but without recognizing hierarchical relationships.

As an example of the transformational approach the classical concept of the carpel as a modified phyllome or leaf-like structure is generally considered well-supported by a variety of evolutionary transformations or "trends" especially in apocarpous angiosperm groups such as the orders Magnoliales and Ranunculales. The evolutionary transformation of a leaf is essentially infolding or conduplication apparently associated with asymmetrical growth so that during development the abaxial region grows more rapidly than the adaxial region (see Stebbins 1974). The acceptance of the phyllome homology for the leaf has led to similar and almost endless discussions on other transformations involving the stigma, marginal and laminal placentation and the homologies of integuments and ovules. Corner (1964) Goethe (1831) and even Theophrastus who employed the word leaf for a petal (according to Arber 1950) make such homologous assumptions. No classification is implied in any of these statements – they are simply general homologies – all of which indicate a love for transformational metamorphosis from an archetype with little regard for taxonomic interrelationships.

The taxic approach provides the alternative to archetypes – morphotypes. Morphotypes are lists of characters which identify the commonality within a group. In much of the higher level taxonomy, in angiosperm systematics at least, while there has been a conscious effort to list characters to find unifying descriptions, or ground plans, there has been little distinction between synapomorphic (grouping-defining homologies) and symplesiomorphic characters (primitive or general features) common to all members of a group but found in wider groups as well).

For those who do not specify the distinction between real and apparent characters, the typical approach is to postulate which of the present living groups are probably closest to the ancestral morphotype. For the angiosperms two major ancestral morphotypes look either like the Apetalae (Engler and Prantl 1897–1915; Wettstein 1901) or the Magnoliidae (Bessey 1897, 1915; Arber and Parkin 1907; Cronquist 1968; Takhtajan 1969; Thorne 1976). For each example the authors constructed an ancestor; for example, among various characters the ancestral Magnolia-like angiosperm would have included vessel-less wood (Bailey 1944; Takhtajan 1969), monosulcate atectate pollen (Walker and Skvarla 1975), an indeterminate strobilus-like flora axis bearing numerous, leaf-like floral parts, including conduplicate carpels and laminar stamens (Sporne 1974). In other words the morphotype is a conjured-up plant description of a "generalized" ancestor. Importantly, it is a taxic approach; it deals with a morphotype and gives a list of characters which can be criticized.

The taxic approach as used by cladists raises a different kind of question, e.g.

for the angiosperms, is the taxon Magnoliidae (1) monophyletic or (2) non-monophyletic – either biphyletic or polyphyletic? (see Burger 1981 for review). To say that they are monophyletic would be to say that there should be one or more synapomorphies (specific group-defining homologies) for the angiosperms. Hill and Crane (1982) tackle this question at great length by examining (1) implied synapomophies from the literature (Takhtajan 1969; Sporne 1974; Niklas *et al.* 1980; Parenti 1980), and (2) synapomorphies from their own observations. They concluded that seven unequivocal synapomorphies unite all angiosperms: (*i*) axially aligned companion cells derived from common mother cells with sieve elements, (*ii*) megaspore walls lacking sporopollenin, (*iii*) 4–16 nuclei in the megaprothallus, (*iv*) "double" fertilization of the embryo and the endosperm, (*v*) pollen wall endexine not laminated under germination apertures as seen in TEM sections, (*vi*) pollen-receptive stigmatic surface present, and borne on an enclosing, or partly enclosing structure which is external to the integument, (*vii*) pollen grains having three nuclei at developmental maturity.

As Hill and Crane (1982) note, six of the characters are concerned with the reproductive system and to hypothesize them as polyphyletic would, on the grounds of parsimony, be extremely cumbersome. To hypothesize the angiosperms as biphyletic or polyphyletic (hypothesis 2) would require characters *i–vii* to have evolved in parallel in two different sister groups. A detailed proposition of parallelism within two groups of gymnosperms has yet to be found in the papers invoking polyphyly (see Melville 1962, 1963; Meeuse 1965, 1972, 1974, 1975, 1977, 1979a, b, 1982; Hughes 1976; Krassilov 1977; Nair 1975). The notion of monophyletic groups in systematics needs to be explored further.

(2) Monophyletic groups

The cladistic approach to character analysis has resulted in the recognition that the only groups worthy of discussion are monophyletic groups. Just as the term homology is ambiguous in the literature so too is the understanding of monophyly (Patterson 1982a). Hennig's (1950, 1966) original method for reconstructing phylogenies used monophyly as an expression of relationship by reference to ancestry. Hennig's original definition was formalized in the following way: in a group of three taxa (Fig. 1), for the sake of argument, the tulip tree (A, representing the Magnoliidae), the pine (B, representing the Pinidae) and the maidenhair tree (C, representing the Ginkoidae), the pine and the tulip tree form a group AB because they share a hypothetical common ancestor, X, which is not shared by the maidenhair tree (C, Fig. 1). Similarly if we were to add the cycad (D, representing the Cycadidae), ABC would form a mono-

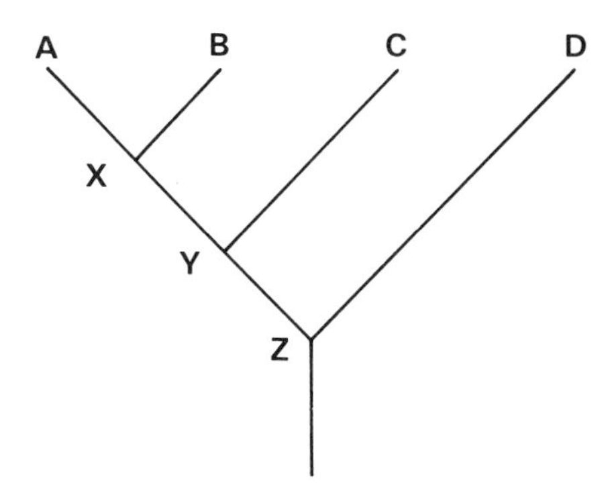

Fig. 1. Cladogram illustrating Hennig's original definition of monophyletic groups. Species A and B share a more recent common ancestor, X, which is not shared by species C or D. Similarly species ABC share a more recent common ancestor, Y, not shared by species D. The most recent common ancestor of the group ABCD is Z.

phyletic group because they share a common ancestor, Y, not shared by the cycad (Fig. 1). Platnick (1977a, b), in summarizing Farris (1974), has restated the definition by specifying that a monophyletic group is one which contains all descendants of a common ancestor or is uniquely derived and unreversed group membership. This common ancestry is indicated by the presence of synapomorphies. Some object to this definition because it includes a notion of common ancestry and because of the implication of group membership based on an abstraction.

Patterson (1982a) shows that these problems arise because we have defined monophyly (hypotheses about groups) in terms of common ancestry which is the same standard by which we judge homologies (hypotheses about characters). He proposes to resolve the problem by re-ordering the dependancies; because empirically we have only organisms and characters it is more obvious to suggest that monophyly is due to common ancestry and that homologous features characterize monophyletic groups. This type of homology can then be referred to as a grouping homology or synapomorphy (*sensu* cladistics).

One difficulty with this solution is the case of loss features conceived as synapomorphies. There is a problem, say, of distinguishing between two sorts of absence. For instance, the absence of petals and other perianth segments in groups such as the subclass Hamamelidae and the absence in genuinely apetalous groups such as the Pinidae. Cronquist (1968:157) says "there is a growing concensus that most or all of the 23 families here assembled in the Hamamelidae do form a natural group" identified as "a loosely knit group of dicots with more or less strongly reduced flowers". Can the loss of petals be treated as an homology and if so with what is it homologous?

Imagine that the Hamamelidae are the only petalless angiosperms and all other groups are petaloid. If that were so it would be possible to remove the Hamamelidae from the angiosperms entirely as a monophyletic, petalless group leaving the remaining angiosperms as a petaloid monophyletic group. According to this type of understanding the petaloid angiosperms would then be a paraphyletic group characterized by a synapomorphy (grouping homology), petals, which means that homologies can then form paraphyletic groups (see p. 349). However, as Platnick (1979) has explained, when one uses characters, such as petals, as a synapomorphy for a set, petalless members are a subset. Petallessness is not a true absence but, in relation to other characters, parsimony requires a hypothesis of secondary loss. Defined in this way the problem with the grouping homology (synapomorphy) defining a non-monophyletic group no longer exists. By contrast, lack of petals in the Pinidae is a primitive absence.

(3) Parsimony

The properties of monophyletic groups, synapomorphies (grouping homologies), have now been qualified and we turn to a criterion for distinguishing between alternative cladograms, the criterion of parsimony. Farris (1983) has recently reviewed comprehensively the subject of parsimony and it is from this publication that much of this discussion is drawn. Farris defines parsimony in the following manner: "most parsimonious genealogical hypotheses are those that minimize requirements for *ad hoc* hypotheses of homoplasy". In other words, when choosing between cladograms one prefers that explanation of character distribution that requires the smallest number of postulated events of parallel or convergent evolution. This is a simple concept, but one that often elicits a response to the effect that parsimony depends on an unrealistic assumption about nature – these individuals believe that evolution is not necessarily parsimonious. Many authors have used "alternative" criteria for determining interrelationship such as: character compatibility, character weighting, maximum likelihood, probability, and phenetic clustering by overall similarity. Farris shows that none of these are valid. The method of character compatibility simply constructs a parsimony tree using only some of the data; character weighting does not eliminate the conflict – it merely shifts the emphasis; maximum likelihood and other probability methods are based on evolutionary models frequently admitted by their authors to conform to unrealistic assumptions;

phenetic clustering by overall similarity has been generally discredited as a method for discerning order in nature, primarily because it assumes evolutionary rates are constant. The only viable alternative is parsimony. However, we do not use parsimony because there is nothing else. Parsimony means avoiding as much as possible multiple origins of homologous features, it does not presuppose the rarity of homoplasy, rather it merely seeks to minimize it. This is the very basis of grouping by synapomorphy, and thus parsimony and grouping by synapomorphy are the same thing.

The primary reason for choosing cladistics is the empirical nature of the method. The empirical basis of cladistics is parsimony, for it is in applying the parsimony criterion that we chose among alternative hypotheses of homology (characters) and alternative hypotheses of relationship (cladograms) to find the most robust cladogram, the one with the greatest explanatory power. To try and divorce grouping by synapomorphy from parsimony is impossible because they are one and the same.

By way of a simple example, consider the five taxa of Fig. 2 – ABCDE – for

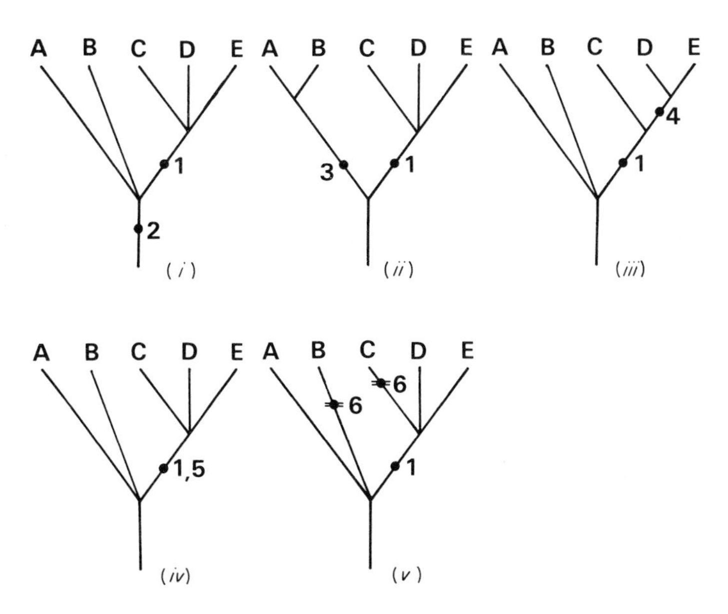

Fig. 2. Cladograms showing five ways (*i–v*) that characters specifying groups (characters 2–6) can relate to another character (character 1 which groups taxa CDE). (*i*) They can define a larger group (ABCDE); (*ii*) they can define a different group (AB); (*iii*) they can define a group within the original group (DE); (*iv*) they can define the same group (CDE); (*v*) they can define a conflicting group (BC).

which we have available six potential synapomorphies (grouping homologies, 1–6). These six characters can relate to one another in different ways. Suppose we select character 1 which identifies the group CDE. Other characters can relate to it in five different ways: they may group CDE into a larger group – ABCDE (as does character 2, Fig. 2*i*); they may group another group entirely – AB (as does character 3, Fig. 2*ii*); they may define a subset DE of CDE (as does character 4, Fig. 2*iii*); they may define the same group – CDE (as does character 5, Fig. 2*iv*); or they may form a conflicting group – BC (as does character 6, Fig. 2*v*). The most informative statements are those shown in Figs 2*iv* and 2*v* showing either full corroboration or complete disagreement with the original potential synapomorphy. (1). In the case of corroboration we fully accept the original grouping because now it is twice as good as before (character 5). In the case of disagreement there is conflict between two characters that group CDE (1, 5) and the one character that groups BC (6). If there is a natural hierarchy in nature, or order vs. disorder (Riedl 1979) not all of these potential synapomorphies can be real synapomorphies (grouping homologies). Figure 3 illustrates the alternatives: (*i*) one potential synapomorphy is found to be false (6) and this necessitates one *ad hoc* explanation of parallel evolution; (*ii*) two potential synapomorphies are found to be false (1, 5) and one must be reversed (6) necessitating three *ad hoc* explanations, two of parallel evolution and one of reversal;

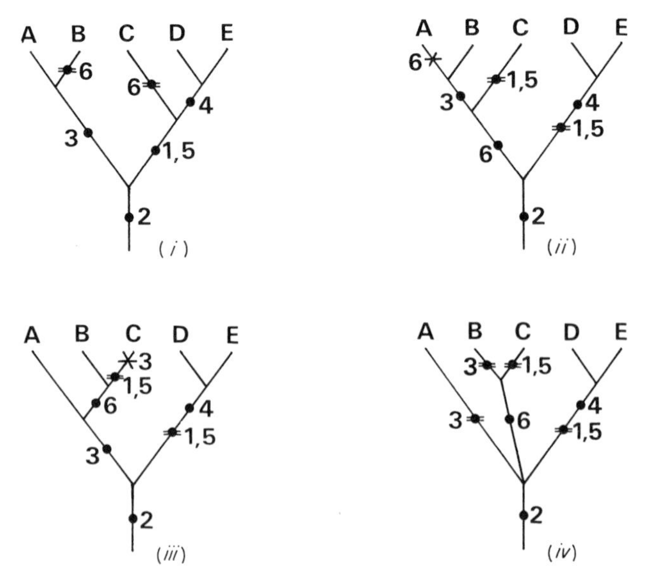

Fig. 3. Four possible cladograms (*i–iv*) that can be constructed from the six characters illustrated in Fig. 2 (see text for explanation).

(*iii*) two potential synapomophies are falsified (1, 5) and one must be reversed (3) necessitating three *ad hoc* explanations, two of parallel evolution and one of reversal; (*iv*) three potential synapomophies are falsified (1, 3, 5) necessitating three *ad hoc* explanations of parallel evolution. Figure 3*i* necessitates the fewest *ad hoc* explanations of parallel evolution and/or reversals and it therefore best represents the data and it is the one hypothesis we accept. Four possible estimates of relationship within this group have been tested and the one with the greatest explanatory power has been chosen – it is this procedure that makes cladistic analysis an empirical method (*sensu* Popper 1968a, b; Gaffney 1979).

CRITERIA FOR ASSESSING APOMORPHIES

Perhaps one of the greatest controversies in comparative biology are the arguments revolving around the initial identification of grouping homologies. In addition to historical comments on the subject (e.g. Hennig 1950, 1966; Ross 1974) several more recent articles have discussed the subject with varying success (Crisci and Stuessy 1980; De Jong 1980; Stevens 1980; Arnold 1981). For our purposes these can be reduced to three main criteria for identifying homologies: by having *ad hoc* criteria like 'common is primitive" (Estabrook 1977), by studying ontogenetic series, and using outgroups by sampling larger taxonomic units than the study group.

(1) Common is primitive

The method of equating frequency of character distribution with relative apomorphy has been shown to be fallacious (Hennig 1966; Wiley 1980; Watrous and Wheeler 1981; Nelson and Platnick 1981). Consequently, one simple example will suffice here. Figure 4 gives a three-taxon problem which

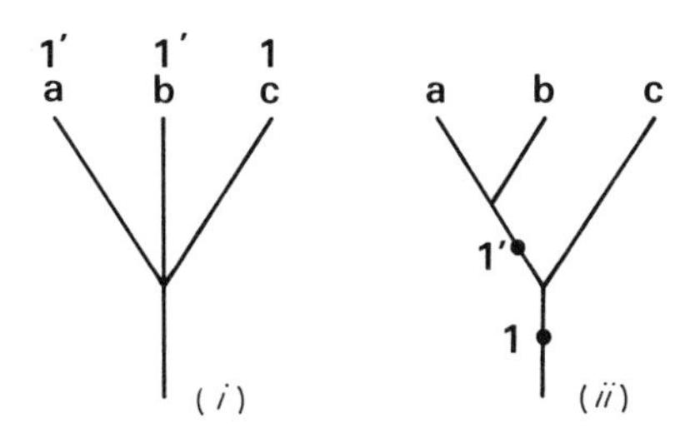

Fig. 4. Two cladograms showing that three taxon statements cannot be solved by using the criterion of "common is primitive". (*i*) At least one character (1′) must be common to 2 out of the 3 taxa to make, (*ii*) a group ab based on that character.

cannot be resolved using the concept of "common is primitive". If one attempts to group two out of the three species, a, b and c by the presence of a synapomorphy (1′) then that character is the most common in a 3-taxon group. An example can be found in the Gramineae. In grasses the common floral characters are 2 lodicules, 3 stamens and 2 stigmas, but careful examination of other monocotyledon outgroups shows that 3 lodicules, 6 stamens and 3 stigmas are the plesiomorphic condition although less frequently found in the family. The only ways of checking "common is primitive" are by outgroups or ontogeny and thus it might as well be abandoned entirely.

(2) Ontogeny

Comparative ontogeny, while a very useful and powerful means of erecting homology hypotheses, has been largely restricted to morphological, rather than phylogenetic, studies in botany. Nelson (1978a) and Wiley (1981) categorize the comparative study of ontogenetic sequences as the direct technique because they are particularly useful in groups with several possible outgroups, in the identification of a sister group and for initial study-group analyses. The key to the relationship between ontogeny and phylogeny is to accept that there is a direct correspondence between ontogenetic sequencies and phylogenies. This idea relies on the fact that, in all ontogenetic sequences, characters are added on to, or modified subsequently, to an initial character or starting condition. In other words, intermediate and terminal characters are always phylogenetically younger features than the initial stage and always follow the primary character and fail to develop if the primary characters are removed (ontogenetic precedence – Hennig 1966). The apomorphic characters of a group will go through stages of development recognizable as embryonic or plesiomorphic in comparison with non-members of the group. This concept of recapitulation is better known as the biogenetic law or the theory of recapitulation (Nelson 1978a). In its present form there are two interpretations, Haeckelian recapitulation and von Baer's law (see Løvtrup 1978; Wiley 1981). The distinction between the two depends on the modification of intermediate (or non-terminal) or terminal characters in an ontogenetic sequence. In an ontogeny of $1 \rightarrow 1' \rightarrow 1''$, $1''$ is the terminal character and 1 and $1'$ the non-terminal characters. If $1''$ occurred as an evolutionary novelty in group X we would expect the sister group Y to have the ontogenetic sequence, $1 \rightarrow 1'$. Thus in Haeckelian recapitulation we would expect group Y to have $1'$ as the terminal character and X to go through the ontogenetic stage that is present in the *adults* of the sister group. A good example of Haeckelian recapitulation can be found in the genus

Montanoa (Compositae). The receptacular bracts in all species of this genus become greatly enlarged by the time the achenes are mature. In some species the veins of the bracts are parallel throughout ontogeny whilst in others the veins are parallel in the immature stages and netted at maturity. In other words, those species with netted veins have passed through the adult stages of parallel-veined species and the net vein condition can be postulated as derived. The main threat to this type of interpretation is neoteny, or character loss, wherein the only real check will be to study the ontogeny or adults of the sister group. Hennig (1966), Lundberg (1973), Nelson and Platnick (1981) Nelson (1978a) and Patterson (1982a) have all discussed the problems of character losses as synapomorphies (see section on homology).

Not all changes in ontogeny are the result of additions of new terminal characters. Von Baer's law (see Nelson 1978a; Løvtrup 1978) indicates that many differences in adult phenotypes are the result of modifications of non-terminal characters of the ontogenetic sequence. Consider the following hypothetical example. In Group A, the sequence is: $1 \rightarrow 1' \rightarrow 1'' \rightarrow 1'''$. Rather than adding on a new character 1^{IV}, in Group B $1''$ may mutate to transcribe a new sequence: $1 \rightarrow 1' \rightarrow 1^{v} \rightarrow 1^{vi}$. Thus, in a comparison of the ontogenies of two groups the only common stages are $1 \rightarrow 1'$ and we would never expect the adults of group B to have or pass through the adult phenotypes. Løvtrup (1978) characterized this phenomenon as follows: "Von Baer generalized his empirical observations in his laws of development which imply that in the course of ontogeny there is a gradual change from the general to the special". Wiley (1981) restates the position to say: "In the course of their ontogeny, the members of two sister groups will follow the same course of recapitulation up to the stage of their divergence into separate taxa". In relating the ontogenetic sequence to phylogeny it is necessary to consider that the unique sequence in the groups to be considered is the unique condition. Thus, in three groups with the following character pattern:

x	$1 \rightarrow 1 \rightarrow ' 1''$
y	$1 \rightarrow 1 \rightarrow ' 1''$
z	$1 \rightarrow 1^{iv} \rightarrow 1^{v}$

given y and z are sister groups, and x is sister to both of them then the character sequence $1^{iv} \rightarrow 1^{v}$ must be considered apomorphic. A recent example is that of *Bougainvillea* (Nyctaginaceae). *Bougainvillea* is a group of some 18 perennial species predominantly in South America (Willis 1966). The family *Nyctaginaceae* exhibits two basic flowering positions, epiphylly which occurs in *Bougainvillea* and pedicellate flowering in groups such as *Boerhaavia* and *Mirabilis*.

Sattler and Perlin (1982) examine the flora ontogenies of all three genera and come to the conclusion that all these have compound dichasial cymes and the epiphyllic flower position in *Bougainvillea spectabilis* is the result of ontogenetic displacement resulting from an intercalary meristem at the base of each floral bract and floral bud. At maturity the pedicellate flowers of *Mirabilis* and *Boerhaavia* have arisen as a result of meristematic activity at the base of the flower and can be designated as a three-step process of development from an axillary primordium and then two distinct bract or flower meristems giving rise to a pedicellate flower and a bract. A three-step process designated as $1 \rightarrow 1' \rightarrow 1''$. In *Bougainvillea* the primordium gives rise to an intercalary meristem at the developing bract and flower. Although its exact origin is difficult to detect it develops first at the base of the bract and then extends gradually below the insertion of the floral meristem. The common intercalary meristem has come about through some mutation giving rise to meristem extension (Dickinson 1978), a pattern designated by $1 \rightarrow 1^{iv} \rightarrow 1^{v}$. The intercalary meristem must be considered apomorphic giving rise to the apomorphic epiphyllous inflorescence of *Bougainvillea*.

(3) Out-group comparison

Perhaps the most useful operational method for assessing homologies, or synapomorphies is out-group comparison (see Hennig 1966; Platnick 1979; Watrous and Wheeler 1981). It is a method which serves two functions; checking the validity of particular characters as homologies or synapomorphies and determining their distribution.

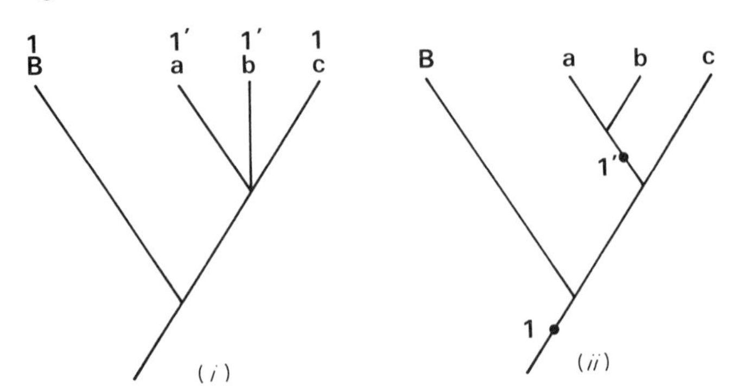

Fig. 5. Out-group comparison. (*i*) A three-taxon problem (the grouping of two species a and b in relation to a third c in genus A; (*ii*) the problem is solved by determining the validity or polarity of the character (1′) by out-group comparison using the sister genus, B as the out-group.

In this method the characters of the study group are examined in a larger group of taxa, which it is hoped contain the sister taxon (a). In Fig. 5 the genus A with three species (a, b, c) can be compared to a related group B. Species a and b have character 1′ while species c has character 1 which occurs also in group B. Character 1 is thus a generalized, or plesiomorphic character because it is found not only in the study group. Character 1′ is synapomorphic for a and b (Fig. 5*i*), and on this amount of information they form a monophyletic group.

Two complications should be mentioned, the problem when the characters of the study group are absent from the out-group and the way of proceeding with an out-group analysis in the absence of a sister group. Figure 6 deals with

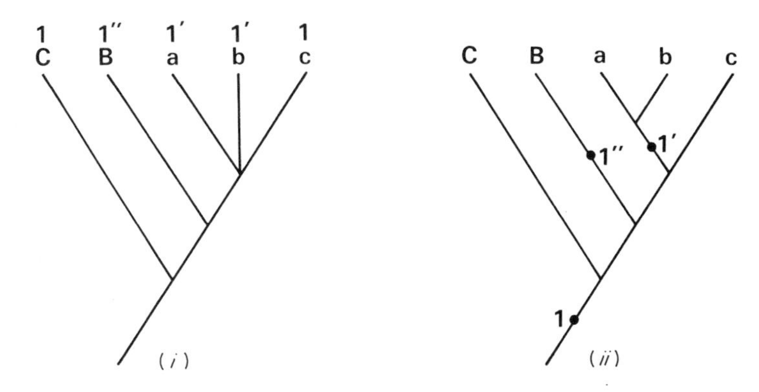

Fig. 6. Out-group comparison (*i*) A three-taxon problem (the grouping of two species a and b in relation to c in genus A) when the first out-group (genus B) has a different character (1″) than either of those (1, 1′) found in genus A; (*ii*) the solution to the three taxon problem by comparison to genus C which shows that the 1″ is an autapomorphy – a unique character for genus B – and that 1′ is a synapomorphy for the two species a and b of genus A.

the first problem. Since the first out-group, genus B, has a different character, 1″, from either of the 1 and 1′ characters in Genus A no conclusion about the relative apomorphy of the characters in A can be achieved. It is necessary to examine a wider sample of taxa, in this case genus C which has character 1 in all of its member taxa (Fig. 6*i*). The interpretation from this result is that character 1″ is autapomorphic for genus B and 1′ is a synapomorphy for species a and b in genus A (Fig. 6*ii*). The same procedure would be followed if both characters 1 and 1″ were initially found in genus B. Should characters 1 and 1′ be both found in genus B and character 1 in genus C (Fig. 7*i*) either 1′, is synapomorphic for part of genus B with genus A species a and b (Fig. 7*ii*) or,

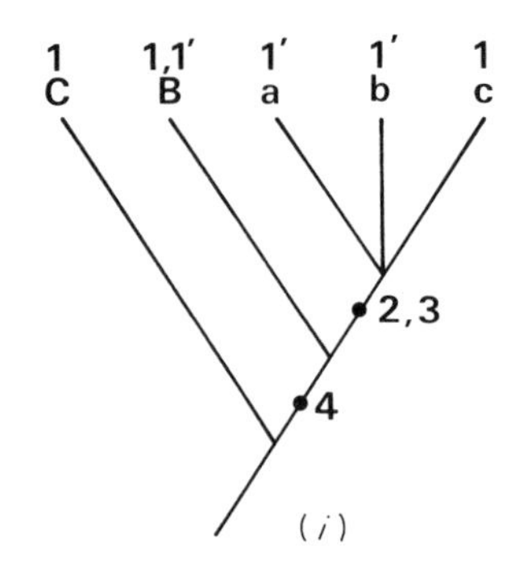

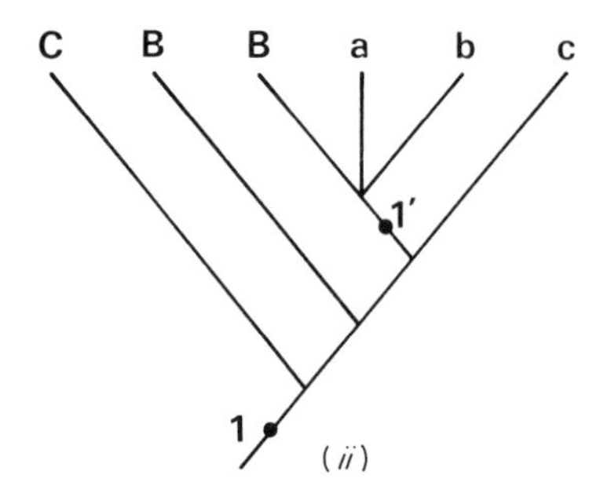

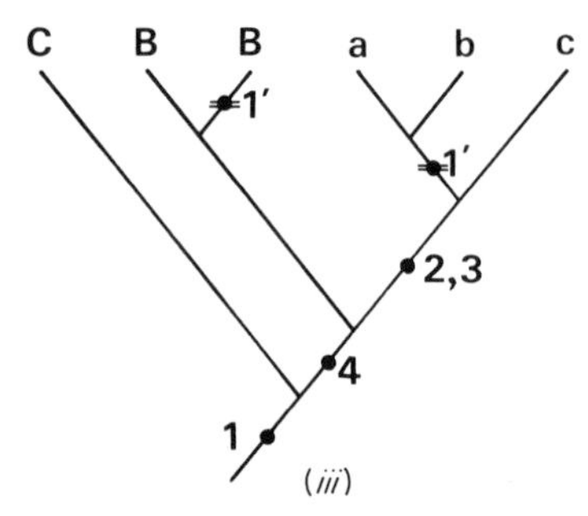

Fig. 7. Out-group comparison (*i*) A three-taxon problem (the grouping of two species a and b, in relation to c in genus A) when the first out-group (genus B) has both of the same characters (1, 1′) found in A; (*ii*) a solution to the three-taxon problem by direct comparison for characters 1 and 1′ in the second out-group (genus C) suggesting that part of genus B should be grouped in with A; (*iii*) a solution to the three taxon problem by comparison with genera B and C in relation to three other characters 2, 3 and 4. It is more parsimonious to assume that character 1′ is homoplasious (i.e. is a result of parallel evolution) for species a and b of genus A and for part of genus B.

on grounds of parsimony and congruence with other characters, 1′ should be considered homoplasious rather than homologous in the two genera (Fig. 7*iii*).

Problems are manifold at the beginning of analysis in the absence of an identifiable sister group to the study group. There are two possibilities for dealing with the problem. One can use a higher level taxonomic out-group, such as a whole family, or tribe, for comparison in the study group or one can try to identify a functional outgroup by some other means. In Fig. 8*i* genera B, C and

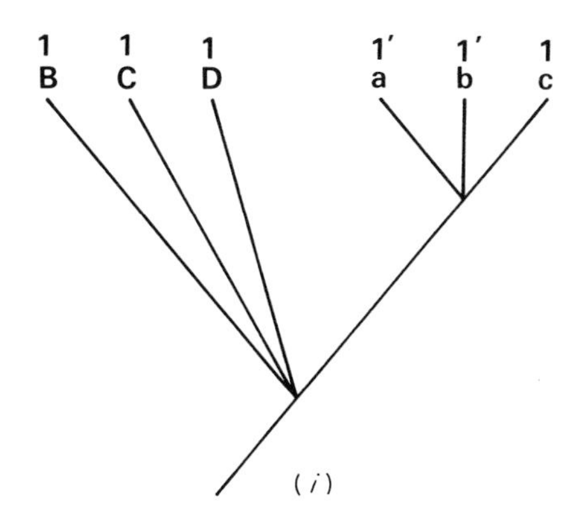

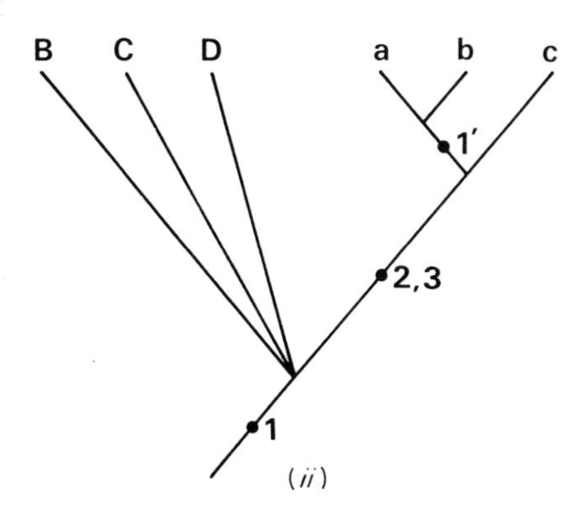

Fig. 8. Out-group comparison. (*i*) A three-taxon problem (the grouping of two species a and b in relation to c in genus A) in the absence of an easily identifiable out-group; (*ii*) A solution to the tree-taxon problem using three genera (BCD) collectively as the out-group.

D are all potential, but unidentifiable, as sister groups and are used as one combined group to solve a homology problem. In other words, there is no character with which to combine one of the genera B, C or D with genus A. In this particular example all three genera, B, C and D show the plesiomorphic character 1 which is sufficient to suggest that character 1′ is synapomorphic for Genus A, species a and b (Fig. 8*ii*). This method is difficult to use, however, in groups exhibiting large amounts of apparent homoplasy or "noisy" character distributions. Alternatively, each potential out-group, B, C or D could be used independently from one another to polarize the characters. The three cladograms produced by this procedure can then be compared with one another by various consensus techniques (Adams 1972; Nelson 1979). Watrous and Wheeler (1981) approached the problem from a different perspective with "functional out group analysis". In this method identifiable monophyletic groups within the study group are used as out-groups of one another. An example is the genus *Montanoa* (Compositae–Heliantheae). *Montanoa* is a very distinct genus that is easily separated from the majority of the Heliantheae tribe by a number of unique characters. However, its uniqueness, coupled with the generally poorly understood higher taxonomy of the tribe (see Robinson 1978) makes the identification of a sister group or a position in a subtribe impossible (Funk 1982). Most potential characters for identifying a sister group repeatedly occur in different genera of the tribe and cannot be hypothesized as homologies using either the entire tribe as an out-group or by ontogenetic studies. However, by using just eleven characters that could be identified as homologies from the entire tribe, or from ontogeny, *Montanoa* was divided up into three monophyletic groups (Fig. 9). To determine the relationships of species within each of these infrageneric groups each group was used as an out-group of the other.

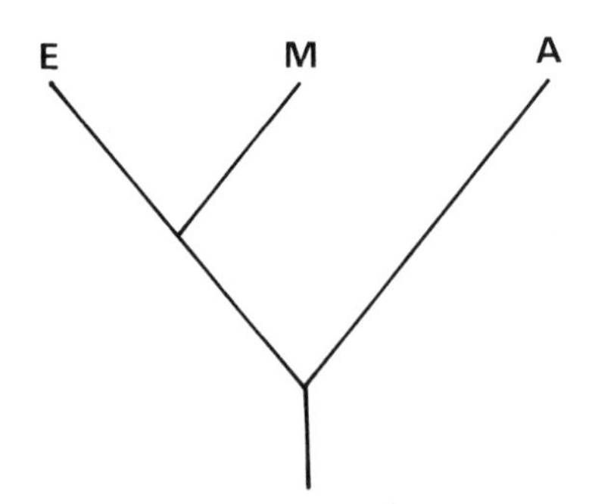

Fig. 9. Three infrageneric groups of *Montanoa* (Compositae–Heliantheae) used as the functional out-groups of one another (see text for explanation).

Montanoa sensu stricto is the out-group of *Echinocephala* and *vice versa* and the two together form the functional out-group of *Acanthocarphae*. The remaining characters could then be analysed with relative ease and a cladogram be constructed for the species of the entire genus (Funk 1982).

HOMOPLASY – PARALLELISM AND CONVERGENCE

Almost all cladistic studies display a degree of character conflict. These conflicting characters are similarities mistaken to be homologies (false synapomorphies). Simpson (1961) called all non-homologous similarity homoplasy and distinguished five sorts: parallelism, convergence, analogy, mimicry and chance similarity. Patterson (1982a) places the last four together under convergence and cites one other term homoiology (homeology) (from Riedl 1979) which refers to "analogies on an homologous base".

As Patterson (1982a) points out there are three tests of homology: similarity (isology), congruence (checking against other possible homology statements) and conjunction (two structures occurring in the same organism). Homology passes all three tests. The different degrees of failure of the tests distinguish between convergence and parallelism (Table I). Convergence passes the conjunction test, i.e. analagous features occur in the same position in unrelated

Table I. Tests of relations specifying groups.

Relation	*Similarity*	*Congruence*	*Conjunction*
Homology	+	+	+
Parallelism	+	–	+
Convergence	–	–	+

– = fail; + = pass; see text for explanation.

groups, but it fails the congruence and similarity tests. Parallelism passes the conjunction and similarity tests but fails the congruence test. A few botanical examples will help to clarify the different possibilities.

Classic examples of convergence include the appearance of superficially similar plants of diverse relationship appearing in widely disjunct habitats of the same general type. The sclerophyllous Zygophyllaceae (*Larrea*, *Bulnesia*) occurring in North American deserts resemble similar looking members of the Rutaceae and Proteaceae in Australia (Stebbins 1974). Here, the leaves and habits look alike, but on the basis of many more floral and fruit characters they fail the congruence test. Careful inspection of leaf anatomy reveals a wide range of differing structures and so they also fail the similarity tests.

The difference between convergence and parallelism is a matter of degree and parallel characters often originate from an homologous base. A good example of parallelism in the land plants is the presence of heterospory; dimorphism for different sized male and female spores, in several major groups. It is notable that when gametophytes are retained in spores, in heterosporus ferns, horsetails (Equisetales) and Lycopods etc., the larger megaspores give rise to female prothalli, and the smaller microspores give rise to male prothalli bearing only antheridia. It is impossible to group together heterosporous taxa on the basis of this one character because of lack of congruence with other group defining characters (see Parenti 1980). It is better to hypothesize that heterospory has arisen on a number of separate occasions.

The problem of comparing loss features cannot be tested in the same way. Losses can only be interpreted by the parsimony criterion in congruence with other characters (p. 330).

The homologies between male and female coniferalean cones as compared with cycad cones provide a good example of a conjunction test. All cycads are dioecious, and in all, except *Cycas*, the ovules are borne in compact cones, ranging in size from 2 cm long in *Zamia pygmaea* to 70 cm or more in *Macrozamia denisonii*, possibly the largest cone in the world (Sporne 1965). The sporophylls are arranged in a spiral axis, ranging in form from the loosely arranged leaf-like appendages in *Cycas* to the peltate appendages of *Zamia*. Although some of the structurally reduced strobili are sometimes considered stachysporous the differences in structure are usually interpreted as a gradually reducing transformation series of phyllosporous appendages arranged on simple, determinate terminal or axillary branches. All the male strobili of cycads are compact cones having spirally arranged appendages with microsporangia on the undersides. Although there is obvious sexual dimorphism, the spirally arranged appendages are considered directly homologous to one another. On the other hand the cones of the Coniferales are very variable in shape and size and degree of fusion of the various parts but, apart from groups such as the Taxales and Podocarpaceae, they are considered to have the same basic structure. The male cones are simple, bearing spirally arranged microsporophylls. By contrast on the main axis of the female cones are two groups of spirally arranged scales – the ovuliferous scales each subtended by an extra structure, the bract scale. Although the male cone can be considered homologous to the cycadian male and female cones, the female coniferalean cone has the extra bract-scale. It is the presence of the bract-scale, an extra structure, which led Florin (1951) to consider, when looking at the fossil Lebachiaceae, the whole male coniferalean cone to be homologous with a single ovuliferous scale of the female. In other words, the female cone is a complex inflorescence and each ovuliferous scale is equivalent to a branch. Hence the female ovuliferous scale (homologous to the male conifer cone) occurs conjointly with the "cone" of the female and they are therefore non-homologous.

It appears then of the three main relations (see Table I), congruence is the most important test because it fails everything except homologies. The only way to implement the congruence test is through a cladistic analysis.

RETICULATE EVOLUTION

Reticulate evolution, or hybridization, is a cause of incongruent, intersecting data. Incongruent characters "destroy" phyletic information and cause a logical inconsistency in a method designed to represent information in hierarchical diagrams. Three major ideas have been proposed to deal with the problem of putative hybridization. Some authors have stressed that multiple branching in cladograms resulting from the probable presence of hybrids is the true reflection of a character pattern and should be left in (Nelson and Platnick 1980; Bremer and Wanntorp 1979b). Others advocate removing the known hybrids at the beginning of the analysis and eventually placing them above the cladograms to indicate putative parents (Wagner 1969). A third group advocates leaving all of the taxa in the first analysis and then closely examining the cladograms for polytomies. Taxa in polytomies that can be identified as hybrids by alternative criteria should be removed and placed above the redrawn cladogram (Funk 1981; Humphries 1981a; Sanders 1981).

There are problems with all three approaches. The first does not reflect accurately the taxonomic relationships and character distributions. The second method assumes that it is possible to identify hybrids and parents prior to analysis. All three approaches assume that most of the time hybrids cause polytomies, although Humphries (1983) and Wiley (*pers. com.*) have shown that this is not necessarily the case.

The problems can be examined by considering hybrids in two different situations; between sister and non-sister taxa. Figure 10 illustrates the situation in sister taxa; species A and B hybridize to give an intermediate H. If there were an equal number of apomorphies in A and B (characters 2 and 3) and if both were found in the hybrid, the result could be expressed as a trichotomy (Fig. 10*ii*). Alternatively, if the incongruent character set is inferred to be the result of hybridization, the hybrid could be placed above the diagram connecting

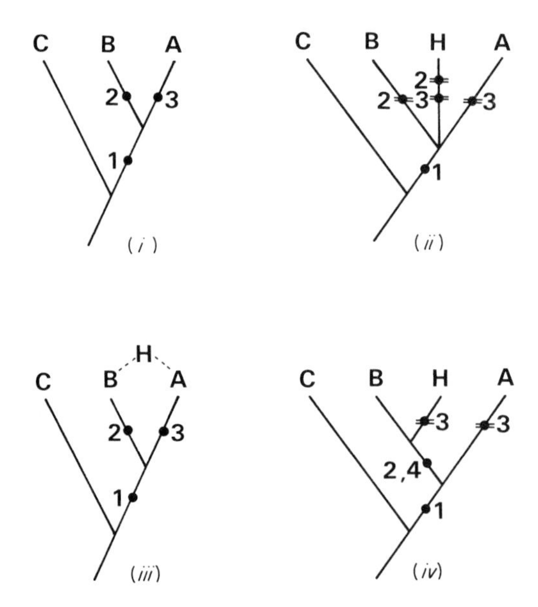

Fig. 10. A hybrid (H) between sister species. (*i*) A three-taxon group, ABC showing that A and B are sister taxa formed on the basis of synapomorphy 1 and each with an autapomorphy (2 and 3); (*ii*) a hybrid (H) between A and B represented as a trichotomy when autapomorphies 2 and 3 are both inherited; (*iii*) The hybrid (H) represented as a reticulation on the cladogram eliminating homoplasies on the cladogram; (*iv*) The hybrid (H) will appear as a resolved sister taxon to B when autoapomorphies (2, 3, and 4) are inherited unevenly.

with both parents (Fig. 10*iii*, after Funk 1981; Humphries 1979; 1981a). However, if one parent taxon had one more autapomorphy than the other, or if the hybrid showed unequal character inheritance, then a fully resolved cladogram results. For instance, if taxon B had two autapomorphies (2 and 4), the parsimony criterion would give the result in Fig. 10*iv*. The problems of hybrids in non-sister taxa are illustrated in Fig. 11. The hypothetical cladogram in Fig. 11*i* shows a minimum number of synapomorphies and autapomorphies for identifying taxa and forming groups in five taxa. There is a hybrid between taxa C and D and this hybrid (H) inherits all of the apomorphies from both parents (1, 3, 5, 6, 7).

Consider the various ways of expressing the relationships and the lengths of the cladograms involved. In Fig. 11*ii* the most parsimonious cladogram places the hybrid (H) as the sister taxon of the parent that involves the least number of

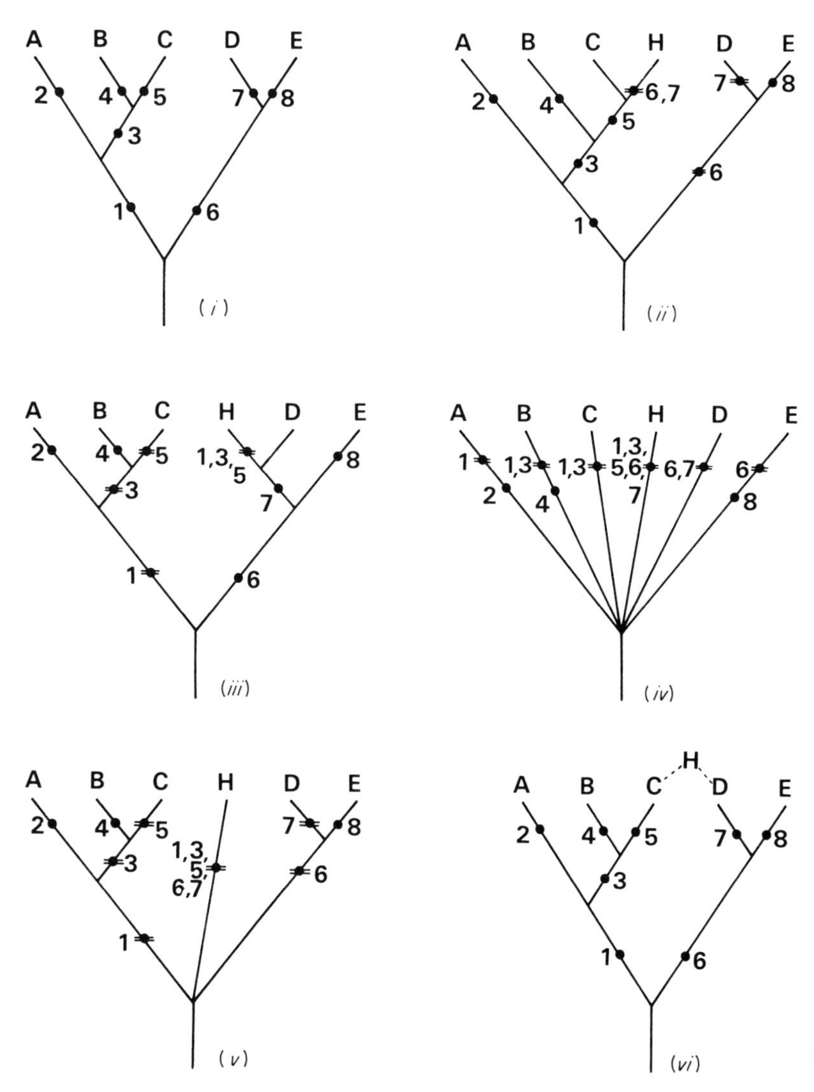

Fig. 11. A hybrid (H) formed between two non-sister species (C, D). (*i*) Cladogram excluding the hybrid species, showing eight character changes; (*ii*) Cladogram grouping the hybrid with the parent (species C) with which it shares the most synapomorphies (ten character changes); (*iii*) cladogram grouping the hybrid with the parent (species D) with which it shares the fewest synapomorphies (eleven character changes). (*iv*) a collapsed cladogram showing seventeen character changes; (*v*) an Adams consensus tree (see Adams 1972) produced by adding together the cladograms *ii* and *iii*; (*vi*) the most parsimonious cladogram (showing eight character changes) achieved by placing the hybrid as a reticulation.

homoplasious events (the number of character changes or length of this cladogram is L = 10). Figure 11*iii* shows that the length would increase if the hybrid were placed with the other parent (L = 11) because there is one more homoplasy event. These data can also be used to show that if intersecting characters should be expressed in resolved cladograms (Nelson and Platnick 1980, 1981) then both the collapsed cladogram (Fig. 11*iv*) and placing the hybrid at the base of the diagram (Fig. 11*v*) produces greater lengths L = 17 and L = 13). The most parsimonious way to construct the cladogram is a way which does reflect the hybrid origin of taxon H. If the hybrid would be identified it would be removed from the cladogram and placed above it giving an even shorter diagram (Fig. 11*vi*) (L = 8). The difficulty comes in assuming hybrid status if a polytomy is not formed.

Nelson (1983) has recently suggested a method for analysing cladograms for possible hybrids. His procedure begins with the most parsimonious cladogram without a reticulation and continues by adding reticulations in a particular order so as to minimize the character conflict at each step. It is based on the idea that when there are two equally parsimonious ways of representing a homology on a cladogram one should investigate the possibility of inserting a reticulation. If the reticulation results in a decrease of apparent homoplasy and if the taxon exhibits character conflict of the "intermediate" type the reticulation can be maintained. Although data containing hybrids may always be somewhat difficult to deal with in a cladistic analysis, these recent papers indicate that progress is being made and that at least in some cases it is now possible to distinguish homologues that appear to be homoplasies as a result of hybridization or from homoplasy.

CLASSIFICATION

For most cladists a cladogram is a classification and the question becomes how best to represent it in a name hierarchy. It seems best at this time to make every effort to maintain as much of the Linnaean system as possible since it is an integral part of our taxonomy although other systems have been suggested (see Wiley 1981 for a review). It seems that the most successful way of accomplishing this is through a method called phyletic sequencing which was developed over a period of years by Nelson (1973), Cracraft (1974), Patterson and Rosen (1977) and Wiley (1979, 1981). Wiley refers to his system, which is by far the most detailed explanation, as the "Annotated Linnaean Hierarchy".

Regardless of the names they are given, all methods operate on the same basic principle: all information from a cladogram is available in a classification and a cladogram can be reconstituted from a classification. In other words, the cladogram and the classification are one and the same.

Monophyletic groups have the same importance in a classification as in a cladogram, and any monophyletic group must be reflected as such in the classification (Wiley 1981). Consider some examples:

(*i*) the Compositae genus *Tetragonotheca* (from Seaman and Funk 1983). A

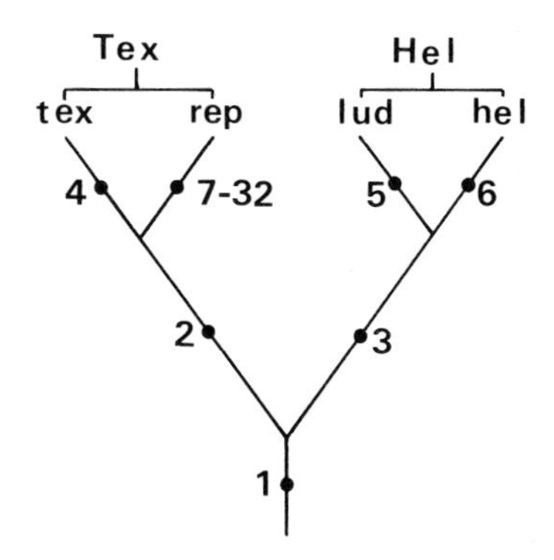

Fig. 12. Cladogram for the four species of *Tetragonotheca* (Compositae–Heliantheae). See text for explanation of characters.

cladogram for four species in this genus is illustrated in Fig. 12 and the sequence is as follows:

Tetragonotheca
  Texana group
    T. texana
    T. repanda
  Helianthoides group
    T. helianthoides
    T. ludoviciana

Each taxon is the sister taxon of all listed below it at the same level of indentation. If one wished to recognize subgeneric groups then one option would be to make both the "Texana group" and the "Helianthoides group" subgenera or sections. The "Texana group" is the sister group of the "Helianthoides group". Then at the next level of indentation *T. texana* and *T. repanda* are sister species as are *T. helianthoides* and *T. ludoviciana*, and the classification can be reconstructed from one another and both contain the same amount of information.

(*ii*) *Anacyclus* (Compositae), a genus with three putative hybrids (Fig. 13; *A.* × *inconstans*, *A.* × *officinarum* and *A.* × *valentinus*; from Humphries 1979).

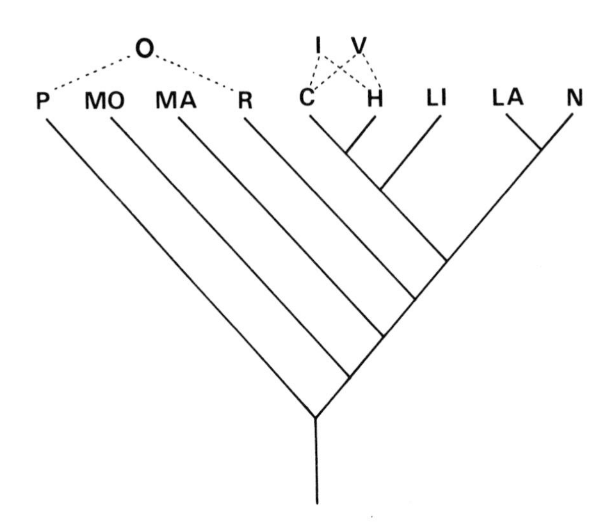

Fig. 13. Cladogram for 11 species of *Anacyclus* (Compositae–Anthemideae) 2 of which are shown as putative hybrids (I, V).

The phyletic sequence without hybrids, sections or species groups is as follows:

Anacyclus
- *A. pyrethrum*
- *A. monanthos*
- *A. maroccanus*
- *A. radiatus*
- *Clavatus* species group
  - *A. linearilobus*
  - *A. clavatus*
  - *A. homogamos*
- *A. latealatus*
- *A. nigellifolius*

The hybrids can be added in several ways; one is to make them the sister taxon of either one of the parents. Using the first parent in the phyletic sequence:

Anacyclus
- Section Pyrethraria
  - *A. pyrethrum*
  - †*A. officinarum sedis mutabilis* (*A. pyrethrum* × *radiatus*)
- Section Anacyclus
  - *A. monanthos*

A. maroccanus
A. radiatus
Clavatus species group
  A. linearilobus
  A. homogamos
  †*A. inconstans sedis mutabilis* (*A. homogamos* × *clavatus*)
  †*A. valentinus sedis mutabilis* (*A. homogamos* × *clavatus*)
  A. clavatus
A. latealatus
A. nigellifolius

The special notations include a dagger (†) for hybrids, the parental species listed after them in parenthesis and the latin words *sedis mutabilis* ("changeable position") which are used to mean a polytomy in the cladogram. Sometimes it may be preferable to list the hybrids as the sister taxa of both parents especially when they are in different subgenera. In any case, the cladogram is recovered in the following manner. Section *Pyrethraria* is the sister group of Section *Anacyclus*. *A. pyrethrum* and *A. officinarum* are sister taxa within section *Pyrethraria* (Fig. 13). However *A. officinarum* is a hybrid and should be placed above the diagram. In succession; *A. monanthos* is the sister taxon of the remaining species; *A. maroccanus* is the sister taxon to the remaining species; *A. radiatus* is the sister taxon to the remaining species (*A. officinarum* can now be linked to both parents); the *clavatus* group is the sister group to *A. latealatus* and *A. nigellifolius* (Fig. 13); finally, within the *clavatus* group, *A. linearilobus* is the sister taxa of *A. clavatus* and *A. homogamos* and their two hybrids; *A. homogamos* is the sister taxon of *A. clavatus*, but forms a polytomy with the two hybrids. The hybrids are placed above the cladogram, as in the original cladogram (Fig. 13).

Despite the logic of this technique, its implementation by botanists is slow. The point of most contention lies in the recognition of monophyletic groups. Consider again the cladogram for *Tetragonotheca* (Fig. 12). Suppose there was one synapomorphy to identify the "Texana group" (2) and one for the "Helianthoides group" (3). Let us suppose also that species *T. texana*, *T. helianthoides* and *T. ludoviciana* have only one autapomorph each (4, 5, 6). *Tetragonotheca repanda*, on the other hand has, let us say, 25 apomorphies (7–32) and is therefore uniquely and vastly different from the other three species. The tendency of some taxonomists is to group the three that have changed very little together in one group, for instance in the "Helianthoides group", because of their "overall similarity". *Tetragonotheca repanda* would then be placed in a satellite group by itself because it is so "different". However, these three species do not represent all of the descendents of a common ancestor and so do not constitute a monophyletic group. Why is this important? Well, to place *T. texana* in a reformed "Helianthoides group" it is necessary to do two things, (1) ignore the synapomorphy that places *T. texana* and *T. repanda* together, and (2) form a paraphyletic group because it does not include all the taxa with common homologies. In other words a group by default, identified by no characters at all.

There has been much written and argued regarding this aspect of cladistics (see Eldredge and Cracraft 1980; Wiley 1979, 1981; Nelson and Platnick 1981; Nelson 1973; Patterson and Rosen 1977), but all of the discussion can be reduced to one consideration – What do we want from a classification? If it is natural groups; if it is prediction; if it is groups useful for further studies of biogeography and ecology, then cladistic groups show the best resolution. As Wiley (1981) pointed out a good classification should be minimally redundant, minimally novel and maximally informative. The best method we have to date, is the representation of the cladogram in the "Annotated Linnaean Hierarchy".

EXAMPLE

By refining the meaning of synapomorphy as a grouping homology, the relation that defines monophyletic groups, the implications of character analysis can be explained with an example. By returning to our rather simple case of the interrelationships of the Cycadidae, the Ginkoidae, the Pinidae and the Angiospermae it is possible to explore in a novel way the character hypotheses, the identification of groups and the validity of existing classifications.

Consider the four taxa labelled as ABCD. Table II gives all the possible character groupings in the three gymnosperm groups and the angiosperms. Table III gives a list of 28 characters based on Hill and Crane (1982). Any character available for these four taxa must fall into one of four categories: (*i*) the character can pertain to only one taxon (Table II, character types 12–15), (*ii*) it can be present in all four taxa (Table II, character type 1) – or it can be informative regarding the groupings of the taxa by either being – (*iii*) present in two taxa (six combinations possible, AB, AC, AD, BC, BD, CD Table II, character types 6–11) or three taxa (four combinations possible ABC, ACD, BDC or ABD Table II, character types 2–5). There are 15 possible fully resolved cladograms for any four taxa (Fig. 14). Neither the characters present in all four taxa (category *i*) nor those present in only one taxon (category *ii*) are of any help in deciding which of the possible cladograms is most likely because they are consistent with all 15 possible cladograms. The remaining ten types (those shared

Table II. Type of possible character distributions.

Taxa	1	2	3	4	5	6	7	8	9	10	11	12	13	14	15
A Angiospermae	+	+	+	+	–	+	+	+	–	–	–	+	–	–	–
B Pinidae	+	+	–	+	+	+	–	–	+	+	–	–	+	–	–
C Ginkgoidae	+	+	+	–	+	–	+	–	+	–	+	–	–	+	–
D Cycadidae	+	–	+	+	+	–	–	+	–	+	+	–	–	–	+
Grouping potential	No	Yes	Y	Y	Y	Y	Y	Y	Y	Y	Y	N	N	N	N
Number of characters	10	3	n.p.	1	2	3	n.p.	n.p.	n.p.	n.p.	n.p.	8	1	n.p.	n.p.
Actual characters (see distribution, Fig. 15)	a–j	k–m	—	n	o, p	q,r,s	—	—	—	—	—	t–z a_1	b_1	—	—

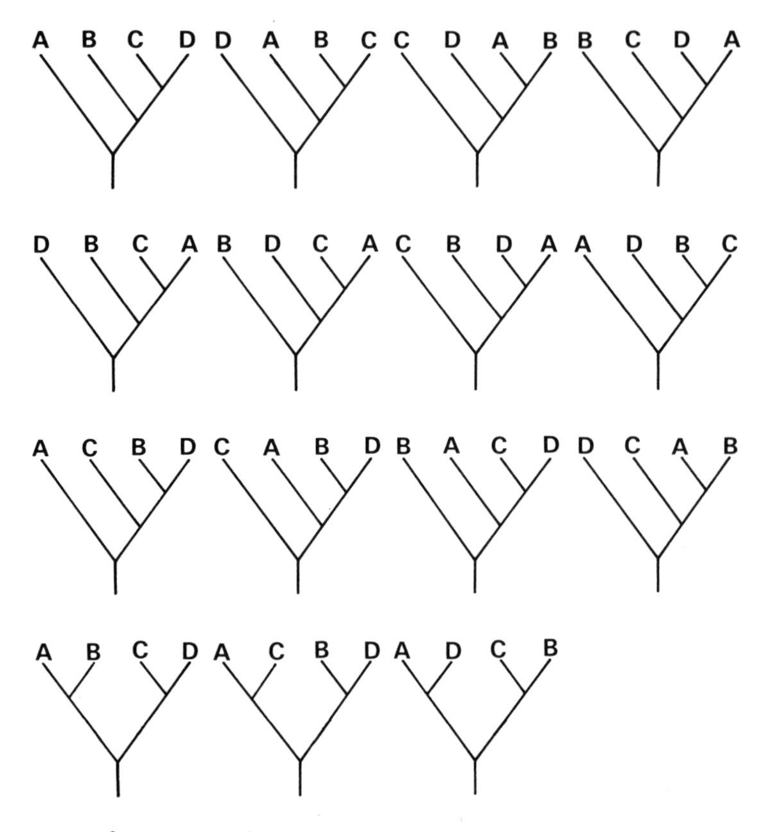

Fig. 14. Fifteen possible fully resolved cladograms for any four taxa (ABCD).

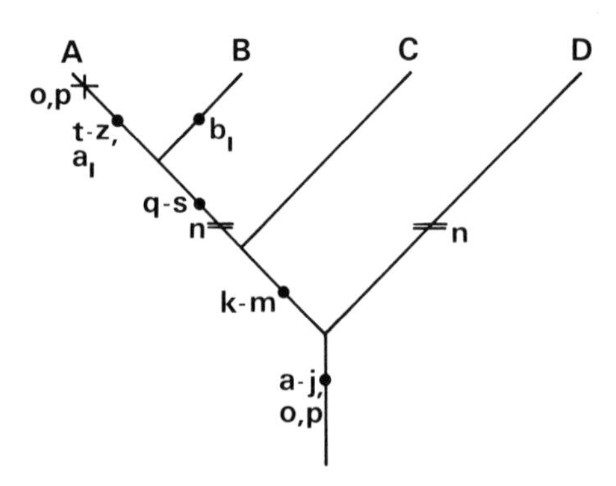

Fig. 15. The most parsimonious cladogram based on the characters (a–z, a_1b_1) of Table III ●, synapomorphies; ═, parallelisms; +, reversals.

by either two or three taxa) have the potential for grouping. The identification of the characters was determined by out-group comparison to ferns, bryophytes and green algae. It is apparent that of the ten character types in Table II, only four (character types 2, 4, 5 and 6) have any grouping potential. Character types 2 and 6 (a total of 6 characters) are congruent with one another and character type 4 (a total of 1 character) and type 5 (a total of 2 characters) are neither congruent with one another nor with character types 2 and/or 6. Because it requires fewer assumptions of false synapomorphy (parallelism or convergence) the most robust or most parsimonious hypothesis is that which agrees with character types 2 and 6 (Fig. 15). A re-examination of some of the grouping potential or the homologies of character types 4 and 5 shows that some statements are less probable and may not actually conflict with the hypothesis in Fig. 15. For example Florin (1951) in a brilliant study showed that the female cones of the Pinidae are "double", in that each ovuliferous scale is subtended by a "sterile" bract-scale which is absent in the cycads (see p. 343). In other words character "n": "sporangia arrangement in strobili or not" is not a synapomorphy (grouping homology) for the cycads and coniferales since the ovuliferous scale of the conifers is a modified shoot emerging from the axil of a bract and thus homologous to a whole cycad strobilus. Furthermore, character "n" is a dubious synapomorphy (grouping homology) for the angiosperms and the coniferales unless the carpel is considered homolgous to both the ovuliferous scale and bract scale. At this point it is probably more appropriate to consider character "n" as a false synapomorphy (non-grouping homology) which should perhaps be rearranged into two autapomorphies (characters unique to a single taxon) and thus like character distributions 12–15 (Table II) and of no value in grouping. A check of the potential out-groups (e.g. ferns) of the study group reveals that the homology statements of characters "o" and "p" yield no change in the hypotheses given in Table III. This means that unless they can be shown to be incorrectly interpreted through ontogenetic studies it is most parsimonious for them to be regarded as a reversal in the angiosperms, to account for their distribution in the study group.

Within the four taxon group some of the groupings above the level of terminal taxa already published in the literature are as follows: BC (Coniferopsida: Sporne 1965; Pinicae: Cronquist 1971); BCD (Gymnospermae: Sporne 1965) and ABCD (subclass Pinatae: Bremer and Wanntorp 1981). Since paraphyletic groups (those that cannot be defined by a synapomorphy or grouping homology without including other non-members) are groups by default, they are based on symplesiomorphies and have no real place in systematics. As has been shown (see section on Classification) only monophyletic groups

Table III. Characters for grouping four taxa (see Table II and Fig. 15).

		−	+
a	Secondary thickening	Absent	Present
b	Shoot meristem	One-celled	Several cells
c	Leaf gaps in shoot vascular system	Absent	Present
d	Thickening of tracheid walls in centrifugal primary xylem with bordered pits	Absent	Present
e	Thickening of tracheid walls of 2° xylem with bordered pits	Absent	Present
f	Megagametophyte retained on sporophyte	No	Yes
g	Spores	Homosporous	Heterosporous
h	Seeds	Absent	Present
i	Microspore pollen		
j	Primary metaxylem	Mesarch/exarch	Endarch
k	Secondary xylem	Manoxylic	Pycnoxylic
l	Thickening of tracheid walls in 2° xylem	Multi-seriate bordered pits	1–2 seriate bordered pits
m	Stamens	Laminate	On stalks
n	Sporangia arrangement	Not in strobili	In strobili
o	Thickening in tracheid walls of protoxylem	Without bordered pits	With bordered pits
p	Embryogenesis early divisions	With cell walls	Free nuclei
q	Male gametes	Motile	Non-motile
r	Pollen tube	Haustorial	Male gamete transfer
s	Maule reaction	Negative	Positive
t	Vessels	Absent	Present
u	Sieve tubes	Absent	Present
v	Companion cells	Absent	Present
w	Strobili	Unisexual	Bisexual
x	Megaprothallus at micropylar end	Cellular	Free nuclear
y	Pollen ektexine	Spongy	Tectate columnate
z	Stigma	Absent	Present
a_1	Seeds	Naked	Enclosed at anthesis
b_1	Strobili	Simple	Compound

are named in cladistics since they are only ones based on synapomorphies (grouping homologies). Thus, groups like the Coniferopsida, Pinicae and Gymnospermae should all be abandoned since they have no value in predicting new character distributions within the study group, and new taxa should be recognized for groups AB, and ABC.

CONCLUSIONS

As so elegantly reviewed by Nelson and Platnick (1981) the three major components of the history of life are form, time and space. As Patterson (1982b) points out the three biological disciplines concerned with these are systematics, which considers the patterns of diversity and their relationships, palaeontology, which can provide information about time, and biogeography, the discipline concerned with space. Cladistics is a rigorous methodological discipline which has called into question traditional attitudes to all three.

As a method of pattern and character analysis cladistics holds a much stronger possibibility for the reconstruction of the history of life in plants than any other proposed method.

We think that it has been apparent in this brief exposition that most other current methods in plant taxonomy, particularly in higher plants, are elementary by comparison. For example, to think that the balloon diagrams of Dahlgren (1975), Cronquist (1968), and Stebbins (1974), so applauded as the best representations of relationship by the Hamburg conference in 1976 (Merxmüller 1977), could show anything beyond the visualization of Bessey (1915) is beyond our comprehension. Traditional systematics of higher plants (e.g. Takhtajan 1969; Thorne 1976; Heywood 1978) has been for the last 75 years a tinkering process with no advancement in character analysis, expression of relationship, or the solving of relationships above the family level. In palaeontology, the role of fossils has yet to be evaluated. In historical biogeography the pioneering works of Croizat, who seems to us to be the only man to examine distribution without prior geological theories, have been systematically ignored by most botanists (see Croizat *et al.* 1974; Nelson and Platnick 1981; Nelson 1978b for reviews). And yet, if we are to have a basis to test the theories of process, we need the best evaluations of pattern. It seems in this respect, to us at least, that cladistics gives something beyond an eclectic choice.

ACKNOWLEDGEMENTS

We thank the following friends and colleagues for reading and commenting on a draft of this manuscript: M. Crisp, P. Ladiges, P. Jaeger and C. Stirton.

We also thank Mr J. R. Press for helping with the illustrations and Ms D. M. Stephenson and Ms L. G. M. Hosking for typing the many drafts of the manuscript. Financial assistance (to V.A.F.) from the Fluid Research Fund of the Secretary of the Smithsonian Institution and from the Director of the Natural History Museum of the same institution, is gratefully acknowledged.

REFERENCES

Adams, E. N. (1972). Consensus techniques and the comparison of taxonomic trees. *Syst. Zool.* **21,** 390–397.

Arber, A. (1950). "The Natural Philosophy of Plant Form". Cambridge University Press, Cambridge.

Arber, A. and Parkin, J. (1907). On the origin of angiosperms. *J. Linn. Soc. Bot.* **38,** 29–80.

Arnold, E. N. (1981). Estimating phylogenies at low levels. *Z. Zool. Syst. Evol. Forsch.* **19,** 1–35.

Bailey, I. W. (1944). The development of vessels in angiosperms and its significance in morphological research. *Ann. J. Bot.* **31,** 421–428.

Baum, B. R. (1975). Cladistic analysis of the diploid and hexaploid oats (*Avena*, Poaceae) using numerical techniques. *Canad. J. Bot.* **53,** 2115–2127.

Baum, B. R. (1977). Assessment of cladograms obtained for fourteen species of *Avena* by two methods of numerical analysis. *Syst. Bot.* **2,** 141–150.

Bessey, L. E. (1897). Phylogeny and Taxonomy of the angiosperms. *Bot. Gaz.* **24,** 145–178.

Bessey, C. E. (1915). The phylogenetic taxonomy of flowering plants. *Ann. Missouri Bot. Gard.* **2,** 109–164.

Bock, W. J. (1974). Philosophical foundations of classical evolutionary taxonomy. *Syst. Zool.* **22,** 375–392.

Bolick, M. R. (1981). A cladistic analysis of *Salmea* DC. (Compositae-Heliantheae). *In* "Advances in Cladistics: Proceedings of the First Meeting of the Willi Hennig Society" (V. A. Funk and D. R. Brooks, eds) pp. 115–127. New York Botanical Garden, New York.

Boulter, D. (1973). Amino acid sequences of cytochrome C and platocyanins in phylogenetic studies of high plants. *Syst. Zool.* **22,** 549–553.

Boulter, D., Ramshaw, J. A. M., Richardson, P. M. and Brown, R. H. (1972). A phylogeny of higher plants based on the amino acid sequences of cytochrome C and its biological implications. *Proc. Roy. Soc. Lond.* (*B*) **181,** 441–455.

Bremer, K. (1976a). The genus *Relhania* (Compositae). *Opera Bot.* **40,** 1–86.

Bremer, K. (1976b). The genus *Rosenia* (Compositae). *Bot. Notiser* **129,** 97–111.

Bremer, K. (1978a). The genus *Leysera* (Compositae). *Bot. Notiser* **131,** 369–383.

Bremer, K. (1978b). *Oreoleysera* and *Antithrixia*, new and old South African genera of the Compositae. *Bot. Notiser* **131,** 449–453.

Bremer, K. and Wanntorp, H. E. (1978). Phylogenetic systematics in Botany. *Taxon* **27,** 317–329.

Bremer, K. and Wanntorp, H. E. (1979a). Geographic populations or biological species in phylogeny reconstruction? *Syst. Zool.* **28,** 220–224.

Bremer, K. and Wanntorp, H. E. (1979b). Hierarchy and reticulation in systematics. *Syst. Zool.* **28,** 624–627.

Bremer, K. and Wanntorp, H. E. (1981). The cladistic approach to plant classification. *In* "Advances in Cladistics: Proceedings of the First Meeting of the Willi Hennig Society" (V. A. Funk and D. R. Brooks, eds) pp. 87–94. New York Botanical Garden, New York.

Brundin, L. (1966). Transantarctic relationships and their significance as evidenced by chironomid midges. *K. Svenska Vetensk-akadamien Handl.* **11(1),** 1–472.

Burger, W. C. (1981). Heresy revived: the monocot theory of angiosperm origin. *Evolutionary Theory* **5,** 189–225.

Camin, J. H. and Sokal, R. R. (1965). A method for deducing branching sequences in phylogeny. *Evolution* **19,** 311–326.

Churchill, S. P. (1981). A phylogenetic analysis, classification and synopsis of the genera of the Grimmiaceae (Musci). *In* "Advances in Cladistics: Proceedings of the First Meeting of the Willi Hennig Society" (V. A. Funk and D. R. Brooks, eds) pp. 127–144. New York Botanical Garden, New York.

Corner, E. J. H. (1964). "The Life of Plants". Weidenfeld and Nicolson, London.

Cracraft, J. (1974). Phylogenetic models and classification. *Syst. Zool.* **23,** 71–90.

Crisci, J. V. and Stuessy, T. F. (1980). Determining primitive character states for phyletic reconstruction. *Syst. Bot.* **5,** 112–135.

Croizat, L., Nelson, G. and Rosen, D. E. (1974). Centers of origin and related concepts. *Syst. Zool.* **23,** 265–287.

Cronquist, A. (1968). "The Evolution and Classification of Flowering Plants". Nelson, London.

Cronquist, A. (1971). "Introducing Botany" (2nd edn). Harper and Row, New York.

Dahlgren, R. (1975). A system of classification of the angiosperms used to demonstrate the distribution of characters. *Bot. Notiser* **128,** 119–147.

De Jong, R. (1980). Some tools for evolutionary and phylogenetic studies. *Z. Zool. Syst. Evol. Forsch.* **18,** 1–23.

Dickinson, T. A. (1978). Epiphylly in angiosperms. *Bot. Rev.* **44,** 181–232.

Duncan, T. (1980). Cladistics for the practicing taxonomist – an eclectic view. *Syst. Bot.* **5,** 136–148.

Duncan, T., Phillips, R. B. and Wagner, W. H. (1980). A comparison of branching diagrams derived by various phenetic and cladistic methods. *Syst. Bot.* **5,** 264–293.

Eldredge, N. (1979). Alternative approaches to evolutionary theory. *Bull. Carnegie Mus. Nat. Hist.* **13,** 7–19.

Eldredge, N. and J. Cracraft (1980). "Phylogenetic patterns and the Evolutionary Process: Method and Theory in comparative Biology". Columbia University Press, New York.

Engler, A. and Prantl, H. (1897–1915). "Die natürlichen Pflanzenfamilien". 20 vols. Leipzig.

Estabrook, G. E. (1977). Does common equal primitive? *Syst. Bot.* **2,** 36–42.

Estabrook, G. F., Johnson, C. S. and McMorris, F. R. (1976a). An algebraic analysis of cladistic characters. *Discrete Mathematics* **16,** 141–147.

Estabrook, G. F., Johnson, C. S. and McMorris, F. R. (1976b). A mathematical foundation for the analysis of cladistic character compatibility. *Mathematical Bioscience* **29,** 181–187.

Estabrook, G. F. and Anderson, W. R. (1978). An estimate of phylogenetic relationships within the genus *Crusea* (Rubiaceae) using character compatibility analysis. *Syst. Bot.* **3,** 179–196.

Farris, J. S. (1970). Methods for computing Wagner trees. *Syst. Zool.* **19,** 83–92.

Farris, J. S. (1974). Formal definitions of paraphyly and polyphyly. *Syst. Zool.* **23,** 548–554.

Farris, J. S. (1983). The empirical basis of Phylogenetic Systematics. *In* "Advances in Cladistics: Proceedings of the Second Meeting of the Willi Hennig Society" (N. Platnick and V. A. Funk, eds) pp. 7–36. Columbia University Press, New York.

Farris, J. S. and Kluge, A. G. (1979). A botanical clique. Cladistics and plant systematics. *Syst. Zool.* **28,** 400–411.

Florin, R. (1951). Evolution in *Cordaites* and conifers. *Acta Horti Bergiani* **15,** 285–388.

Funk, V. A. (1981). Special concerns in estimating plant phylogenies. *In* "Advances in Cladistics: Proceedings of the First Meeting of the Willi Hennig Society" (V. A. Funk and D. R. Brooks, eds) pp. 73–86. New York Botanical Garden, New York.

Funk, V. A. (1982). The systematics of *Montanoa* (Asteraceae: Heleanthene). *Mem. N.Y. Bot. Gard.* **36,** 1–133.

Funk, V. A. and Stuessy, T. F. (1978). Cladistics for the practicing plant taxonomist. *Syst. Bot.* **3,** 159–178.

Gaffney, E. S. (1979). An introduction to the logic of phylogenetic reconstruction. *In* "Phylogenetic Analysis and Palentology" (J. Cracraft and N. Eldredge, eds) pp. 79–111. Columbia University Press, New York.

Gardner, R. C. and LaDuke, J. C. (1978). Phyletic and cladistic relationships in *Lipochaeta* (Compositae). *Syst. Bot.* **3,** 197–207.

Goethe, J. W. von (1831). "Versuch über die Metamorphose der Pflanzen". Übersetzt von F. Soret, nebst geschichtlichen Nachträgen. Stuttgart.

Haeckel, E. (1866). "Generelle Morpholigie der Organismen", Vol. II. Georg Reiner, Berlin.

Hennig, W. (1950). "Grundzüge einer Theorie der Phylogenetischen Systematik". Deutscher Zentralverlag, Berlin.

Hennig, W. (1966). "Phylogenetic Systematics". University of Illinois Press, Urbana.

Heywood, V. H. (ed.) (1978). "Flowering Plants of the World". Oxford University Press, Oxford.

Hill, C. R. and Crane, P. R. (1982). Evolutionary cladistics and the origin of angiosperms. *In* "Problems of Phylogenetic Reconstruction" (K. A. Joysey and A. E. Friday, eds) pp. 269–361. Academic Press, London, New York.

Hughes, N. F. (1976). "Palaeobiology of angiosperm origins". Cambridge University Press, Cambridge.

Humphries, C. J. (1979). A revision of the genus *Anacyclus* (Compositae: Anthemideae). *Bull. Brit. Mus. Nat. Hist.* (*Bot.*) **7,** 83–142.

Humphries, C. J. (1981a). Cytogenetic studies in *Anacyclus* (Compositae: Anthemideae). *Nordic J. Bot.* **1,** 83–96.

Humphries, C. J. (1981b). Biogeographical methods and the southern beeches (Fagaceae:

Nothofagus). *In* "Advances in Cladistics: Proceedings of the First Meeting of the Willi Hennig Society" (V. A. Funk and D. R. Brooks, eds.) pp. 177–207. New York Botanical Garden, New York.

Humphries, C. J. (1983). Primary data in hybrid analysis. *In* "Advances in Cladistics: Proceedings of the Second Meeting of the Willi Hennig Society" (N. Platnick and V. A. Funk, eds) pp. 89–103. Columbia University Press, New York.

Humphries, C. J. and Richardson, P. M. (1980). Hennig's methods and phytochemistry. *In* "Chemosystematics: Principles and Practice" (F. A. Bisby, J. G. Vaughan and C. A. Wright, eds) pp. 353–378. Academic Press, London, New York.

Jensen, R. J. (1981). Wagner networks and Wagner trees: a presentation of methods for estimating most parsimonious solutions. *Taxon* **30,** 576–590.

Kiriakoff, S. G. (1959). Phylogenetic systematics versus typology. *Syst. Zool.* **8,** 117–118.

Kiriakoff, S. G. (1962). On the neo-Adansonian school. *Syst. Zool.* **11,** 180–185.

Kluge, A. G. and Farris, J. S. (1969). Quantitative phyletics and the evolution of Anurans. *Syst. Zool.* **18,** 1–32.

Koponen, T. (1968). Generic revision of Mniaceae Mitt. (Bryophyta). *Annls bot. Fenn.* **5,** 117–151.

Koponen, T. (1973). *Rhizomnium* (Mniaceae) in North America. *Annls bot. Fenn.* **10,** 1–26.

Krassilov, V. A. (1977). The origin of angiosperms. *Bot. Rev.* **43,** 143–176.

Kruskal, J. B. (1956). On the shortest spanning subtree of a graph and the travelling salesman problem. *Proc. Amer. Math. Soc.* **7,** 48–50.

Løvtrup, S. (1978). On von Baerian and Haeckelian recapitulation. *Syst. Zool.* **27,** 343–352.

Lundberg, J. C. (1973). More on primitiveness, higher level phylogenies and ontogenetic transformations. *Syst. Zool.* **22,** 327–329.

Meacham, C. A. (1980). Phylogeny of the Berberidaceae with an evaluation of classifications. *Syst. Bot.* **5(2),** 149–172.

Meeuse, A. D. J. (1965). Angiosperms – past and present. *Advg. Front. Pl. Sci.* **11,** 1–228.

Meeuse, A. D. J. (1972). Sixty-five years of theories of the multiaxial flower. *Acta biotheor.* **21,** 167–202.

Meeuse, A. D. J. (1974). Flora evolution and emended anthocorm theory. *In* "International Bio-Sci. Monograph I" (T. M. Varghese, ed.) pp. 1–188. Hissar, India.

Meeuse, A. D. J. (1975). Origin of angiosperms – problem or inaptitude? *Phytomorphology* **25,** 373–379.

Meeuse, A. D. J. (1977). Coincidence of characters and angiosperm phylogeny. *Phytomorphology* **27,** 314–322. Issued 1978.

Meeuse, A. D. J. (1979a). Why were the early angiosperms so successful? A morphological, ecological and phylogenetic approach (Parts I and II). *Proc. K. ned. Akad. Wet.* (*C*) **82,** 343–369.

Meeuse, A. D. J. (1979b). 5. The significance of the Gnetatae in connection with the early evolution of the angiosperms. *In* "Glympses in Plant Research" (P. K. K. Nair, ed.), Vol. 4, pp. 62–73. Vikas Publishing House, New Delhi.

Meeuse, A. D. J. (1982). Once again: cladistics in botany. *Nordic J. Bot.* **2,** 189–190.

Melville, R. (1962). A new theory of the angiosperm flower. I. The gynoecium. *Kew Bull.* **16,** 1–50.

Melville, R. (1963). A new theory of the angiosperm flower. II. The androecium. *Kew Bull.* **17,** 1–63.

Merxmüller, H. (1977). Summary Lecture. *In* "Flowering Plants: Evolution and Classification of higher Categories. Plant Systematics and Evolution" Suppl. 1 (K. Kubitzki, ed.) pp. 397–405. Springer Verlag, Wien, New York.

Nair, P. K. K. (1979). The palynological basis for the triphyletic theory of angiosperms. *Grana palynol.* **18,** 141–144.

Nelson, C. H. and Van Horn, G. T. (1975). A new simplified method for constructing Wagner networks and the cladistics of *Pentachaeta* (Compositae: Asteraceae). *Brittonia* **27,** 362–372.

Nelson, G. J. (1969). Gill arches and the phylogeny of fishes with notes on the classification of vertebrates. *Bull. Amer. Mus. Nat. Hist.* **141,** 475–552.

Nelson, G. (1973). Classification as an expression of phylogenetic relationships. *Syst. Zool.* **22,** 344–359.

Nelson, G. (1978a). Ontogeny, phylogeny and the biogenetic law. *Syst. Zool.* **27,** 324–345.

Nelson, G. J. (1978b). From Candolle to Croizat. Comments on the history of biogeography. *J. Hist. Biol.* **11,** 269–305.

Nelson, G. J. (1979). Cladistic analysis and synthesis: Principles and definitions, with a historical note on Adanson's *Familles des Plantes* (1763–1767). *Syst. Zool.* **28,** 1–21.

Nelson, G. (1983). Reticulation in cladograms. *In* "Advances in Cladistics: Proceedings of the Second Meeting of the Willi Hennig Society" (N. Platnick and V. A. Funk, eds) pp. 105–111. Columbia University Press, New York.

Nelson, G. and Platnick, N. I. (1980). Multiple branching in cladograms: two interpretations. *Syst. Zool.* **28,** 86–91.

Nelson, G. and Platnick, N. I. (1981). "Systematics and Biogeography: Cladistics and Vicariance". Columbia University Press, New York.

Niklas, K. J., Tiffney, B. H. and Knoll, A. H. (1980). Apparent changes in the diversity of fossil plants: a preliminary assessment. *Evol. Biol.* **12,** 1–89.

Parenti, L. R. (1980). A phylogenetic analysis of the land plants. *Biol. J. Linn. Soc.* **31,** 224–242.

Patterson, C. (1981). Methods of Palaeobiogeography. *In* "Vicariance Biogeography; a Critique" (G. Nelson and D. E. Rosen, eds) pp. 446–489. Columbia University Press, New York.

Patterson, C. (1982a). Morphological characters and homology. *In* "Problems of Phylogenetic Construction" (K. A. Joysey and A. E. Friday, eds) pp. 21–74. Academic Press, London, New York.

Patterson, C. (1982b). Cladistics and classification. *New Scientist* **94** (1303), 303–306.

Patterson, C. and Rosen, D. E. (1977). Review of ichthyodectiform and other mesozoic teleost fishes and the theory and practice of classifying fossils. *Bull. Am. Mus. Natur. Hist.* **158,** 81–172.

Platnick, N. I. (1977a). Paraphyletic and polyphyletic groups. *Syst. Zool.* **26,** 195–200.

Platnick, N. I. (1977b). Cladograms, phylogenetic trees, and hypothesis testing. *Syst. Zool.* **26,** 438–442.

Platnick, N. I. (1979). Philosophy and the transformation of cladistics. *Syst. Zool.* **28,** 537–546.

Popper, K. R. (1968a). "The Logic of Scientific Discovery". Harper Torchbooks, New York.

Popper, K. R. (1968b). "Conjectures and Refutations". Harper Torchbooks, New York.

Prim, R. C. (1957). Shortest connection networks and some generalizations. *Bell. Syst. Tech. J.* **36,** 1389–1401.

Riedl, R. (1979). "Order in Living Organisms. A Systems Analysis of Evolution". John Wiley, London, New York.

Robinson, H. R. (1978). Studies in the Heliantheae (Asteraceae). XV. Various new species and new combinations. *Phytologia* **41,** 33–44.

Ross, H. H. (1974). "Biological Systematics". Addison-Wesley, Reading.

Sanders, R. W. (1981). Cladistic analysis of Agastache (Lamiacede). *In* "Advances in Cladistics: Proceedings of the First Meeting of the Willi Hennig Society" (V. A. Funk and D. R. Brooks, eds) pp. 95–114. New York Botanical Garden, New York.

Sattler, R. and Perlin, L. (1982). Floral development of *Bougainvillea spectabilis Willd., Boerhaavia diffusa* L. *Mirabilis jalapa* L. (Nyctaginaceae). *Bot. J. Linn. Soc.* **84,** 161–182.

Seaman, F. C. and Funk, V. A. (1983). Cladistic analysis of complex natural products: developing transformation series from sesquiterpene lactone data. *Taxon* **32,** 1–27.

Simpson, G. G. (1959). Anatomy and morphology: classification and evolution: 1859 and 1959. *Proc. Am. phil. Soc.* **103,** 286–306.

Simpson, G. G. (1961). "Principles of Animal Taxonomy". Columbia University Press, New York.

Sneath, P. H. A. and Sokal R. R. (1973). "Numerical Taxonomy". W. H. Freeman, San Francisco.

Sporne, K. R. (1965). "The Morphology of Gymnosperms". Hutchinson University Library, London.

Sporne, K. R. (1974). "The Morphology of Angiosperms". Hutchinson University Library, London.

Stebbins, G. L. (1974). "Flowering Plants: Evolution above the Species Level". Edward Arnold, London.

Stevens, P. (1980). Evolutionary polarity of character states. *Ann. Rev. Ecol. Syst.* **11,** 333–358.

Takhtajan, A. (1969). "Flowering Plants; Origin and Dispersal". Oliver and Boyd, Edinburgh.

Thorne, R. F. (1976). A phylogenetic classification of the Angiospermae. *Evol. Biol.* **9,** 35–106.

Wagner, W. H. (1961). Problems in the classification of ferns. *In* "Recent Advances in Botany", pp. 841–844. University of Toronto Press, Toronto.

Wagner, W. H. (1969). The role and taxonomic treatment of hybrids. *Bioscience* **19,** 785–789.

Wagner, W. H. (1980). Origin and philosophy of the groundplan-divergence method of cladistics. *Syst. Bot.* **5,** 173–193.

Walker, J. W. and Skvarla, J. J. (1975). Primitively columellaless pollen: a new concept in the evolutionary morphology of angiosperms. *Science, N.Y.* **187,** 445–447.

Watrous, L. and Wheeler, Q. (1981). The outgroup method of phylogeny reconstruction. *Syst. Zool.* **30,** 1–21.

Wiley, E. O. (1979). An annotated Linnaean hierarchy with comments on natural taxa and competing systems. *Syst. Zool.* **28,** 308–337.

Wiley, E. O. (1980). Phylogenetic systematics and vicariance biogeography. *Syst. Bot.* **5(2),** 194–220.

Wiley, E. O. (1981). "Phylogenetics: the Theory and Practice of Phylogenetic Systematics". Wiley Interscience, New York, Chichester.

Willis, J. C., revised by Shaw, H. K. (1966). "A Dictionary of Flowering Plants and Ferns". Cambridge University Press, Cambridge.

Wettstein, R. (1901). "Handbuch der systematischen Botanik". Franz Deuticke, Leipzig, Wien.

Taxonomic Priorities

18 | Completing the Inventory

G. T. PRANCE

The New York Botanical Garden, Bronx, USA

Abstract: The botanical inventory of world vegetation is far from complete. It has been carried out unevenly with greater focus on some areas. Recent international reports point out the inadequacy of inventory in the tropics where the vegetation is most species diverse.

Inventory is not just the collection and identification of herbarium specimens from a certain area. In order to understand the dynamics of the vegetation it should include data on biological relationships such as pollination and dispersal, as well as quantitative phytosociological studies. Basic forest inventory of tropical forest types is most inadequate and data are not readily comparable because of the use of different measurements in different places. There is a need to decide on basic parameters such as 10 cm tree diameter as the minimum tree size in forest inventories. To be useful these inventories must also be based on accurate botanical identification and not just the use of local names as has been the practice in many cases.

The urgency for further inventory is great because of the rate of deforestation. Some areas such as the rain forests of eastern Brazil have been destroyed before even a basic inventory was made. Inventory must be designed to provide logical reasons for conservation and the inclusion of quantitative vegetation studies and biological studies furnishes data useful for preservation of the habitats. The identification of threatened and endangered species and habitats has unfortunately become a most important part of inventory, because of the rapid destruction of plant habitats.

The written inventory is still far from complete, and serious gaps in the Floras of the world exist in both temperate and tropical regions.

The deposition of specimens resulting from increased inventory is also becoming a major problem as the cost of herbarium maintenance increases. Many of the world's herbaria, especially in the tropics, are inadequately curated and protected. Laws in many countries require deposition of specimens in local herbaria that are badly infested with insects, do not mount their specimens and have inadequate storage cabinets and staff.

Systematics Association Special Volume No. 25, "Current Concepts in Plant Taxonomy", edited by V. H. Heywood and D. M. Moore, 1984. Academic Press, London and Orlando.
ISBN 0 12 347060 9

Accelerated inventory must also include added assistance and care of the places in which the results are deposited. Herbaria of the world are also faced with the problem of when to limit accession of collections from botanically well-known areas. Herbaria can no longer afford to accession everything and must be more selective. There is a need for greater specialization and agreements between herbaria to avoid too much repetition and for collections by specialists who can collect with discretion. Data are provided to show that there is a tendency to collect common species repetitively and little material of the rare ones.

It is concluded that botanists of the last two decades of this century should make a major effort to complete the inventory of the tropics before it is too late. Future inventory should include the collection of data suitable for the formulation of conservation policies in which we should play a greater part.

INTRODUCTION

The topic before us is vast, and so I can only touch on a few of the most important issues concerning botanical inventory. In a previous paper (Prance 1977, 1978), I outlined the status of floristic inventory of the tropics. That paper summarized the current situation of collecting and of the publication of Floras around the tropics. The situation has not changed much since then, and so I do not need to repeat the data that are readily available. I will present a few additions to update that paper and concentrate on some of the other important aspects that are an integral part of modern botanical inventory, as well as try to provide some references as to where further information can be found.

The fact that many recent reports, particularly about tropical areas, have repeatedly called for an increase in the rate of inventory would indicate that at present it is far from complete. For example, the report of the National Research Council (1980) of the US National Academy of Sciences on "Research Priorities in Tropical Biology" recommended a *greatly accelerated* pace of biological inventory in the tropics. It suggested a combination of regional studies of relatively well-known groups of organisms or those of economic importance with detailed local inventories of other areas. The "World Conservation Strategy" of the IUCN, UNEP and WWF (1980) picked out inventory as one of the three aspects of research needed to improve the capacity to manage the earth. That report points to the need for the inventory to include research on the distribution of ecosystems and species in each country. A report to the United States President by a US Interagency Task Force on Tropical Forests (1980) also stated as one of its goals "Initiation of an international program to inventory, evaluate, classify and catalogue unique forest, plant and animal types". All these reports and many others like them, assert that there is much

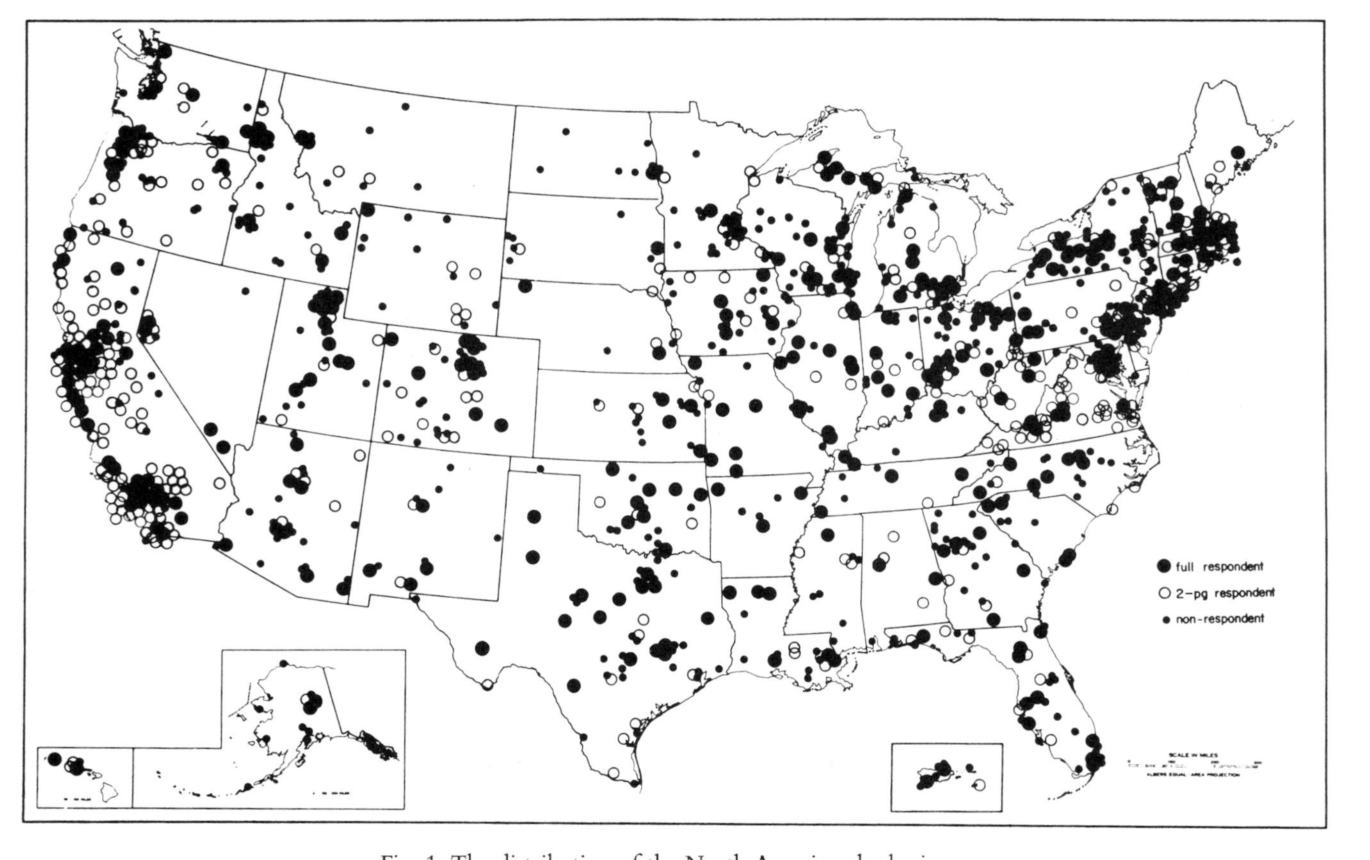

Fig. 1. The distribution of the North American herbaria.

more inventory still to be performed, and would indicate that "completing the inventory" is still an important current topic in plant taxonomy.

The inventory of the world's Flora is very uneven. The reports cited above all refer to the tropics with good reason, because the temperate region Floras are much better known. For example, the recent completion of "Flora Europaea" (Tutin *et al.* 1980) has brought our knowledge of that region up to an excellent level. In Great Britain every species is accurately mapped on a 10 km^2 grid and is recorded in a computerized system (Perring and Walters 1962; Perring and Sell 1968). In contrast it is often impossible to map tropical species even on a 100 km^2 grid because they are so poorly known. Other major temperate regions such as the USSR have their Floras (Komarov 1968; Schischkin 1972, etc.), and efforts are now being renewed to produce a Flora of North America. Even though they still lack a country wide Flora in the United States there are many important regional Floras in progress such as the "Intermountain Flora" (Cronquist *et al.* 1977) and the soon to be published "Great Plains Flora" (Great Plains Flora Association, 1977, and in prep.).

These Floras in the United States are based on material deposited in 1127 herbaria distributed throughout the country (see Fig. 1). In Britain alone there were 267 institutional herbaria listed in Kent *et al.* (1957). In the temperate region, temperate South America is probably the least known with Argentina's 35 herbaria containing 2 615 795 specimens (Zardini 1980), and Chile's 8, only 211 845 specimens (Holmgren *et al.* 1981). Although there are still many interesting and important things to be done for the inventory of the temperate regions of the world, I will concentrate the rest of this paper on the tropics where the need for basic inventory is far greater, and also where a much larger number of the world's species are being threatened with extinction. Rather than repeat a list of the status of inventory around the tropics as was done in Prance (1977), I will concentrate on some aspects and figures that might provide some hints as to where and how we should go about future inventory, leaving out some of the important aspects of inventory that are treated in other papers in this volume such as data storage and handling (Bisby, Chapter 16) and Floras (Heywood, Chapter 19).

THE PROBLEM OF RARE SPECIES IN THE TROPICS

One of the most important problems of inventory was touched upon by Kubitzki (1977) in a paper entitled "The problem of rare and of frequent species". In this paper he used data from recent Flora Neotropica monographs to show the difference in collecting frequency between common species and rare species. The data included among other genera an analysis of *Licania* (Chrysobalanaceae) from my own monograph (Prance 1972a). In the monograph 25 of the 152 species recognized were represented by only one collection, and another 19 by only two. One hundred and four of the species were known from ten or fewer collections and were documented by a total of only 373 of the 2961 specimens consulted. This means that 12.6% of the known material refers to 68.4% of all described species. The twelve most common species of *Licania*, with over 50 collections, were represented by 1886 collections, that is 7.9% of the species represented by 63.7% of the material. Kubitzki used these data and others from *Caraipa*, *Doliocarpus*, *Davilla* and *Eperua* to show the uneven rate of collecting of neotropical species, and to suggest the use of this type of analysis as a way of indicating rare and endangered species.

It is interesting to compare these data with data from Chrysobalanaceae which have accrued over the ten year period since the monograph. I have compiled similar data to that of Kubitzki's for the genera *Hirtella* and *Licania*, both from the monograph (Prance 1972a), and for the past ten years (1973–1982). The data are summarized in Tables I–V.

At first sight the statistics look impressive. In *Licania* there has been a 49.7% increase of available material (1426 collections) and in *Hirtella* a 41.5% increase (1182 collections). This looks good in comparison with Kubitzki's (1977)

Table I. Collection history of *Licania* and *Hirtella* (Chrysobalanaceae) in Prance's (1972a) monograph and subsequently.

	Licania	Hirtella
No. of species in 1972 monograph	152	88
Total no. of collections consulted in 1972	2961	2849
Average no. of specimens/species	19.5	32.37
Dates of specimen collecting up to 1900	14.8%	
1901–1950	50.6%	
1951–1976	34.5%	
No. of monographed species collected 1972–82	109 (72.18%)	66 (73.33%)
No. of collections consulted 1972–82 (previously described species)	1426	1182
No. of new species since 1972	24	11
No. of collections of new species	46	20
Increase in new collections 1972–82	49.71%	41.49%
New average No. of specimens/species (including new species)	24.9	40.71

Table II. Number of new collections of species of *Licania* and *Hirtella* with $\geqslant$ 30 collections since 1972.

No.	Licania	*Distribution*	No.	Hirtella	*Distribution*
116	*L. apetala* (2 vars.)	Amazonia	55	*H. bicornis* (2 vars.)	Amazonia
32	*L. canescens*	Amazonia	48	*H. elongata*	Amazonia
30	*L. egleri*	Amazonia	31	*H. eriandra*	Amazonia
190	*L. heteromorpha* (2 vars.)	Amazonia	47	*H. glandulosa*	Planalto
60	*L. hypoleuca*	Amazonia & Central America	37	*H. gracilipes*	Planalto
			52	*H. hispidula*	Amazonia
30	*L. leptostachya*	Amazonia	37	*H. paniculata*	Amazonia
36	*L. michauxii*	Florida, USA	30	*H. physophora*	Amazonia
32	*L. micrantha*	Amazonia			
80	*L. octandra* (3 subsp.)	Amazonia			
31	*L. pallida*	Amazonia			
38	*L. parvifolia*	Amazonia			
675			*780*		

675 collections of 11 species = 45.68% of total; 780 collections of 10 species = 65.99% of total.

Table III. Number of new collections (1973–82) of the species monographed in 1972.

Collections	*No. of species* Licania	*No. of species* Hirtella
1	15	6
2	18	7
3–10	37	32
10	39	21
	109	*66*

Table IV. Number of collections of new species of Chrysobalanaceae described 1973–82.

Collections	Licania	Hirtella
1	14	7
2	7	2
3	1	—
4	1	1
5		1
10	1	—
	24	*11*

Table V. Poorly-known species of *Licania* and *Hirtella* (either 5 collections total, or no new collections made 1972–82). *Indicate species where vegetation is being destroyed.

Monograph no.	*Species*	*No. cols. 1972*	*No. new cols. 1972–82*	*Distribution*
2	*L. boliviensis*	3	0	Bolivia
3	*L. maritima*	3	0	Colombia-Chocó
7	*L. klugii*	1	1	Peru: Loreto
*9	*L. retifolia*	2	1	Mexico: Guerrero
10	*L. longipedicellata*	2	2	Brazil: Amazonas
14	*L. angustata*	2	0	Brazil: Amazonas
*16	*L. gonzalezii*	2	2	Mexico: Nayarit
*19	*L. maranhensis*	1	0	Brazil: Maranhão
20	*L. fritschii*	4	0	Brazil: Amazonas
29	*L. maguirei*	3	0	Brazil: Mato Grosso
31	*L. cuspidata*	1	0	Colombia: Santa Marta
34	*L. calvescens*	2	1	Colombia: Valle
35	*L. persuadii*	9	0	Guyana
38	*L. albiflora*	2	0	Guyana, Suriname
40	*L. fuchsii*	1	0	Colombia: Chocó
42	*L. foveolata*	2	0	Guyana
46	*L. velata*	1	0	Colombia: Valle
47	*L. subarachnophylla*	2	1	Colombia: Boyacá
48	*L. salicifolia*	1	0	Colombia: Antioquia
49	*L. arenosa*	4	0	Brazil: Goiás
50	*L. silvatica*	1	0	Brazil: Espírito Santo
51	*L. chocoensis*	3	0	Colombia: Chocó

(*cont.*)

Table V (*cont.*)

Monograph no.	Species	No. cols. 1972	No. new cols. 1972–82	Distribution
53	*L. hirsuta*	2	2	Brazil: Amazonas
*54	*L. costaricensis*	1	0	Costa Rica
58	*L. minuscula*	1	1	Colombia: Valle
59	*L. operculipetala*	2	0	Costa Rica
61	*L. arachnoidea*	2	1	Guyana, Brazil: Amazonas
65	*L. latistipula*	4	1	Venezuela: Delta Amacuro
70	*L. glazioviana*	2	1	Brazil: Guanabara
73	*L. irwinii*	3	0	Suriname: Fr. Guiana
74	*L. cyathodes*	4	0	French Guiana
76	*L. silvae*	3	2	Brazil: Amazonas, Pará
84	*L. cymosa*	3	2	Brazil: Pará, Bahia
87	*L. piresii*	3	0	Brazil: Amapá
90	*L. buxifolia*	3	2	Guyana
91	*L. orbicularis*	7	0	Venezuela: Amazonas
92	*L. niloi*	1	0	Brazil: Rondônia
96	*L. glauca*	3	2	Colombia: Valle, Nariño
100	*L. couepifolia*	1	1	Guyana
101	*L. trigonioides*	1	0	Peru: Loreto
102	*L. cordata*	5	0	Venezuela: Amazonas
103	*L. foldatsii*	8	0	Venezuela: Amazonas
104	*L. hebantha*	3	2	Colombia: Amazonas
106	*L. subrotundata*	7	0	Venezuela: Dist. Federal
107	*L. crassivenia*	1	0	Brazil: Amazonas
110	*L. hitchcockii*	2	0	Venezuela: Bolívar
111	*L. sandwithii*	1	0	Guyana
115	*L. bellingtonii*	1	0	Brazil: Rondônia
116	*L. compacta*	1	0	Guyana
118	*L. caldasiana*	3	0	Colombia:(?)
120	*L. microphylla*	1	0	Guyana
123	*L. apiculata*	3	0	Brazil: Amazonas
125	*L. pruinosa*	5	0	Fr. Guiana, Braz. Amapá
127	*L. riedelii*	4	0	Brazil: Planalto
132	*L. spicata*	8	0	Brazil: Rio de Janeiro
137	*L. vaupesiana*	3	1	Colombia: Vaupés
138	*L. bahiensis*	1	1	Brazil: Bahia
139	*L. maxima*	1	0	Brazil: Amapá
143	*L. indurata*	2	0	Brazil: São Paulo
149	*L. amapaensis*	1	2	Brazil: Amapá

(*cont.*)

Table V (*cont.*)

Monograph no.	*Species*	*No. cols. 1972*	*No. new cols. 1972–82*	*Distribution*
150	*L. tepuiensis*	1	0	Venezuela: Bolívar
151	*L. obtusifolia*	1	0	Fr. Guiana
152	*L. roraimensis*	1	0	Guyana
*1a	*L. naviculistipula*	1	0	Brazil: Espírito Santo
3	*H. vesiculosa*	2	0	Colombia: Guainia
4	*H. dorvalii*	1	4	Brazil: Roraima
11	*H. araguariensis*	1	3	Brazil: Amapá
12	*H. cordifolia*	2	0	Venezuela: Amazonas
14	*H. tocantina*	3	0	Brazil: Pará
17	*H. subglanduligera*	1	0	Peru: Madre de Dios
26	*H. deflexa*	1	4	Venezuela: Bolívar
*31	*H. corymbosa*	4	0	Brazil: Espírito Santo
32	*H. pendula*	2	1	Lesser Antilles
33	*H. barrosoi*	8	0	Brazil: Rio de Janeiro
34	*H. leonotis*	1	4	Venezuela: Dist. Federal
37	*H. bahiensis*	3	2	Brazil: Bahia
39	*H. suffulta*	6	0	Brazil: Amazonas, Pará
44	*H. orbicularis*	3	0	Venezuela: Amazonas
45	*H. guyanensis*	8	0	Guyana, Venezuela: Bolívar
46	*H. lightioides*	3	0	Bolivia
47	*H. aramangensis*	1	0	Peru: Amazonas
48	*H. rasa*	2	0	Peru: San Martín
51	*H. angustissima*	8	0	Guyana
57	*H. juruensis*	3	0	Brazil: Mato Grosso
58	*H. kuhlmanii*	2	0	Brazil: Mato Grosso
60	*H. longifolia*	1	0	Brazil: Amazonas
71	*H. pimichina*	4	1	Venezuela: Amazonas
72	*H. subscandens*	1	1	Venezuela: Amazonas
76	*H. adenophora*	3	0	Colombia: Meta
78	*H. fasciculata*	2	0	Brazil: Amazonas
79	*H. couepiflora*	2	2	Brazil: Amapá
80	*H. tubiflora*	1	2	Colombia: Valle
81	*H. floribunda*	8	0	Brazil: Minas Gerais
84	*H. scaberula*	1	0	Brazil: Amazonas
86	*H. enneandra*	1	0	Colombia: Valle
*87	*H. pauciflora*	1	2	Ecuador: Los Rios
88	*H. glaziovii*	7	0	Brazil: Rio de Janeiro, São Paulo

figures of only a 34.5% increase in material of *Licania* in the 25 years from 1951–1976 or a 50.6% increase in the fifty year period 1901–1950. There has, therefore, been a considerable acceleration in the number of collections entering the herbarium during the last ten years. Collecting during that period has also produced a large number of new species from a wide range of places throughout the region (Table IV), 24 new species in *Licania* (15.8% increase), and 11 or 12.5% in *Hirtella*. One hundred and nine (= 72.2%) of the species of *Licania* monographed in 1972 have been re-collected in the intervening period and 66

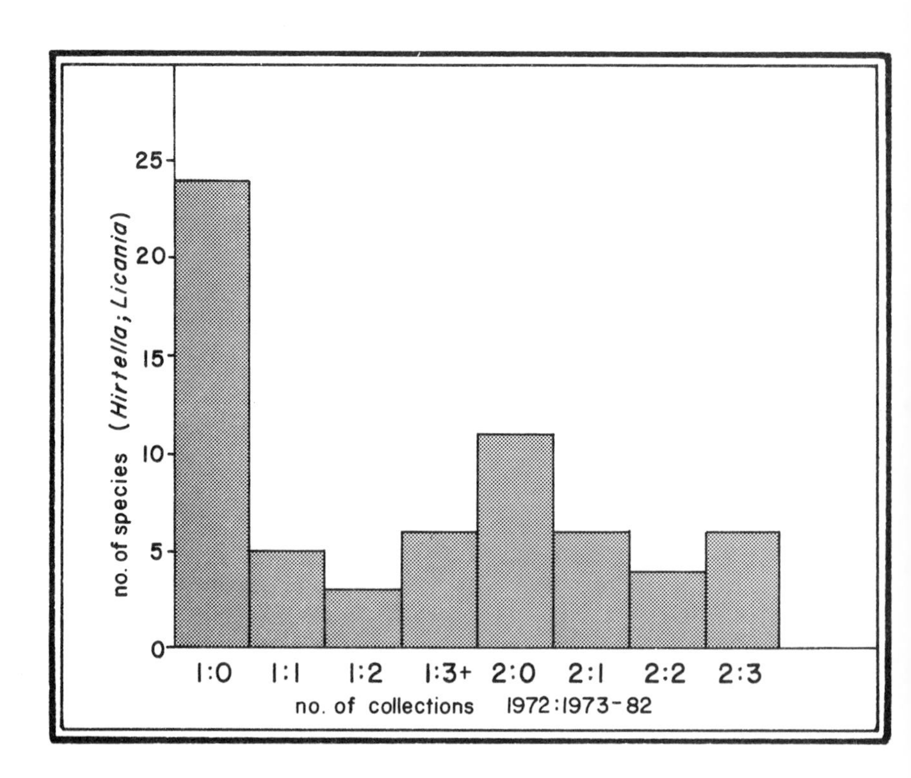

Fig. 2. Collection frequency of poorly-known species of *Licania* and *Hirtella*. The vertical axis represents total number of species and the horizontal axis different combinations of collecting for all species known by 1 or 2 collections in 1972. The first digit is the number of specimens consulted for the monograph of Chrysobalanaceae (Prance 1972a), and the second is the number of collections made since 1972. For example, the left-hand bar represents the 24 species known by a single collection in 1972 for which no new collections have been made and the right-hand bar species known by 2 collections in 1972 and 3 new ones since.

of *Hirtella* (73.3%). Also the average number of collections per species increased from 19.5 to 24.9 in *Licania* and from 32.4 to 40.7 in *Hirtella*.

These averages can be misleading, however. In *Hirtella* 780 collections or 66% of the total made between 1972–1982 belong to ten common species such as the ubiquitous *H. racemosa* Lam. (Table II), and 675 or 46% of the collections of *Licania* belong to only 11 common species. Figure 2 and Table V show the persisting problem of rare species that are not being adequately touched by current collecting. For example, 24 of the 38 species of the two genera which were represented by a single type collection in 1972, remain in the same situation today, and 11 of the 27 species known from two collections have not been re-collected. The data for these and other rare species are given in Table V together with their geographic distribution. This would indicate that some of the key areas for future collecting of lowland species (Table VI) include Pacific Coastal Colombia, Bolivia, Guyana, Amazonian Brazil and Venezuela, and Atlantic Coastal Brazil. The new species described since 1972 are also mostly local rare ones represented by few collections (21 by one, 9 by two and only 5 by more than 2, see Table IV).

Table VI. Geographical breakdown of poorly-known species of Chrysobalanaceae (see Table V), in order of frequency.

Brazil: Amazonas	12
Guyana	12
Brazil: Atlantic Coast States	11
Colombia: Chocó, Valle, Nariño	9
Venezuela: Amazonas	8
Brazil: Amapá	6
Colombia: North & Central	5
French Guiana	4
Venezuela: Bolívar	4
Brazil: Pará	4
Brazil: Mato Grosso	3
Brazil: Planalto states	3
Bolivia	2
Brazil: Rondônia	2
Colombia: Amazonian states	2
Costa Rica	2
Mexico	2
Peru: Loreto	2
Suriname	2
Venezuela: Dist. Federal	2

Most of the species that have not been collected since 1972 are of extremely restricted geographical range. However, this does not necessarily mean that they are very rare in the areas where they do occur. For example, *Hirtella dorvalii* Prance, described from a single collection in 1972, is an extremely abundant species in a very small transition zone of caatinga forest around Caracarai in the sandy area between the Roraima savannas and the Amazon rain forest. *Hirtella tocantina* Ducke, represented by three collections in 1972, turned out to be the commonest Chrysobalanaceae on a three hectare survey plot in the Xingu River region which I made in 1980 (Prance and Daly unpublished). The numerous sterile voucher collections for that inventory are not entered into the statistics in Table V. *Hirtella tocantina* is restricted to the area of the Xingu and Tocantins river basins, where it appears to be very common. The same observations about rarity can be true at the generic level. The least known of all genera of Chrysobalanaceae is *Acioa*, in which three species were known from only 10 specimens in 1972. Since that time six new collections have been made of two of the species. However, *A. schultesii* Maguire, known only from the type in 1972, turns out to be an abundant species on the white sand caatinga forest around San Carlos on the Rio Negro where it has turned up frequently on inventory plots of the ecological survey of that region (Howard Clark *pers. comm.*).

Table VII. Rare species of Chrysobalanaceae in 1972 for which much new material has been collected in the succeeding 10 years.

Species	*to 1972*	*1972–82*	*Distribution*
Licania macrocarpa Cuatr.	2	5	Colombia: Chocó, Valle
L. lanceolata Prance	2	17	Venezuela: Amazonas
L. belemii Prance	1	5	Brazil: Bahia
Hirtella insignis Briq. ex Prance	2	6	Brazil: Bahia
H. adderleyi Prance	2	14	Venezuela: Amazonas
H. standleyi Baehne & Macbr.	2	7	Peru: Loreto
H. lemsii Wms. & Prance	1	7	Costa Rica

The conclusion about differential collecting of rare and common species is borne out by comparing the Chrysobalanaceae activity given above with that of a much rarer family with few common species, the Dichapetalaceae. This was also monographed by Prance (1972b). In the same ten-year period the 884 collections consulted have grown by a further 217, i.e. a 24.5% increase which is considerably less than the 49.7% in *Licania*. The rarer plants are the ones in need of attention.

The conclusion from these data taken from Chrysobalanaceae is that further collection must be carefully planned to include areas where there are high concentrations of the poorly-collected species. At the same time we are reaching a point where a certain discretion must be used about the collection of very common species. The 365 collections of *Hirtella racemosa* Lam. made in the last ten years have added very few new records or interesting facts about this species that is common from Mexico to Rio de Janeiro. Common species such as *Abuta grandifolia* (Mart.) Sandw. and *Hirtella racemosa* cannot continue to fill the limited storage space of our herbaria, or occupy the processing time they take that diverts our attention from the need for new collections of rarer species. This in general means that we are reaching the point where there is a need for specialists or at least well-trained general collectors who are familiar with the local Flora and who can collect with some discretion to do the field work. The areas that are chosen for collection based on data from one large plant family may not be appropriate for another. The danger of this approach is that numerous collections may still be needed for common but highly variable species (polymorphic ochlospecies of White 1962).

One further important aspect of these data is given in Table V, that is species that are in areas where the vegetation is being devastated. Some of the most obvious are indicated by asterisks in the table. In these cases part of the reason for the lack of recent collections is probably the disappearance of the native habitat of these species. Areas where this is true are of prime importance for rapid further collection to learn as much about these rare species to help avoid their becoming extinct, and to provide the data which are needed to plan conservation areas for them.

THE PROBLEM OF NEGLECTED AREAS

The same kinds of discrepancies found for collections of rare and common species are seen in the comparison of collecting in geographical regions. Vink (1981) recently updated the collectors indices for some parts of the Flora Malesian area. These density indices, showing the number of herbarium collections per 100 km^2, were referred to and discussed in Prance (1977, 1978) where similar indices for Central America were also provided. The updated Malesian figures are given in Table VIII. They illustrate one of the points that also came from the data on Neotropical Chrysobalanaceae, i.e. that in the tropics there are many local inequalities in collecting. The density index for the island of

Borneo is at present 32 specimens per 100 km², but this apparently high number is obtained because of intensive collecting in Sarawak, Brunei and Sabah which together have a density index of 87. These small territories occupy only 26.5% of the total area of Borneo. The larger part of the island, Indonesian Borneo,

Table VIII. Density index for selected areas in Malesia, number of herbarium specimens per 100 km² (from Vink 1981).

	Area km[1]	*1950*	*1972*	*1980*	*% increase 1972–80*
Sumatra	479 513	18 (88 000)	21 (97 000)	22 (106 000)	7
Celebes	182 870	18 (33 000)	19 (34 000)	23 (42 000)	23.5
Borneo (whole island)	739 175	12 (92 000)	26 (194 000)	32 (228 000)	17.5
Kalimantan	542 815	7.3 (52 000)	9.2 (145 000)	10.5 (171 000)	17.9
Sarawak, Brunei Sabah	196 360	26 (39 000)	73 (49 000)	87 (57 000)	16.3

or Kalimantan, has a density index of only 10.5. This shows dramatically where there is a need for further inventory. Collecting expeditions should concentrate on Kalimantan and ignore Sarawak, Brunei and Sabah until the Kalimantan density index is up to 87. On an island with a Flora of an estimated 10 000 species a density index of 10.5 is a far from adequate inventory.

The collecting in Malesia over the past eight years has contributed 7–18% more material depending on the area. This indicates considerable botanical activity, but rather less than the neotropical data where *Licania* specimens increased by 50% in ten years. It is interesting that the highest increase was in the Celebes, one of the areas pointed out as in need of much further inventory in Prance (1977). In the neotropics Bolivia remains in a similar uncollected state. However, recent progress there includes the establishment of a collecting program by Missouri Botanical Garden, and of an ecological survey by the University of Göttingen.

WHAT IS INVENTORY?

The considerations above deal solely with the gathering of conventional herbarium material, and traditionally botanists have tended to regard this as the only aspect of inventory. However, there are many other important aspects that need to be included. For example, the National Research Council report (1980) called for botanists to "emphasize collecting specimens in part by unusual methods, such as freeze drying or preservation in fluids". It also called for an emphasis on biosystematics and evolutionary biology and collecting for seed banks and botanical gardens.

(1) Special collections

With the impending extinction of many tropical species, the need for these special collections is even more urgent. It would have been impossible to resolve the taxonomy of the Neotropical Lecythidaceae (Prance and Mori 1980) without the aid of the vast collection of pickled flowers which we amassed over a ten-year period. Many species of Lecythidaceae produce one large fleshy flower per inflorescence at any one time. The squashed pressed androecia do not revert to their original shape when boiled. However, androecium shape is critical for the taxonomy and for the understanding of their floral biology. Many other groups such as the Orchidaceae, Lentibulariaceae and the saprophytic Gentianaceae are very difficult to study without liquid preserved material. There is also a great need for fixed buds for chromosome studies, for wood samples, for material for leaf anatomy and nodal anatomy studies and especially for sufficient extra material for chemical studies. These are all an integral part of the inventory and their importance should be stressed to our students and colleagues who are active in the field.

2) Biological data

Data on the various biological relationships of plants are a particularly important part of inventory as they often furnish some of the best arguments for conservation. Botanists involved in the completion of the inventory should be aware of the complexity of interrelationships between organisms and gather data on pollination and floral biology, dispersal of diaspores, predator defense mechanisms and mycorrhizal relationships of the plants which they collect. For example, the Brazil nut (*Bertholletia excelsa* Humb. and Bonpl.) is dependent upon its euglossine bee pollinators and its agouti dispersal agent for survival in the forests of Amazonia. In the course of basic inventory work it is often possible to furnish useful biological data about insect visitors to a flower, flower scent, the phenology of flower production, the presence or absence of nectar, etc. Detailed studies, such as those from Costa Rica of Bawa (1974) on breeding systems of forest trees, and of Frankie (1975) and Frankie *et al.* (1974)

on pollination and phenology are needed. The extensive work on biological relationships carried out in Costa Rica by many investigators needs to be extended around the tropics as the inventory is completed.

(3) Quantitative inventory

Much of the early quantitative inventory of tropical forests was carried out by foresters and international agencies. For example, in Amazonian Brazil the basic inventories upon which many decisions were made, are those of Heinsdijk, who coordinated an FAO inventory that covered an area of 15 000 000 hectares (Heinsdijk 1960; Heinsdijk and Miranda Bastos 1965). This work relied heavily on local names given by field workers and so in many large woody plant families the results bear little relationship to botanical species. For example *matamata* is the local name for over fifty species of *Eschweilera* (Lecythidaceae), *abiu* for many species of *Pouteria* (Sapotaceae), etc. In addition only 260 of the 348 local names provided were identified to species. Heinsdijk (1960) himself pointed out these limitations and called for greater collaboration with botanists. Recently many more field botanists have followed the example of Cain *et al.* (1956) who made an early numerical inventory in the region, collecting much herbarium material so that more accurate identifications were possible and the sterile material could be sorted into morphological species.

This type of quantitative inventory provides many useful data about the interrelationships of the different species and about forest diversity. In the last few years there have been many useful inventories around the tropics, but there is a need for many more based on accurate botanical identification and identical numerical parameters. A few examples of successful inventories by botanists are Ashton (1964) in Brunei, Black *et al.* (1950) and Pires *et al.* (1953) in Belém, Brazil, Jones (1955, 1956) in Nigeria, Grubb *et al.* (1963) in Ecuador, Knight (1975) in Panama, Paijmans (1969) in New Guinea, Prance *et al.* (1976) in Manaus, Brazil, and Schulz (1960) in Suriname. Gentry (1982) gave a review of many previous inventories as well as many new data from his own inventories. His data are presented in the framework of a discussion of species diversity and he shows a strong correlation between rainfall and tropical rain forest species richness. This correlation is backed up by my own data. For example, the hectare in Manaus studied by Prance *et al.* (1976) contained 179 species $>$15 cm diameter per hectare, that which we studied in the Xingu River region (Prance and Daly unpublished) had 118 species $\geqslant$10 cm diameter per hectare, and that of the much drier region south of Amazonia (Prance and Schaller 1982) contained only 36 species $\geqslant$15 cm diameter. The species diversity is greater in the areas of greater rainfall. Quantitative inventories furnish important data about diversity, one of the subjects that is most important for conservation. For example, two well-established diversity gradients in tropical vegetation have been established from recent research: the first, referred to above, greater diversity with increasing precipitation (Gentry 1978, 1981, 1982), and the second, decreasing species diversity with increased altitude (Gentry 1979, 1981). However, Janzen *et al.* (1976) and Scott (1976) suggested a rather different picture with a "mid-altitude bulge" in diversity of insects and leaf litter herps respectively.

For a continuation of this type of inventory, one of the most important aspects is standardization of data. The minimum diameter in past inventories has varied greatly because many have been interested only in the larger trees of commercial timber size. For example, Aubréville (1961) and Takeuchi (1962) and many others used trees >10 cm diameter; Prance *et al.* (1976), 15 cm; Ayres and Milton (1981), 19.1 cm (60 cm circumference); Letouzey (1968), 20 cm; Heinsdijk (1960), 25 cm; the Centre Technique Forestier Tropical in Gabon, 60 cm; and the Brazilian Projeto RADAM, 100 cm/π! In the lower diameter classes many interesting and important species are included that do not appear in inventories of 25 cm diameter or more. Most recent inventories have used the 10 cm minimum diameter criterion. This should be the standard for future inventories of this type. Also the theoretical bases for the various approaches to quantitative inventory of this type have varied considerably, and are quite muddled.

A useful aspect of quantitative inventory is that it can and should be used to set up permanent study plots in different tropical vegetation types, especially in rain forest. A plot with a good basic inventory of properly identified and permanently labelled trees can lead to many further important biological studies. For example, the hectare of rain forest which we inventoried near Manaus, Brazil (Prance *et al.* 1976) was subsequently used for many other studies such as a survey of the mycorrhizal associations of the trees (St. John 1980), sand fly ecology and diaspore dispersal. Those carrying out the inventory should set up plots which also lend themselves as study areas for thesis projects. Many interesting data will be generated because of the comparisons that can be made when research in many different disciplines is carried out in the same location.

In addition to this quantitative study of selected hectares or transects, another useful contribution to the recent botanical literature has been that of the publication of various mini-Floras. The detailed survey of Barro Colorado Island (Croat 1979), an area of 15.6 km^2 with 1316 species, and of the Río Palenque Science Center (Dodson and Gentry 1978), 167 hectares with 1100 species, have provided much useful data that are being widely used. The "Flora of Avila" by Steyermark and Huber (1978) is another well-produced Flora of a limited area. Such detailed studies furnish data from which broad extrapolations can be made. Both the quantitative inventories and the mini-Floras provide the type of data that are essential for the formulation of conservation policy. Completion of the inventory should include many more Floras of such small areas.

These intensive local Florulas based on field work are producing many useful data such as the recognition of clearly related sympatric but biologically distinct species that are often overlooked or united by large-scale monographers, as for example, the species of Meliaceae in the much-studied Flora of La Selva, Costa Rica. A well-studied area also furnishes advantages similar to specialist collecting, in that every species is known and anything unusual can be picked out for special study.

Another important aspect for complete inventory is the need for year-round collecting. This can be done most effectively by either local botanists or visitors who spend extended periods in, rather than flying visits to, the tropics. Year-long residence enables one to get material of plants that flower for extremely short periods (the big-bang phenology of Gentry 1974) and for other phenological specialists such as trees that flower only in some years and not in others. This extended work can be combined with work on local Florulas. It is often as important to find out what species are not in a certain region as those that are present. Only extensive study can show what is absent from a region.

Some of the collecting techniques that have become more important in recent years include tree bicycles, telescopic lightweight clipper poles and the use of professional tree climbing people. All three aspects are now used by many collectors, but by no means universally. Tree climbers are a rare commodity in many regions and we need to find ways to make such a work a viable livelihood.

(4) Plant groups to be collected

The last section dealt mainly with the collection of trees in forest inventories. However, it is most important to maintain a balance and collect all types of plants from all habitats and in all synusiae. Since all plants in any ecosystem are closely connected through their interactions, their history in a region and the future effect upon it, it is essential that inventory take them into account. The inventory has been uneven and there are various neglected groups and habitats.

All synusiae in a forest are important, yet the trees have received far more attention than the shrubs, herbs, epiphytes, climbers, saprophyttes or parasites. Gentry (1982), in the study of forest diversity cited above, included lianas and was careful to stress their importance as components of tropical forests. This is one of the few studies that has included lianas, a much neglected group. Rollet (1974) compiled a list of 191 local names of lianas in the Guyana region of Venezuela and was able to identify only 42 to species, 64 to genus and 112 only to family. The lianas are often neglected in inventory as are the epiphytes, another important component of the forest. Epiphytic ferns, mosses, hepatics, and lichens are an important part of the tropical vegetation. Holdridge (1972) also emphasized the importance of lianas when he pointed out that vegetation type could be indicated by densities of epiphytes and lianas. How many botanists bother to collect the epiphyllous members of these groups? When you do it it can get you into trouble. I was once accused by the Curator of a small South American herbarium of leaving only rubbish in his herbarium. He had picked up the package of leaves with epiphyllous plants in them and assumed the leaves were the specimens of trees left for his herbarium! Studies of the non-vascular plants and fungi have been far from adequate in the tropics. The reason is often that specialist knowledge is required to make adequate specimens of these groups. Our training of field botanists needs to include more details about how to look for and how to prepare specimens of these groups.

It is significant that many large tropical families are the greatest problems in inventory work. It is hard to obtain accurate and rapid identifications in such families as Annonaceae, Burseraceae, Euphorbiaceae, Lauraceae, Myrtaceae, Rubiaceae, Sapindaceae and Sapotaceae. A greater effort needs to be made to train people who can cope with these families and not leave the undetermined material to accumulate for many years in the basements of herbaria. Difficult-to-collect plant families such as Palms, Araceae, Bromeliaceae or Pandanaceae have been severely neglected, and badly collected. We need to encourage future collectors to make adequate collections and sufficient field notes when they gather palms.

A careful monitoring needs to be kept on the inventory from different habitats. Some habitats are neglected and others are studied carefully. The need for work along altitudinal gradients in the tropics is one of the most important. Most collectors rush from the forested lowlands to the top of the mountains to find the high altitude endemics (the cloud forest syndrome described by Edgar Anderson 1952!). The slopes in between have a rich Flora with most important records in terms of the history of the region. Far more work needs to be done on these slopes. The inventory that is complete will have included material from all possible habitats from the benthos to the mountain snow line.

Small habitat islands within a larger vegetation type often tend to be overlooked. For example, small patches of forest on white sand in Amazonian Peru have yielded the greatest number of new species in recent collecting, even near Iquitos, the centre for all collecting in upper Amazonia (Gentry *pers. comm.*). Many collectors do not even recognize the different types of indundated forest (Prance 1979b) which have quite different species complements. For future collecting there is a need for this to be done by people who understand the local ecology. The different types of tropical forest often look the same to inexperienced botanists who have been trained in the temperate region.

(5) Remote sensing

The potential of remote sensing has not been fully realized as part of the botanical inventory of the world, in spite of the publication of various volumes dealing with its application to biological sciences (e.g. National Research Council 1970; Billings *et al.* 1976). One of the crucial facts that must be known for conservation is the extent of deforestation of any particular habitat. In the case of tropical rain forest there have been wild guesses of greatly varied figures, often based on the bias of the person or organization producing the statistics. For example, a recent Brazilian government claim that only 1.5 % of the Amazon forest had been cut by 1978 must be regarded as utterly false. Yet it is reputedly based on satellite data (Tardin *et al.* 1979, 1980). The problem is the inability to distinguish between primary forest and recently regrowing secondary forest with the present grid size of satellite photographs. This problem was discussed by Fearnside (1982) who pointed out the actual rapid rate of deforestation in Amazonia. A review of the situation is given in Myers (1980). The importance of remote sensing is not only to determine the amount of deforestation, but especially to monitor the rate at which it is occurring. Conventional aerial photographs have sometimes been used for this purpose, but there are few places in the tropics where aerial photographs have been taken more than once. In addition cloud cover is a constant problem. One notable study using aerial photographs was the FAO study of the Azuero Peninsula in Panama (FAO 1972) compared in 1956 and 1964 and showing that in the eight-year period 26% of the forest had been cleared by agricultural settlers.

Side-looking air-borne radar (SLAR) has furnished another important technique which has the advantage of passing through cloud cover. It can certainly be used to distinguish major vegetation types as has been shown by SLAR projects in Brazil and Colombia (Projeto RADAM, 1973–1975, Projeto RADAMBRASIL 1975–1978 and Proyecto Radagrametrico del Amazonas

1979) where detailed vegetation maps have been produced from their radar images.

Satellite data from the ERTS and Landsat satellites have been used since they orbit the entire earth in a few days. Cloud cover has hampered the production of comparative data in many areas and the resolution of a ground area of 70 m has limited its use. However, later Landsat satellites will considerably reduce the size of the interpretable area and increase the importance of satellite data for botanical inventory. There is a need for a review to be prepared that documents the resources available to biologists in the area of remote sensing and its application to the inventory of tropical forests. Remote sensing will almost certainly become an increasingly important component of inventory in the future.

Myers and Benson (1981) and Myers (1982) emphasized the importance of large-scale, colour arerial photographs as a useful aid to botanists working in the species-rich tropical rain forest. Aerial colour slides at a scale of 1:2000 were found to be extremely reliable for the identification of over 50% of the tree species in plots of the Queensland rain forest. It is an additional method which should prove applicable in many small-scale sample plots used for ecological research. Further details of this technique are also found in Aldrich *et al.* (1959) and Spencer (1974).

The use of remote sensing techniques as part of botanical inventory will require more botanists trained in these techniques.

THE WRITTEN INVENTORY AND DATA PROCESSING

The completion of inventory obviously involves adequate presentation and dissemination of the results. This is far from complete and there are many large gaps in the floristic and monographic record of the inventory. Space does not permit a detailed review of this aspect of the inventory, but another paper in this volume deals with Floras (Heywood, Chapter 19). Also I have reviewed the major tropical Floras in Prance (1977, 1979a). A notable addition since these reports is the inauguration of the "Flora Mesoamericana" project as a joint effort of Mexican, Central American, North American and British botanists. It will cover an important part of the Neotropics that is now adequately collected for such a project. Furthermore, it will provide a major Flora in the regional language, Spanish, which is another important step forward.

Another most important aspect of completion of the inventory is the role of electronic data processing. This will certainly play an increasing role in facilitating the management of data. Since this is the subject of another paper in this volume (Bisby, Chapter 16), I will mention only its importance. Many herbaria are using word processing equipment extensively, and a few herbaria have their herbarium label data fully computerized including that of Bogotá, Colombia (Forero and Pereira 1976). This will become an increasingly important part of the inventory as techniques improve and equipment becomes cheaper. A future problem will be the inter-compatibility of the various data storage systems being developed independently.

THE DEPOSITION OF SPECIMENS

The continuation of the inventory will result in the collection of many more herbarium specimens and, it is hoped, associated liquid preserved collections, wood samples, chromosome fixations, etc. As any herbarium curator can attest, these specimens are bulky and costly to maintain, and curatorial staffs are small. Planning for the completion of the inventory must also include the preparation of adequate housing for the inventory. Each country needs large herbaria to house that part of the inventory in which it is involved. A recent trend in the United States has been for some small university herbaria to close and deposit their collections with one of the large ones. For example, recently at the New York Botanical Garden we have received the mycological collections of the Carnegie Museum and of the University of Utah and the herbarium of Wesleyan University. The large herbaria must be ready to receive this sort of accession as well as the recently collected material.

The data from the Chrysobalanaceaean genera *Hirtella* and *Licania* given above indicate that, even in the greatly undercollected Neotropics, we must begin to be selective in collections and concentrate on the rarer species. This is even more true in better collected areas. There will be a need to work out better specialization programs between different herbaria on an international basis to avoid too much duplication but to ensure enough replication so that material is housed in enough places to minimize the effect of a disaster in one herbarium and to make material available in the major regions where study is underway. The agreement between the two major herbaria in London, Kew and the British Museum, is a good example of what can and should be done. Their arrangement has divided the responsibility for herbarium material of the world both by botanical groups and geographic areas based on the historical specialities at both institutions. This is a good start that should be copied elsewhere. The uniting of the herbarium material of the Gray Herbarium and the Arnold Arboretum of Harvard University, and of the California Academy of Sciences herbarium with that of Stanford University are examples of steps in the right direction, and Solem and Burger (1982) outlined specialization policy at the Field Museum in Chicago. The Field Museum in Chicago and the Missouri Botanical Garden in St. Louis came to a specialization agreement resulting in the depositing of their joint holdings of mosses at Missouri and liverworts at Chicago. Missouri has also sent all their algal, fungal and lichen collections to herbaria active in these areas. Such specialization is leading to more defined research programs at each institution but also has the obvious disadvantage that it is cutting future research possibilities of the institution rather drastically. However, the principal message of this report is that the inventory is far from complete, and so in addition to consolidation, we must prepare space and provide funds to house the important new specimens that are yet to be collected.

The recent tendency to consolidation of herbaria dealing largely with tropical material in several countries would indicate that tropical inventory can be much more efficiently carried out from a few large herbaria that are fully equipped to process and identify the collections rather than from the smaller and middle-sized herbaria. If the tropical vegetation is to be studied sufficiently before it disappears, there is a need to work with the greatest efficiency possible. Countries involved in inventory of the tropics would do well to support a few centres of excellence in inventory and specimen storage and processing rather than fracture the effort into a lot of minor efforts that are not properly focussed. At the same time this should be backed up by interdisciplinary ecologically orientated studies, such as those carried out in the Manu National Park in Peru, and by the World Wildlife Fund near Manaus, Brazil on the minimum critical size of conservation areas, in which participants from the smaller institutions will be an essential part.

There is a special problem in many of the tropical countries where herbaria are inadequately curated or protected largely through the lack of funds available in developing areas. The work of the UNESCO Committee that produced a "Manual for Tropical Herbaria" (Fosberg and Sachet 1965) has been helpful. It is a pity that their work stopped at that point. Laws in many tropical countries, quite rightly, require the deposition of specimens in local herbaria to make them available for study by local botanists. I realized how difficult it is to carry out a good systematic revision in Brazil when I began to train Brazilian graduate students in systematics. Their theses suffered because of the lack of type specimens that are all in the historic herbaria of Europe and the USA. The problem is not the demand that material be left in the country of origin, but what happens to the material after the collector leaves the country. It is depressing to see half of one's collections remain in an insect infested herbarium or to be left unmounted and undistributed for many years. This means that outside collaborators must be involved in the improvement of the local herbaria where they work. For example, the small rat and roach infested herbarium of the Instituto Nacional de Pesquisas da Amazônia (INPA) in Manaus, that existed when I first arrived in that city in 1964, is now housed in its own modern air-conditioned building with a regular fumigation program. This involved much time coaxing directors and talking to influential people in the capital, Brasília, but the result is a herbarium where there is no reluctance to leave specimens. Even there the duplicates of one expedition were burned in a fire in a separate drying room. The fire was caused by members of a North American expedition, not Brazilians!

The herbarium of the Amazonian University of Peru in Iquitos is another good example of the collaboration between foreign botanists and local institutions. The development has been rapid from a university biology library of six books and a single half herbarium case of unidentified plants to a modern well-housed, well-curated herbarium of 30 000 specimens with a good taxonomic library. All this in a period of only six years.

The accelerated inventory of the tropics must also include assistance to the local regional herbaria that help us so much, and in which much of the results of the inventory are deposited. Some good guidelines for working in Neotropical countries were recently produced by Mori and Holm-Nielsen (1981), but they were not concerned with the curatorial problems of the tropical herbaria.

A recent publication that updates the manual of Fosberg and Sachet (1965) is a FAO publication on plant collecting and herbarium development by Womersley (1981). This book contains much elementary information about collecting and herbarium technique since it is directed towards anthropological, ecological and other non-taxonomic collectors and updates another paper by Fosberg (1960) with the same purpose. However, it also contains much information of great use to all tropical systematists such as special instructions for the collecting of various difficult-to-collect groups such as *Pandanus*, bamboos, Araceae and aquatic plants. It also includes instructions for making many of the special collections recommended by the National Research Council (1980) report, such as liquid preserved collections, wood specimens and the use of chemical preservatives like paraformaldehyde. Womersley's book will be useful material for those concerned with improving conditions in tropical herbaria and will encourage more systematists to collect better specimens and thus become a more important part of the task of inventory. For further references on plant collection and herbarium curation, the bibliography of Hicks and Hicks (1978) provided many useful references.

The curatorial situation of the tropical herbaria is poor, but even in the major herbaria of the world there is a great shortage of curatorial staff to cope with the processing of the inventory (Table IX). Curators range from 1–13 per million specimens, with an overall average of 5.3 which seems a very low

Table IX. Data on the curatorial staff of the world's 21 largest herbaria, taken from "Index Herbariorum" (Holmgren *et al.* 1981).

Institution	*Date founded*	*No. of specimens* (10^6)	*No. of Curators*	*Curators/ Million*
Muséum National d'Histoire Naturelle, Phanérogamie and Cryptogamie, Paris	1635	10.5	38	3.6
Royal Botanic Gardens, Kew	1853	5	47	9.4
Komarov Botanical Institute, Leningrad	1823	5	31	6.2
Conservatoire et Jardin botaniques, Geneva	1817	5	12	2.4
Harvard University, Cambridge	1858	4.5	19	4.2
New York Botanical Garden, Bronx	1891	4.3	12	2.8
US National Herbarium, Smithsonian Institution, Washington, DC	1868	4.1	17	4.1
British Museum (Natural History), London	1753	4	25	6.3
Institut de Botanique, Montpellier	1890	4	23	5.8
Naturhistoriska riksmuseet, Stockholm	1758	4	12	3.0
Université de Lyon, Villeurbanne	1925	3.8	*	*
Università de Firenze, Florence	1842	3.5	10	3.1
Naturhistorisches Museum, Vienna	1807	3.5	4	1.1
The Manchester Museum, Manchester	1821	3	*	*
Missouri Botanical Garden, St. Louis	1859	2.9	10	3.4
Rijksherbarium, Leiden	1575	2.5	23	9.2
Friedrich-Schiller Universität, Jena	1895	2.5	8	3.2
University of Helsinki, Helsinki	1750	2.4	10	4.2
Field Museum of Natural History, Chicago	1893	2.4	10	4.2
Botanical Museum, Lund	1770	2.3	30	13.0
University of Uppsala, Uppsala	1785	2.2	24	10.9
			Average	*5.3*

* information not available.

number in light of the task that is before us and the number of specimens which a person can handle in a year. There is certainly a need for those responsible for the continuation of the inventory to work towards a more adequate staffing of the herbaria that will receive the specimens.

CONSERVATION AND INVENTORY

I have alluded to conservation in several of the previous sections because it is progressively becoming a more important part of inventory. Inventory work is now involved in conservation in two ways: (1) the salvage of specimens from areas that are about to be destroyed, and (2) the collection of the right sort of data that can be used for conservation programs and save species and habitats from falling into the first category. Conservation is the theme of Chapter 20 and so I do not need to treat it in detail. However, no contemporary discussion on inventory would be complete without mentioning conservation.

The Global 2000 Report (Barney 1980) stated that "of the 3–10 million species now present on earth, at least 500 000–600 000 will be extinguished during the next two decades". The state of the inventory is such that many of these soon to be lost species are not even named. With tropical moist forest disappearing at 20 hectares per minute around the world (Myers 1980), the role of the botanical collector in conservation activity is vital and urgent. One of the current needs is for a program to establish a list of priority areas for immediate concentration of effort and resources where tropical rain forest and other habitats are in immediate danger of total extinction, such as the Atlantic coastal rain forests of Brazil. Some of the areas where collecting efforts are needed are not necessarily the areas where the collection sample is most inadequate, because many of the poorly-known areas are under little threat. It is much more urgent to work in areas about to be clear cut for agriculture, flooded by a hydroelectric project or turned into an oil field, as is some of Ecuadorian Amazonia.

The collectors responsible for the continuation of the botanical inventory of the globe need to be aware of the current conservation and destruction issues of their region. When there is an area about to be flooded by a hydroelectric project, every effort should be made to discover as much as possible about the biology of the species in the area and to rescue as many as possible. In this case inventory will involve the transplanting of living material to botanic gardens and other germplasm collections. Such a project has recently taken place in Amazonian Brazil in the Tocantins River region where the Tucuruí dam is being built, but, alas, it was started too late to be really effective. A three year survey by biologists of the Museu Goeldi in Belém and INPA in Manaus has revealed a lot about the plants and animals of the region. We have also been involved in two botanical expeditions to the area as well as in an environmental impact study.

Much basic conservation information such as the compilation of threatened and endangered plant species lists is largely based on data from the herbarium sheets of a region. Thus, the herbarium collections furnish many useful data to the conservationists. However, inventory that also includes observations on interrelationships such as pollination, dispersal, animal protection of plants or mycorrhizal associations, provides many more data for the conservation planner. Knowledge about the interrelated ecosystem gives the National Park planner a much stronger case than just a list of endangered species. The quantitative data from forest inventories are also an essential part of this process.

I will give just one example of this type of inventory work from Atlantic Coastal Brazil where the rain forest has completely disappeared (Mori *et al.* 1981; Sick and Teixeira 1979; da Vinha *et al.* 1976). This forest once occupied about one million square kilometres, and extended from Rio Grande do Norte to Rio Grande do Sul in a strip ranging from a few to 160 kilometres wide. This forest has been reduced to a small remnant (perhaps less than 5% of the original) by sugar, coffee and cocoa plantations. After two years of general herbarium collecting in this region my colleague Scott Mori felt the need for more detailed inventory in order to try to convince the authorities of the necessity to conserve as much of the remnant as possible. In a program financed by the World Wildlife Fund, US, we first analysed the distribution of 127 tree species with at least part of their range in the coastal rain forest. Data for this were taken from the "Flora Neotropica" monographs. The results showed that 53% of these trees are endemic to the coastal forest, 11.8% endemic to the region plus part of the adjacent Planalto of Central Brazil, 7.8% disjunct with Amazonian rain forest and 26% widespread. This high endemism, in view of the rapid destruction of the area, emphasized further the need for increased preservation of the remaining areas. This was followed by a field trip to carry out forest inventories of selected patches of forest to compile quantitative data. Interestingly, we found 180 species per hectare of $\geqslant$10 cm diameter which is almost identical to that of Manaus, 179 species/ha reported in Prance *et al.* (1976). These data are now being combined with those of ornithologists (Sick and Teixeria 1979) and primate specialists to make a compelling case for conservation. This type of botanical inventory will be increasingly needed in the future and it is imperative that systematic botanists furnish the data that can help to preserve the plants that they study rather than just collect specimens.

In conclusion it is obvious that the greatest need for completing the inventory is extensive and accelerated collection of the tropics, especially in areas that are under threat of destruction. Contemporary inventory will provide material suitable for the application of modern methods such as chemotaxonomy, electronic microscope work and chromosome studies. It will also be deeply involved in conservation issues and provide the data necessary for the formulation of conservation policy.

ACKNOWLEDGEMENTS

I thank Douglas Daly, Al Gentry and Patricia K. Holmgren for a critical reading of an earlier draft of this paper and Frances Maroncelli for typing. Knowledge about botanical inventory was made possible by various grants from the National Science Foundation for field work in the Neotropics.

REFERENCES

Aldrich, R. C., Bailey, W. F. and Heller, R. C. (1959). Large scale 70 mm color photography techniques and equipment and their application to a forest sampling problem. *Photogramm. Engny* **25,** 747–754.

Anderson, E. (1952). "Plants, Man and Life", 251 pp. Little, Brown & Co, Boston.

Ashton, P. (1964). Ecological studies in mixed dipterocarp forests of Brunei State. *Oxford Forest Memoirs* **25,** 110 pp.

Aubréville, A. (1961). "Etude écologique des principales formations végétales du Bresil et contribution à la connaissance des fôrets de l'Amazonie Brésilienne", 268 pp. Centre technique forestier tropical (CTFT), Nogent-Sur-Marne.

Ayres, J. M. and Milton, K. (1981). Levantamento de primatos e habitat no Rio Tapajós. *Bol. Mus. Paraense Emílio Goeldi. Zool.* **111,** 1–11.

Barney, G. O. (1980). "The Global 2000 Report to the President of the U.S.". Pergamon Press, New York.

Bawa, K. (1974). Breeding systems of tree species of a lowland tropical community. *Evolution* **28,** 85–92.

Billings, W. D., Golley, F. B., Lange, O. L. and Olson, J. S. (eds) (1976). "Remote Sensing for Environmental Sciences", 367 pp. Springer Verlag, Berlin, New York.

Black, G. A., Dobzhansky, T. and Pavan, C. (1950). Some attempts to estimate species diversity and population density of trees in Amazonian forests. *Bot. Gaz.* **111,** 413–425.

Cain, S. A., Castro, G. M. O., Pires, J. M. and da Silva, N. T. (1956). Application of some phytosociological techniques to Brazilian rain forest. *Amer. J. Bot.* **43,** 911–941.

Croat, T. B. (1979). "Flora of Barro Colorado Island". Stanford University Press, California.

Cronquist, A., Holmgren, A. H., Holmgren, N. H., Reveal, J. L. and Holmgren, P. K. (1977). "Intermountain Flora", Vol. 6, 584 pp. Columbia University Press, New York.

Department of State (1980). "The World's Tropical Forests: A Policy, Strategy and Program for the United States. Report to the President by a U.S. Interagency Task Force on Tropical Forests". Dept. of State publication 9117.

Dodson, C. and Gentry, A. H. (1978). Flora of the Río Palenque Science Center. *Selbyana* **4,** 1–638.

FAO (1972). "Reconocimiento general de los bosques e inventario detallado de Azuero,

por Centre technique forestier tropical, FO: SF/PAN 6". Informe tecnico FAO 12, Vol. 4, documentos, cartográficos.
Fearnside, P. M. (1982). Deforestation in the Brazilian Amazon: How fast is it occurring? *Interciencia* **7,** 82–88.
Forero, E. and Pereira, F. J. (1976). EDP-IR in the national herbarium of Colombia. *Taxon* **25,** 85–94.
Fosberg, F. R. (1960). Plant collecting as an anthropological field method. *El Palacio* **67**(**4**).
Fosberg, F. R. and Sachet, M.-H. (1965). "Manual for Tropical Herbaria". *Regnum vegetabile* No. 39, 132 pp. Internat. Bureau for Plant Taxonomy and Nomenclature, Utrecht, Netherlands.
Frankie, G. W. (1975). Tropical forest phenology and pollinator plant coevolution. *In* "Coevolution of Animals and Plants" (L. Gilbert and P. H. Raven, eds) pp. 192–209. University of Texas Press, Austin.
Frankie, G. W., Baker, H. G. and Opler, P. A. (1974). Comparative phenological studies of trees in tropical wet and dry forests in the lowland of Costa Rica. *J. Ecol.* **62,** 881–919.
Gentry, A. H. (1974). Coevolutionary patterns in Central American Bignoniaceae. *Ann. Missouri Bot. Gard.* **61,** 728–759.
Gentry, A. H. (1978). Floristic knowledge and needs in Pacific Tropical America. *Brittonia* **30,** 134–153.
Gentry, A. H. (1979). Extinction and conservation of plant species in Tropical America: a phytogeographical perspective. *In* "Systematic Botany, Plant Utilization and Biosphere Conservation" (I. Hedberg, ed.) pp. 110–120. Almquist and Wiksell, Stockholm.
Gentry, A. H. (1981). Distributional patterns and an additional species of *Passiflora vitifolia* complex: Amazonian species diversity due to edaphically differentiated communities. *Pl. Syst. Evol.* **137,** 95–105.
Gentry, A. H. (1982). Patterns of neotropical plant species diversity. *Evol. Biol.* **15,** 1–84.
Great Plains Flora Association (1977). "Atlas of the Flora", 600 pp. Iowa State University Press.
Great Plains Flora Association. "Great Plains Flora". (In prep.)
Grubb, P. J., Lloyd, J. R., Pennington, T. D. and Whitmore, T. C. (1963). A comparison of montane and lowland rain forest in Ecuador. *J. Ecol.* **51,** 567–601.
Heinsdijk, D. (1960). "Interim Report to the Government of Brazil on the Dry Land Forests on the Tertiary and Quarternary South of the Amazon River". FAO Report No. 1284, Rome.
Heinsdijk, D. and Miranda Bastos, A. (1965). "Report to the Government of Brazil on Forest Inventories in the Amazon". FAO Report No. 2080, Rome.
Hicks, A. J. and Hicks, R. M. (1978). A selected bibliography of plant collection and herbarium curation. *Taxon* **27,** 63–99.
Holdridge, L. R. (1972). "Forest Environments in Tropical Life Zones; a Pilot Study", 747 pp. Pergamon Press, Oxford, New York.
Holmgren, P. K., Keuken, W. and Schofield, E. K. (1981). "Index Herbariorum Part I. The Herbaria of the World", 452 pp. Bohn, Scheltema & Holkema, Utrecht, Antwerp.

IUCN, UNEP, WWF. (1980). "World Conservation Strategy. Living Resource Conservation for Sustainable Development". IUCN, Geneva.

Janzen, D. H., Ataraff, M., Jariñas, M., Reyes, S., Rincon, N., Soler, A., Soriano, P. and Vera, M. (1976). Changes in the arthropod community along an elevation transect in Venezuelan Andes. *Biotropica* **8,** 193–203.

Jones, E. W. (1955). Ecological studies on the rain forests of southern Nigeria. IV. The plateau forest of Okomu forest reserve. *J. Ecol.* **43,** 564–594.

Jones, E. W. (1956). Ecological studies on the rain forest of southern Nigeria. IV. The plateau forest of the Okomu forest reserve. *J. Ecol.* **44,** 83–117.

Kent, D. H., Bangerter, E. B. and Lousley, J. E. (1957). "British Herbaria. An Index to the Location of Herbaria of British Vascular Plants with Biographical References to their Collectors", 101 pp. Bot. Soc. British Isles, London.

Knight, D. H. (1975). A phytosociological analysis of species-rich tropical forest on Barro Colorado Island, Panama. *Ecol. Monog.* **45,** 259–284.

Komarov, V. L. (1968). "Flora of the U.S.S.R.", Vol. I. Israel Program for Scientific Translations, Jerusalem.

Kubitzki, K. (1977). The problem of rare and frequent species: the monographers view. *In* "Extinction is Forever" (G. T. Prance and T. S. Elias, eds) pp. 331–336. New York Botanical Garden, New York.

Letouzey, R. (1968). "Étude phytogéographique du Cameroun", 508 pp. Lechevalier, Paris.

Mori, S. A., Boom, B. M. and Prance, G. T. (1981). Distribution patterns and conservation of eastern Brazilian coastal forest tree species. *Brittonia* **33,** 233–245.

Mori, S. A. and Holm-Nielsen, L. B. (1981). Recommendations for botanists visiting neotropical countries. *Taxon* **30,** 87–89.

Myers, B. J. (1982). Large-scale color aerial photographs – a useful tool for tropical biologists. *Biotropica* **14,** 156–157.

Myers, B. J. and Benson, M. L. (1981). Rainforest species on large-scale color photos. *Photogramm. Engng and Rem. Sens.* **47,** 505–513.

Myers, N. (1980). "Conversion of Tropical Moist Forests", 205 pp. National Research Council of National Academy of Sciences, Washington, D.C.

National Research Council (1970). "Remote Sensing with Special Reference to Agriculture and Forestry", 424 pp. National Academy of Sciences, Washington, D.C.

National Research Council (1980). "Research Priorities in Tropical Biology", 116 pp. National Academy of Sciences, Washington, D.C.

Paijmans, K. (1969). An analysis of four tropical rain forest sites in New Guinea. *J. Ecol.* **58,** 76–101.

Perring, F. H. and Sell, P. D. (eds) (1968). "Critical Supplement to the Atlas of the British Flora", 159 pp. Nelson, London.

Perring, F. H. and Walters, S. M. (eds) (1962). "Atlas of the British Flora", 432 pp. Nelson, London.

Pires, J. M., Dobzhansky, T. H. and Black, G. A. (1953). An estimate of species of trees in Amazonian forest community. *Bot. Gaz.* **114,** 467–477.

Prance, G. T. (1972a). Monograph of Chrysobalanaceae. *Flora Neotropica* **9,** 1–410.

Prance, G. T. (1972b). Monograph of Dichapetalaceae. *Flora Neotropica* **10,** 1–84.

Prance, G. T. (1977). Floristic inventory of the tropics: Where do we stand? *Ann. Missouri Bot. Gard.* **64,** 659–684.

Prance, G. T. (1978). Floristic inventory in the tropics: a correction. *Ann. Missouri Bot. Gard.* **65,** i. ii.

Prance, G. T. (1979a). South America. *In* "Systematic Botany, Plant Utilization and Biosphere Conservation" (I. Hedberg, ed.) pp. 55–70. Almquist and Wiksell, Stockholm.

Prance, G. T. (1979b). Notes on the vegetation of Amazonia III. The terminology of Amazonian forest types subject to inundation. *Brittonia* **31,** 26–38.

Prance, G. T. and Mori, S. A. (1980). Monograph of Lecythidaceae. *Flora Neotropica* **21,** 1–270.

Prance, G. T. and Schaller, G. B. (1982). A preliminary study of some vegetation types of the Pantanal, Mato Grosso, Brazil. *Brittonia* **34,** 228–251.

Prance, G. T., Rodriques, W. A. and da Silva, M. F. (1976). Inventário florestal de um hectare de mata de terra firme, km 30 da Estrada Manaus-Itacoatiara. *Acta Amazônica* **6,** 9–35.

Projeto RADAM (1973–75). "Levantamento de recursos naturais", Vols. 1–7. Ministério das minas e energia, Rio de Janeiro.

Projeto RADAMBRASIL (1975). "Levantamento de recursos naturais", Vols. 8–18. Ministério das minas e energia, Rio de Janeiro.

Proyecto Radargrametrico del Amazonas (1979). "La Amazonia Colombiana y sus recursos", Vols. 1–5. Proradam, Bogotá.

Rollett, B. (1974). "L'architecture des forêts denses humides sempervirents de plaine", 298 pp. Centre technique forestier tropical (CTFT), Nogent-sur-Marne.

St. John, T. V. (1980). Root size and mycorrhizal infection; A re-examination of Baylis's hypothesis with tropical trees. *New Phytol.* **84,** 483–487.

Schulz, J. P. (1960). Ecological studies on rain forests in northern Suriname. *Verh. Konik. Nederl. Akad. Wetensch. Afd. Natuurk.* **53,** 1–267.

Scott, N. J. (1976). The abundance and diversity of herpetofaunas of tropical forest letter. *Biotropica* **8,** 41–58.

Schischkin, B. K. (1972). "Flora of the U.S.S.R.", Vol. 24. Israel Program for Scientific Translation, Jerusalem.

Sick, H. and Teixeira, D. M. (1979). Notas sobre aves brasilieros raras ou ameaçada de extinção. *Publ. Avulsas do Museu Nacional-Universidade Federal de Rio de Janeiro* **62,** 1–39.

Solem, A. and Burger, W. C. (1982). The Tyranny and opportunity of numbers. *ASC Newsletter* **10(1),** 1–5.

Spencer, R. D. (1974). Supplementary aerial photography with 70 mm and 35 mm cameras. *Austral. For.* **37,** 115–125.

Steyermark, J. A. and Huber, O. (1978). "Flora del Avila". Ministério del ambiente y de los recursos naturales renovables, Caracas.

Takeuchi, M. (1962). The structure of the Amazonian vegetation. *J. Fac. Sci. Univ. Tokyo Sect. III Bot.* **8,** 1–26, 27–35.

Tardin, A. T., dos Santos, A. P., Lee, D. C. L. *et al.* (1979). "Levantamento de áreas de desmatamento na Amazônia legal através de imagens de Satelite Landsat", 9 pp. INPE-COM 3/NTE, C.D.U. 621.38SR. São José do Campos, São Paulo.

Tardin, A. T., Lee, D. C. L., Santos, R. J. R. *et al.* (1980). "Subprojeto desmatamento, convenio IBDF/CNPq-INPE 1979", 44 pp. Inst. Nac. de Pesq. Esp. (INPE) Relatorio No. INPE-1649 RPE/103. São José dos Campos, São Paulo.

Tutin, T. G., Heywood, V. H., Burges, N. A., Moore, D. M., Valentine, D. H., Walters, S. M. and Webb, D. A. (1980). "Flora Europea", Vol. 5, 452 pp. Cambridge University Press, Cambridge.

US Interagency Task Force on Tropical Forests (1980). "The World's Tropical Forests: A Policy, Strategy, and Program for the United States", 53 pp. Dept. of State Publication 9117, Washington, D.C.

Vinha, S. G. da, Jesus Soares Ramos, T. de and Hori, M. (1976). Inventário florestal, pp. 20–212. *In* "Diagnóstico socioeconômico da região cacaueira, recursos florestais". Vol. 7. Comissão Executiva do Plano da Lavousa Cacaueira and the Instituto Interamericano de Ciências Agrícolas-OEA. Ilhéus, Bahia, Brazil.

Vink, W. (1981). Density indexes updated. *Fl. Malesiana Bull.* **34,** 3567–3568.

White, F. (1962). Geographic variation and speciation in Africa with particular reference to *Diospyros*. *Syst. Assoc. Publ.* **4,** 71–103.

Zardini, E. M. (1980). Index of Argentinan Herbaria. *Taxon* **29,** 731–741.

Womersley, J. S. (1981). "Plant collecting and herbarium development", 137 pp. FAO Plant Production and Protection Paper 33, Rome.

19 | Designing Floras for the Future

V. H. HEYWOOD

Department of Botany, University of Reading, England

Abstract: Different kinds of Floras are needed for different purposes so that in considering the design of Floras for the future, two inter-related aspects have to be considered: (1) the purpose of the Flora and the audience aimed at, (2) technical aspects of data presentation in a computer-orientated world. Also important is the distinction between Floras as basically practical tools containing, primarily, information related to their identification role, and Floras as repositories of taxonomic information which can be used for a wide range of purposes. A brief survey is given of the major kinds of Flora-programme currently in progress across the world, ranging from conventional, institutional, monographic Floras to those heavily dependent on the use of EDP systems.

INTRODUCTION

At the final Flora Europaea Symposium held in 1977 at Cambridge, I commented how surprising it was that little had been written about the design and nature of Floras since the classic accounts of J. D. Hooker (1855), Bentham (1874) and Auguste de Candolle (1880), all published 100 years ago or more. This century, one of the few substantial contributions to the subject was the characteristically pungent essay by van Steenis (1954), presented at the 1954 International Botanical Congress at a symposium entitled "General Principles in the Design of Floras". Apart from those given by van Steenis, few general principles emerged from this symposium and one important decision was that it was not then appropriate to write a Flora of Europe, a viewpoint shared by many taxonomists working at the national institutions in Europe. The basis for

Systematics Association Special Volume No. 25, "Current Concepts in Plant Taxonomy", edited by V. H. Heywood and D. M. Moore, 1984. Academic Press, London and Orlando.
ISBN 0 12 347060 9

this honestly held viewpoint was a conviction that the amount of revision necessary to present a Flora of adequate quality had not yet been undertaken and also that there was no likelihood of institutional resources being diverted to achieve such an end. Such an attitude, which largely ignores the need or demand for a Flora by the various consumers, was widespread and is still prevalent today.

As is well known, the conviction of a group of British botanists that a Flora of Europe was both timely and necessary was so strongly held that a decision to go ahead and plan one was taken at an informal meeting held after the symposium just referred to. This conviction was based on the experience of a number of University taxonomists whose assessment of the academic, didactic and practical benefits of a continental European Flora was shared by many other University taxonomists in Europe who were beginning to re-establish cooperative links in the post-war years. It was a reflection too of the European movement which led in 1957 to the signing of the Treaty of Rome and the creation of the European Economic Community, and later to institutions such as the Council of Europe devoted to social and cultural cooperation and much later to the European Science Foundation. Plant taxonomists have been in the forefront of European cooperation since the end of the last world war.

It may be a characteristic of the great periods of Flora-writing that there has to be some socio-political-economic climate or pressure that acts as the underlying stimulus: in the nineteenth century, world exploration (and exploitation) and building of empires, by the great European powers leading to the great Colonial Floras and creation of Botanic Gardens (Heywood 1983); in the recent post-war period just alluded to, the European movement stemming from a desire for reconciliation and cultural integration and leading to "Flora Europaea" which in turn served as a stimulus for other parts of the world such as North America and, after many false starts, Australia; and in the last ten to fifteen years, a remarkable intensification of Flora-writing that is attributable in large part to the conservation movement highlighting the need for basic floristic inventorying assessments of many parts of the world not so far provided with Floras as a basis for resource management, conservation and other biological activities.

Later there was a series of papers and guides arising out of the *Flora Europaea* project (notably the so-called "Green Book" – Heywood 1958, 1960, 1964), including the introduction of the "Basic" and "Standard Floras" concept. Finally, critical assessments of various aspects of the nature, purpose and design of Floras were published by Shetler (1971), Fisher (1968), Heywood (1973, 1976), Davis and Heywood (1963), Watson (1971), Jacobs (1969, 1980), Taylor

(1971), and Frodin (1976, 1983) (several of these arising out of the "Flora North America" project, Shetler (1971).

Recent criticisms of the outmoded format of "Institutional large-scale handbook Floras" (what I have called hard-core Floras!) by Jacobs (1977), Heywood (1978, 1980) and others and the need to relate much more clearly to the needs of the consumer were in fact first enunciated by Corner in 1946 and by Symington in 1943. Corner in fact regarded most tropical Floras as fundamentally unsuitable and useless for the local user, whether agronomist, forester, teacher or student. This raises the need to establish at the outset the purpose of a Flora and the potential consumer being aimed at, which is returned to later in this paper.

The classic concept of a Flora was expounded by Bentham (1861) in the following terms "The principal object of the Flora of a country is to afford the means of determining (i.e. ascertaining the name of) any plant growing in it, whether for the purpose of ulterior study or of intellectual exercise". He also had much to say about the style and technicalities to be adopted. This philosophy was essentially similar to that expressed by W. J. Hooker and was adopted in most of the British Colonial Floras that emanated from Kew. Such Floras became widely influential and were regarded as a model or standard to be followed in many English-speaking countries. This type of approach has aptly been described as "phytographical utilitarianism" by Frodin (1984) in a detailed historical review of floristic works. Similar approaches were adopted in continental Europe and in North America. The essential features were the use of concise but reasonably detailed descriptions of the species included and the provision of analytical or diagnostic keys, together with due regard for accurate summaries of geographical distribution reflecting an increasing interest in geographical botany and patterns of variation. The underlying philosophy of this type of Flora is a combination of various factors: (1) it is essentialist in the sense that it reduces a complex set of essentially morphological variables into a set of basic statements which represent the whole; there is also an element of Aristotelean or Linnaean essentialism involved in that the characters selected for description were regarded as the essential ones in a functional sense, either explicitly or implicitly; (2) it is pragmatic in that it used the minimal number of statements to provide a recognizable plant portrait, in the semi-telegraphic stylized form of presentation, and in the use of analytical keys to facilitate identification. The emphasis moved away from codification in the Linnaean sense (cf. Heywood 1980, 1983b) to descriptions, but descriptions concise enough to serve as a means of confirming the identification indicated by use of the keys, as well as conveying a picture of the plant at the same time.

A quite different tendency was the detailed descriptive Flora which became prevalent in central Europe. Here the aim was to provide comprehensive, meticulously detailed descriptions, extensive synonymy, discussion material and, often, extensive illustration. The idea of the Flora as a repository of all kinds of systematic information was the inspiration for this approach, and little attention seems to have been paid to practical matters such as ease of identification, comparability of descriptions, etc. These Floras were a part of systematics rather than practical taxonomy in the strict sense. And the tradition persists today in Floras that seem to be more acts of academic scholarship rather than pragmatically-orientated works of identification and inventory. Quite clearly this dichotomy between the Flora as a practical work and as a detailed source of comparative data is a long-standing one.

These concepts were updated by Shetler (1971) in connexion with the "Flora North America" project when he wrote that a Flora is a:

time-honoured information retrieval system. It is a physical repository of descriptive data about plants which are organized and formatted, usually in book form, so as to answer a time-tested series of prescribed questions. It is as old as taxonomy itself and equally indispensable.

He also viewed the Flora of the future as:

a standardized data bank. It will be open-ended, dynamic and ever-growing. The specialist of the future doubtless will store his revisions and monographs in the computer, not on the printed page. Thus the Flora of the future will become a huge memory or series of linked memories available on-line to all users at any place and time.

There are in fact several different types of Flora today (Davis and Heywood 1963; Frodin 1976, 1983): (1) the monographic or research Flora, (2) the concise or field Flora/manual, (3) the excursion Flora, (4) the illustrated Flora, and (5) the enumeration and the checklist, each of which occupies a separate role. None of these has shown much evolution in terms of presentation or content apart from the inclusion of newer types of information such as chromosome number, information on variation of a taxonomic or biosystematic nature, phytosociological data, chemical constituents, etc. In addition, there has been a tendency to use fewer infraspecific categories – often now reduced to either the variety or subspecies – which has come with a realization that the detailed nature of the variation pattern cannot easily be forced into the straightjacket of the conventional categories of subspecies, variety and form (Mayr 1959; Heywood 1963).

A recent development has been the increasing use of detailed guidelines for the use of contributors to multi-author Floras, a good example being the "Green Book" (and Supplement) produced in connexion with the "Flora Europaea" project (Heywood 1958, 1960). Such guidelines are not only necessary for collaborative works so that the various contributors can appreciate the underlying principles and detailed style and format to be adopted, but are of considerable value to the editors too in focussing attention on matters that normally receive little consideration.

FLORAS AS FRONT-LINE PUBLICATIONS

There are many misconceptions about Floras which have arisen in part from the conventions and mystique surrounding them that have built up over the years. These have tended to obscure some of the more important functions of the Flora such as decision-making. A Flora is not just a set of facts and data. It is also a collection of opinions – some of them original and well-founded, some arrived at after only superficial study, others perhaps adopted from other taxonomists without further consideration. In this sense the Flora can be regarded as a front-line synthesis: in the overall context of publication in taxonomy, it occupies a basic role in the collection, collation and revision of taxonomic information (including new taxa, nomenclatural changes, distributional information, revision and monographs) which has been published in a wide array of books and journals; this is then reviewed and assessed and an "authoritative" opinion is expressed and the results are then presented in a ready-digested form for a wide range of users. It provides a rapid means of processing the raw data and publishing the results. There is no equivalent for non-floristic data.

This raises the question of what McNeill (1971) has called "authority". In a conventional printed Flora, the decisions or opinions are attributed to a particular author or group of authors even though the full data-base on which the decision is founded is not presented. Floras (despite what many non-taxonomists believe) do not necessarily record much of the information used by the taxonomist in the decision-making process (Heywood 1973). In that sense Floras are like icebergs, only the tip of the information-base shows. Unfortunately the data-base is usually thrown away (or at least not recorded) after the decisions have been made and the Flora published. It cannot be stressed too often that Floras are *abstracts* of knowledge prepared for practical convenience. Floras, like classifications, as McNeill (1979) has recently pointed out, are simplifications. The question of authority in "computerized" Floras is discussed below.

Generally the Flora is an in-depth survey at local/regional level as regards actual material examined or reviewed. The monograph, in contrast, is in-depth in a different sort of way – more material is examined in total for each species covered, but from a much wider geographical range and the coverage of any particular part of the individual species range may be poor (certainly poorer than that which a Flora of the area concerned would attain).

The *balance* of treatment between a Flora and monograph is usually quite different – by definition, a complete monograph should cover all the taxa/species concerned in the genus and then provide an overall perspective, while in a Flora (unless it is a major regional Flora) the viewpoint is often (usually) more biased by being selective or partial. The distinction is blurred by the existence of partial monographs on the one hand and monographic Floras (e.g. "Flora Neotropica") on the other.

The reality about monographs is often quite otherwise! The sampling of material examined is often very uneven or restricted, often unrepresentative. This was especially true of the early "Pflanzenreich" monographs which were based mainly in the collections in the Berlin herbarium, and other examples by leading systematists could be cited which were similarly restricted. There was, indeed, a tendency to base major taxonomic works very largely on the in-house collections until well into this century, especially in the larger national taxonomic institutions.

The practice of large-scale borrowing of herbarium material for the writing of monographs is a relatively recent one, botanists having relied previously on the system of exchange of herbarium material to provide an adequate range of specimens in their own herbarium. Little attention has been focussed until recently in herbarium costs (cf. Payne 1979) and it is clear that the value of specimens and the curating process have been under-valued or underestimated (the Advisory Committee for Systematic Resources in Botany of the American Society of Plant Taxonomists suggests in a recent report that each mounted, identified herbarium specimen should be valued at seven dollars (Payne 1979)). The costs of writing a Flora or monograph must include a large element for the consultation and borrowing of herbarium material, often running into thousands of dollars in terms of handling a loan, postage or transport in both directions, and depreciation. Not surprisingly, with reduced budgets and staffing, loans are made less readily than previously, as mentioned by Cullen (Chapter 2), and this could have the most serious consequences on the practice of floristic and revisionary research. This, coupled with decreasing travel budgets, will force the professional taxonomic community to rethink some of the present strategies adopted in these kinds of taxonomic activity.

FLORAS AS DATA-BASES

As we noted earlier, many Flora writers have adopted a different philosophy to the practical Benthamian view of the Flora, and regarded it as having an archival or encyclopaedic role ("Flora of Panama", Martius' "Flora Brasiliensis", "Flore de l'Afrique Centrale"). In recent times there has often been a failure to distinguish between the archival and practical functions of Floras. The word "archival" can be applied to a Flora containing data which are not strictly necessary for the practical functions of identification (keys and descriptions) and naming of the plants growing in a specified region together with information about the distribution of the species and other taxa within this region.

Archival Floras do not necessarily (or even often) constitute comparative data-bases. They may contain much more extensive descriptions than concise Floras but these are not strictly comparable across the genera, thus leading Watson (1971) to comment that:

> perusal of the average taxonomic-descriptive work usually reveals that as a source of comparative data it is hopeless. I suggest that the chaotic state of the descriptive literature and the continued lack of organization over descriptive practice are the main reasons why major plant groups are so daunting and why the relatively few attempts to tackle them are at present doomed to be superficial and unsatisfactory.

This is, however, a reflection of a more general problem noted by Crovello (1976):

> Except for the literature which is easily available (and easily verifiable!) to commercial organizations, botanical data are scattered, of varying degrees of accuracy and precision, often not compatible or comparable in their content, and rarely ever recorded in a way that would be suitable for incorporation into a botanical data bank.

So-called archival Floras are simply not organized in such a way as to constitute a data-base and what archival information that is contained in them is perhaps better achieved by other kinds of publication or by storage and retrieval through computerized data banks or other non-print media, rather than the printed book as Watson (1971) comments.

It is, perhaps, unfortunate and certainly ironic that the archival or encyclopaedic approach to Flora-writing has been applied most often to areas of the world where there is greatest floristic diversity and consequently a need for rapid techniques so as to produce usable results within a reasonable time-scale. Thus many Floras of tropical countries are conceived on such a grandiose scale –

many volumes in large format, often with more regard for prestigious appearance than for economy of space. As a consequence, relatively few Floras of tropical countries have been completed so that organized knowledge of the floristics of those parts of the world that are in greatest danger of extinction is woefully incomplete.

A remarkable instance of a European encyclopaedic Flora is Hegi's "Illustrierte Flora von Mitteleuropas". An acknowledged classic, now in its third edition, it is in fact not designed for easy use or access to the mass of information which is encompassed in its large format densely printed pages. The text contains a mass of non-taxonomic information in addition to the very extensive descriptions, thus detracting from its practical role.

FLORAS AND THE CONSUMER

It will be evident from the above discussion that many recent Floras seem to have been written with little explicit regard to questions of design, utility, potential consumers and so forth.

Possibly such questions may have been raised at some stage in their conception, although a study of many Floras would make one doubt this. Many Floras follow the ethic of a former age and are no longer relevant to today's conditions. Even serial, research Floras, such as "Flora Malesiana", "Flora Zambesiaca", "Flora Iranica", which are not markedly encyclopaedic or archival, are planned in such a way that they are not completed within a reasonably short timescale. Consequently they are more like a series of monographs (deliberately so in the case of "Flora Neotropica") and of limited value except to other taxonomists because of their incompleteness; moreover there is also the likelihood that the early parts will become out of date well before the final parts are ready for publication. In practice the materials and information available for writing a Flora in any particular case vary so considerably that it is difficult to make useful generalizations as to the procedure to be adopted. The most neglected factor is, however, usually the consumer – the potential public for the Flora. This is symptomatic of the general alienation of taxonomy from the public (cf. Jacobs 1980).

The needs of the potential consumer should largely determine the style, form and detailed content of a Flora. Floras often seem to be aimed at a quite unspecified general audience. I think one has to question very closely this concept – is there such a thing as a general audience? In some cases it is quite clear that there is not a body of other taxonomists, ecologists, foresters, horticulturists, agronomists, administrators, etc. who would constitute an audience. In other words, apart from some other taxonomists, there may be few sufficiently interested people who would acquire or use the Flora. In some countries there may be no local interest developed in the Flora and one is in the position of saying, in effect, we are going to write a Flora like it or not. This goes against van Steenis's (1954) precept that a Flora should live in the area for which it is prepared, although there may indeed be situations where the consumer will necessarily be other taxonomists only. In some cases the justification for writing the Flora may be purely scientific.

If a "general" audience *can* be identified for a Flora, then more effort should be made to ascertain the needs and desires of that audience. Taxonomy as a whole has in the last 100 years or so gradually become a more and more esoteric science and its vehicles of communication less and less manageable by non-taxonomists. The design of Floras has hardly changed, and what one might call the "hard core" professional Flora often includes an appalling accumulation of technical jargon, unnecessarily complex abbreviations, etc. As Jacobs (1977) comments:

> Taxonomists certainly have to unlearn a lot of indifference towards the consumers of their work. Terminology should be in more plain English and consistency strived for. Nomenclatural tidbits like *quae est*, *pro maj. parte*, *quoad specim.*, should be in English, not in Latin however desirable it is that every botanist should be well versed in that language.

The difficulties experienced by non-taxonomist biologists in using Floras are well known even though they are ostensibly part of the audience aimed at. Admittedly some concessions have been made in some Floras to the non-specialist consumer – use of clearer, simpler language, multiple access or alternative keys, statements in prose about problem taxa, ample illustrations, etc. but too many Floras, in my experience, are uncompromising in their approach to the technicalities of taxonomy. Amongst the points one has to query, for example, are the citation of place of publication of generic names, citation of old, seldom-used synonyms; even the value of citing specimens examined, of no use to anyone other than a professional taxonomist with access to a large herbarium, has to be carefully assessed.

Evidently there is bound to be some conflict of interest: what the professional taxonomist needs or demands is not the same as what the non-taxonomic biologist needs, and the wisdom of attempting both in the same work should be questioned. One unfortunate disadvantage of the "hard core" professional Flora in the hands of non-specialists is that it misleads him into thinking that some of the purely bibliographic/citation information is important to him, rather than the names, keys, descriptions and distributions, quite apart from being put off by it all.

FLORAS AND THE NEW TECHNOLOGY

The application of modern technology in the preparation of Floras covers a wide range of possibilities. There are several approaches.

(1) The first is where an editing typewriter or word-processor is used in the typing of the manuscript of the Flora, so that no matter what the content and style adopted, revised versions of the typescript can be easily produced without retyping and proof reading. Such an approach is so obvious that the only surprising thing is that it is not adopted more frequently.

(2) The second approach involves gradually building up the materials for a Flora from label data – identifications, geography, ecology, etc. supplemented by morphological data when prepared. This is essentially the approach followed by the "Flora of Veracruz" project (Gómez-Pompa and Nevling 1973; Gómez-Pompa *et al.* 1984) in a floristic study of the Mexican State of Veracruz.

(3) The third approach is characterized by the construction of a full-scale descriptive data-base, as was envisaged in the original "Flora North America" project (Shetler and Krauss 1971), and has been the subject of extensive research (e.g. Pankhurst 1984). Because of the acute problems of devising schemes for describing the wide range of morphological features involved in a Flora, this approach has so far been applied only on a pilot scale. Computer-generated keys are much more widely employed than descriptive floristic data-bases and the subject was reviewed in an earlier Systematics Association volume (Pankhurst 1975).

A recent development that could reduce the costs of publishing Floras substantially is the automatic typesetting of text from computer-generated tapes or discs. This is the logical extension of the use of word-processors mentioned above and eliminates the need to reset from the keyboard for the final version to be printed as well as the need to check the proofs in detail.

A note of caution has to be sounded here. The idea that we can use computer technology to allow us to update and print out annually (or at other regular intervals) via computer-linked typesetting, new editions of Floras and other descriptive works is somewhat naive or at least belongs to Bisby's dream world (Chapter 16). The economics of the operations have to be considered carefully: Floras usually sell fewer copies than one realizes unless they are student-orientated. Certainly the larger, serial Floras are frequently produced in small editions and even fewer copies sold.

One possibility that deserves further consideration is the publication of Floras in loose-leaf form with ring-binders or some other easily removed binding procedure. This would allow pages to be substituted as corrected versions become available without having to reprint the whole volume at great expense.

Alongside the new technology, there has to be developed a new approach to the basic procedures of Flora-writing, in particular the possible standardization of descriptions (as well as descriptive terminology). Although descriptions and a Flora are supposed to be original, based on the material studied by the author, all too often they involve a large amount of judicious (or even uncritical) copying from previous descriptions. There is a large element of duplicated effort in description writing and there is no way of knowing what is original and which description is better or more accurate so that each author starts afresh.

In due course I would hope that information systems such as the European Taxonomic, Floristic and Biosystematic Documentation System, which we are setting up at Reading, will become available as a means of providing much of the initial information – nomenclatural, bibliographic, geographical, etc. – that is involved in writing a Flora or monograph. There is no real justification today for having to spend time and money in the repeated but individual accumulation of factual data for each new Flora, revision or whatever kind of taxonomic publication.

The use of data-bases in taxonomic procedure will raise problems such as that of "authority" mentioned above. Unlike conventionally produced Floras which, as we have seen, are collections of opinions, the information in a data-base is initially unprocessed in the sense that no opinion is expressed as to the validity or acceptability of the new data. As McNeill (1971) notes:

> new does not necessarily mean better; indeed it often means worse. A computerized data base can store alternative classifications or opinions and unless it is specifically edited in whole or in part, the user will be presented with a choice and have to make his decisions or express his own opinions.

The taxonomist is, however, used to making such decisions; the non-taxonomist user of data bases would need to be very sophisticated to be able to handle such diversity of opinion and the intermediacy of the taxonomist in editing the data-base would normally be needed. Eventually we can look forward to a situation when alternative, edited, equally authoritative classifications or opinions are made available in the data-base according to the interests or needs of the user – conventional phenetic classifications for the general user, genecological or genetically-based classifications for the plant breeder or agronomist, etc.

CONCLUSIONS

By way of conclusion, I shall list a series of recommendations that might usefully be taken into account by those embarking on writing a Flora.

(1) The first need is to set out clearly the objectives of the Flora, separating out the different roles that can be occupied and deciding which shall receive priority.

(2) The needs of the potential consumer then have to be ascertained and the Flora tailored as far as possible to meet these.

(3) The time-scale and the manpower needs of the Flora should be worked out carefully in advance. Floras with an open-ended time-scale tend not to be completed. Because of the complexity of taxonomy today, the accumulation of synonymy, the wealth of material to be consulted and the many other occupations of most taxonomists, productivity as measured by the number of species written for a "research" Flora per annum has declined from the 250 or so per taxonomists in the nineteenth century to 50 (De Wolf 1964) or even 15–20 (van Steenis 1979).

(4) Whenever possible, standardization of procedures and formats should be aimed at with a view to using modern technology and to save editorial time and effort. A guide for contributors to the Flora (if it is a multi-author work) should be prepared and firmly adhered to.

(5) For large-scale works, the details and implications of editorial responsibility *vis-à-vis* the contributors should be thoroughly explored and the advantages of collegialism as practised in the "Flora Europaea" project should be considered (cf. Heywood 1964, 1978; Webb 1978).

(6) The more extensive use of illustrations should be considered especially for poorly known areas. An excellent example is the "Flora of the Rio Palenque Science Centre" by Dodson and Gentry (1978).

Finally, I believe we ought to think out much more clearly priorities in terms of areas for which Floras are most needed, and then foster international and inter-institutional cooperation to achieve them. Unless we undertake such action soon, the chances of achieving Floras of many areas of the world are rapidly diminishing.

REFERENCES

Bentham, G. (1861). "Flora Hongkongensis", pp. i–xxxvi. Reeve, London.

Bentham, G. (1874). On the recent process and present state of knowledge of systematic botany. *Reports Brit. Ass. Adv. Sci.* (1874), 27–54.

Candolle, A. de (1880). "La phytographie". Masson, Paris.
Corner, E. J. H. (1946). Suggestions for botanical progress. *New Phytol.* **45,** 185–192.
Crovello, T. J. (1976). Botanical data banking. *In* "Proc. Fifth International CODATA Conference", pp. 44–46. Pergamon, New York.
Davis, P. H. and Heywood, V. H. (1963). "Principles of Angiosperm Taxonomy". Oliver and Boyd, Edinburgh.
De Wolf, G. P. (1964). On the sizes of floras. *Taxon* **13,** 149–153.
Dodson, C. H. and Gentry, A. H. (1978). Flora of the Rio Palenque Science Center. *Selbyana* **4** (1–6).
Fisher, F. J. F. (1968). The role of geographical and ecological studies in taxonomy. *In* "Modern Methods in Plant Taxonomy" (V. H. Heywood, ed.) pp. 241–259. Academic Press, London, New York.
Frodin, D. G. (1976). On the style of floras: some general considerations. *Gardens' Bull. Singapore* **29,** 239–250.
Frodin, D. G. (1983). "Guide to Standard Floras of the World". Cambridge University Press, Cambridge.
Gómez-Pompa, A., Allkin, R., Moreno, N., Sosa, V. and Gama, L. (1984). Flora of Veracruz: progress and prospects. *In* "Databases in Systematics" (R. Allkin and F. A. Bisby, eds) pp. 165–174. Academic Press, London, New York.
Gómez-Pompa, A. and Nevling, L. I. (1973). The use of electronic data processing methods in the flora of Veracruz program. *Contr. Gray Herb.* **203,** 49–64.
Heywood, V. H. (1958). "The Presentation of Taxonomic Information: A Short Guide to Contributors to 'Flora Europaea' ". Leicester University Press, Leicester.
Heywood, V H. (1960). "The Presentation of Taxonomic Information Supplement". Flora Europaea Organization, Alcobaca.
Heywood, V. H. (1963). Biosystematics and classification. *Rep. Scott. Pl. Breed. Stn.* **1963,** 1–7.
Heywood, V. H. (1964). *Flora Europaea* and the problems of the organization of floras. *Taxon* **13,** 48–51.
Heywood, V. H. (1973). Ecological data in practical taxonomy. *In* "Taxonomy and Ecology" (V. H. Heywood, ed.) pp. 329–347. Academic Press, London, New York.
Heywood, V. H. (1976). Contemporary objectives in systematics. *In* "Proceedings Eighth International Conference on Numerical Taxonomy" (G. F. Estabrook, ed.) pp. 258–283. W. H. Freeman, San Francisco.
Heywood, V. H. (1978). European floristics: past, present and future. *In* "Essays in Plant Taxonomy" (H. E. Street, ed.) pp. 275–289. Academic Press, London, New York.
Heywood, V. H. (1980). The impact of Linnaeus on botanical taxonomy – past, present and future. *Veröff. Joachim Jungius-Ges. Wiss. Hamburg* **43,** 97–115.
Heywood, V. H. (1983a). Botanic gardens and taxonomy – their economic role. *Bull. Bot. Survey, India* (In press).
Heywood, V. H. (1983b). Linnaeus – the conflict between science and scholasticism. *In* "Linnaeus Symposium". University of Texas, Austin (In press).
Hooker, J. D. (1855). Introductory essay to J. D. Hooker and T. Thomson, "Flora Indica". London.
Jacobs, M. (1969). Large families – not alone! *Taxon* **18,** 253–262.
Jacobs, M. (1977). (Editorial) *Flora Malesiana Bulletin* **30,** 2733–2736.

Jacobs, M. (1980). Revolutions in plant description. *Misc. Papers Landbouwhogeschool Wageningen* **19,** 155–181.
McNeill, J. (1971). "Data Processing in Biology and Geology" (J. L. Cutbill, ed.) p. 93. Academic Press, London and New York.
McNeill, J. (1979). Structural value – a concept used in the construction of taxonomic classifications. *Taxon* **28,** 481–504.
Mayr, E. (1959). Trends in avian systematics. *Ibis* **101,** 293–302.
Pankhurst, R. J. (ed.) (1975). "Biological Identification with Computers". Academic Press, London, New York.
Pankhurst, R. J. (1984). A review of herbaria catalogues. *In* "Databases in Systematics" (R. Allkin and F. A. Bisby, eds) pp. 155–164. Academic Press, London, New York.
Payne, W.W. (1979). "Systematic Botany Resources in America. Part II. The Costs of Services". New York Botanical Garden, New York.
Shetler, S. G. (1971). Flora North America – an information system. *BioScience* **21,** 524–532.
Shetler, S. G. and Krauss, H. M. (1971). Flora North America: a comprehensive program of biological research, information systems development, and data banking concerned with the vascular plants of North America north of Mexico. *Flora North America* Report **61,** 1–124.
Steenis, C. G. G. J. van (1954). General principles in the design of Floras. "8me Congr. Int. Bot. Paris Rapp. et Comm." sects. 2, 4, 5 et 6, 59–66. SEDES, Paris.
Steenis, C. G. G. J. van (1979). The Rijksherbarium and its contribution to the knowledge of the tropical Asiatic flora. *Blumea* **25,** 57–77.
Symington, C. F. (1943). The future of colonial forest botany. *Empire For. J.* **22,** 11–23.
Taylor, R. A. (1971). The "Flora North America" project. *BioScience* **21,** 521–523.
Watson, L. (1971). Basic taxonomic data: the need for organisation over presentation and accumulation. *Taxon* **20,** 131–136.
Webb, D. A. (1978). Flora Europaea – a retrospect. *Taxon* **27,** 3–14.

20 Taxonomic Problems Relating to Endangered Plant Species

E. S. AYENSU

*Office of Biological Conservation,
Smithsonian Institution, Washington, USA*

In the United States, a potentially major taxonomic problem was created by the definition of the term "species" in the Endangered Species Act of 1973, which persisted through the amendments of November 1978 and up to the present time. Section 3(16) of the Act defines "species" to include "any subspecies of fish or wildlife or plants, and any distinct population segment of any species of vertebrate, fish or wildlife which interbreeds when mature". This wording does not provide for plant *varieties* to be considered for official listing, even though the botanical variety is biologically significant and is often interpreted to be the equivalent of the subspecies, a term favoured by the zoologists. The Smithsonian Institution (Ayensu and DeFilipps 1978) pointed out this anomaly, or unequal treatment of plants *vis-à-vis* animals, but fortunately in actual practice varieties of plants have been officially listed by the Department of the Interior. The first variety to be listed as endangered and given protection under the Act was *Erysimum capitatum* var. *angustatum* (Fig. 1), a Californian crucifer known as the Contra Costa wallflower, listed on April 26, 1978, and published concomitantly with several subspecies of Californian *Oenothera* (Onagraceae) species (e.g. Fig. 2).

The successive procedures which must be completed by government agencies and individuals, in order to list a species as endangered, extend over a period of time during which the name of a species may change, either because the original name is found to be illegitimate or else because taxonomists have decided that

---

Systematics Association Special Volume No. 25, "Current Concepts in Plant Taxonomy", edited by V. H. Heywood and D. M. Moore, 1984. Academic Press, London and Orlando.
ISBN 0 12 347060 9

Fig. 1. *Erysimum capitatum* var. *angustatum* (Cruciferae), the Contra Costa wallflower, is the first plant variety officially listed as endangered pursuant to the Endangered Species Act of 1973. Photo by P. Opler.

Fig. 2. *Oenothera deltoides* subsp. *howellii* (Onagraceae), the Antioch Dunes evening primrose, is one of the first officially listed endangered plant subspecies. Photo by R. F. Thorne.

Fig. 3. *Dudleya traskiae* (Crassulaceae), the Santa Barbara Island live-forever from California, is an officially listed endangered plant species. Photo by R. Moran.

the organism should be named at a different rank or within a different superordinate taxon. Changes of name, fortunately, have been accepted as posing no obstacle to the legal process of listing.

Among the 61 plants currently officially listed as endangered or threatened, for explanatory purposes it has been necessary to give, for 27 of them, 44 synonyms which have been applied to them during the course of regulatory proceedings. Five of the synonyms have ultimate epithets (four at the same rank) different from those of the names finally listed; one is a *nomen nudum*; and one of them was a misapplication. Another of the mentioned names replaces a homonym originally suggested as a threatened species.

The Cactaceae dominate the set of 61 listed and protected plants; among the others is the succulent *Dudleya traskiae* (Crassulaceae) (Fig. 3). Moreover, other commercially exploited plants comprise a large proportion of the list, though

Fig. 4. *Pedicularis furbishiae* (Scrophulariaceae), the Furbish lousewort, is an endangered species occurring in Maine and New Brunswick, Canada. Photo by E. S. Ayensu.

Fig. 5. Yellow flowers of *Sarracenia oreophila* (Sarraceniaceae), the endangered green pitcher plant from Alabama and Georgia. Photo by G. W. Folkerts.

the species which have been publicized by journalists, such as the Furbish lousewort (*Pedicularis furbishiae*, Scrophulariaceae) (Fig. 4) are endangered inadvertently by the potential destruction by man-made development projects such as hydroelectric dams, rather than by exploitation for their ornamental qualities. It is among the attractive, exploited, widely-known plants that memorable taxonomic changes and controversies have been concentrated.

The carnivorous pitcher plant genus *Sarracenia* includes several races which variously are, (1) exploited, (2) of doubtful identity or affinity, and (3) possibly not seriously threatened in spite of drastic human interference with their habitats. One of the sarracenias, finally found to be endangered, is listed under the name *S. oreophila* (Fig. 5), without synonyms (US Department of the Interior 1980). Three subspecies, not including the type, of *S. rubra* are currently under review; they have otherwise been known as *S. alabamensis* (Alabama canebrake pitcher plant) and its subsp. *S. wherryi*, and *S. jonesii* (Figs 6–10). Found not to be currently in danger were *S. psittacina* (parrot pitcher plant) and typical *S. rubra*. It is debatable whether some of the pitcher plant taxa now under review actually designate taxonomically separable populations, but in the main they appear to be distinct entities.

Some sarracenias are among the species for which further action has had to be postponed by the Interior Department, because deadlines have passed, until new information justifies further consideration. Such new information is available, however, for most taxa now under review by the US Fish and Wildlife Service, and the sarracenias could be advanced to the status of "proposed rulemaking", if the Service were to be authorized to propose new names for official listing, and if the information were found to justify their proposal.

The sarracenias grow principally in the southereastern part of the United States, although a few species, e.g. *S. purpurea*, grow in the northeastern states. Several species have never been thought endangered or threatened. Among their habitats are bogs, where human visitation quickly converts the sphagnum surface into an aquatic surface, and where grow also many plant species sensitive to disturbance. But some of the sarracenias have been claimed to act almost as pioneers in disturbed areas, and even to propagate themselves more vigorously than usual as a result of the removal of individuals of their own species. Nevertheless, some members of the genus are in trouble.

The growth-habit of a sarracenia, as reflected by leaf-shape and size, is sensitive to environmental influence, so detailed statistical study is required to find out which morphological characters reflect genetic differences. Such study turns out to be possible with herbarium material, in spite of the gross distortion of the leaves' tubular shape when a specimen is pressed. And so some closely-related

Fig. 6. Flower of *Sarracenia alabamensis* subsp. *alabamensis* (Sarraceniaceae), the Alabama canebrake pitcher plant, subjected to recent taxonomic confusion within the *S. rubra* complex. Photo by G. W. Folkerts.

Fig. 7. Leaves of *Sarracenia alabamensis* subsp. *wherryi* (Sarraceniaceae), a component of the *S. rubra* species complex in Alabama. Photo by G. W. Folkerts.

Fig. 8. Flowers of *Sarracenia alabamensis* subsp. *wherryi*. Photo by G. W. Folkerts.

Fig. 9 (*left*). Leaf of *Sarracenia jonesii* (Sarraceniaceae), a component of the *S. rubra* species complex growing in Etowah, North Carolina. Photo by J. D. Pittillo.

Fig. 10 (*below*). Flowers of *Sarracenia jonesii* in Etowah, North Carolina. Photo by J. D. Pittillo.

races have been found constantly separable, and even to inhabit ranges disjunct from each other, though they occur in company with other representatives of the genus which are not so nearly related to them (Case and Case 1974, 1976; Schnell 1977).

What is significant for us here is that a consensus on *Sarracenia* taxonomy is just now appearing, during the time when the well-being of the several taxa is being evaluated. Fortunately, these and other carnivorous plant genera attract scientific investigation, and not only exploitation.

The question of rarity will always present itself in considerations of status. A rare species is one that has a small population in its range. When a rare species, subspecies or variety appears on a list of endangered, threatened or possibly extinct plants, it becomes subject to two critical questions (Ayensu 1981):

(1) Is the "extinct" species truly extinct? Perhaps it actually does occur in small numbers or is more common or widespread but has not been observed or collected because, (*a*) it is small or inconspicuous, (*b*) it produces flowers rarely, briefly, or at unusual times, germinates infrequently, or has long dormant periods, or (*c*) it occurs only in difficult or inaccessible terrain or habitats rarely visited by collectors.

(2) Has the taxon been correctly categorized as a natural entity? Perhaps it is a taxonomically untenable segregate of a more common species; it could be an unusual phenotype owing to some unusual edaphic condition or environmental stress. Or, it may be a rarely produced polyploid, aberrant, mutant, or hybrid.

Under the US Endangered Species Act, hybrids have no status and cannot be afforded conservation measures. That is why they are absent from the Smithsonian recommended lists of endangered and threatened species. But this situation does not peclude hybrids from being given protection by individual states. One example is a plant protected as endangered by North Carolina state law, the Tennessee bladder fern (*Cystopteris* × *tennesseensis*). It occurs in two coastal plain counties in North Carolina as a disjunct separated from the normal range in Tennessee, 400 miles away. It is thought to have arisen as a hybrid between *C. bulbifera* and *C. protrusa*; neither of the putative parental species is found on the coastal plain. Another remarkable disjunct protected in North Carolina is Wright's cliff brake fern (*Pellaea* × *wrightiana*), which occurs 800 to 1000 miles away in Oklahoma and Texas (Hardin *et al.* 1977). Is protection of such rare disjuncts justifiable, even if they are hybrids? Obviously so, in the state of North Carolina. Yet, the very rare *Panicum shastense* from California was found to be a first generation sterile hybrid between *P. pacificum* and *P. scribnerianum* (Spellenberg 1970), and summarily dropped from further consideration by state and federal authorities.

The existence of taxonomic confusion in certain plant groups is advantageous to unscrupulous importers and exporters, who have tried to trade in endangered plants by labelling them with obscure synonyms, in order to avoid detection by inspectors who rely on one commonly accepted name only for a species.

In the opinion of George Argus (1978), with regard to the orchids and cacti on the CITES Appendices:

> the fact that many of these plants are traded in a condition in which positive identification is difficult, if not impossible (i.e. lacking flowers), would make it possible for Canada, or any other country, to be used as a "country of convenience" for the re-export of endangered exotic species if we (Canada) did not monitor trade of all species in these families.

Incidentally, it may be mentioned that the taxonomy of *Cannabis*, or marijuana, has been contested in a number of court cases in the United States over the past few years. *Cannabis sativa* is the illegal or controlled substance, but sometimes material of *C. indica* is involved. Some botanists have testified their opinions that the two entities are, at least anatomically, distinct. On the other hand Small and Cronquist (1976) believe *Cannabis* consists of a single highly variable species, *C. sativa*, with subsp. *sativa* and subsp. *indica*. Within each subspecies are two parallel phases (discernible groups recognized as named varieties): a wild (weedy, naturalized or indigenous) phase and a domesticated (cultivated or spontaneous) phase (see also Small 1979). In many cases it is the judge who decides in court which taxonomic opinion as to the consituent elements of *Cannabis* is valid for a particular lawsuit.

"Parochial taxonomy" is a term used to describe the lack of communication that resulted in, for example, nine different names being given to one species of shrub, *Cestrum megalophyllum* (Solanaceae), in nine different countries, a nomenclaturally monstrous situation reported by D'Arcy (1977). Multiple names in this case created an illusion of many endemic species, when in fact the species is only one entity and is widely distributed. But, as Fernández-Pérez (1977) has noted, certain populations of the same species (ecotypes) can become adapted to different ecological conditions, presenting major obstacles in saving the species from extinction by transplanting to different areas.

By means of selecting endemic species as the ones for first consideration for protection, the Smithsonian Institution/Threatened Plants Committee Latin American Project has been able to find from taxonomic specialists that some taxa listed as endemic in the standard Floras are now obsolete or otherwise questionable. Examples include nine taxonomically unreliable endemic Phoradendrons (Loranthaceae) from Guatemala; four Ipomoeas (Convolvulaceae) from Guatemala; five Meliaceae now considered synonyms, *formae*, or based on too fragmentary material for determination of status, from Guatemala; and the fact that there is a taxonomic problem with *Gouania hypoglanca* (Rhamnaceae) in Costa Rica, which is probably *G. polygama* (Jacq.) Urban *sensu lato*.

For these data we thank Drs J. Kuijt, D. Austin, T. D. Pennington, and M. C. Johnston, and others who are helping to make the selection of species for conservation a more exact exercise by weeding out the confusing and problematical elements.

Dr Richard A. Howard, editor of the "Flora of the Lesser Antilles" series from the Arnold Arboretum of Harvard University, has observed with regard to the taxonomic studies and previous Floras of the Caribbean, that "the nationalist and single-island approach has led to the over-description of species, as recent monographs clearly show. In all such studies, more endemic species are being reduced than new species described" (Howard 1977). Some of the problems here relate to old collections with incomplete or misleading data, including plants grown in botanical gardens but originating in countries elsewhere, for which only the name of the city in which the garden is located is given on the collector's label. Howard also details the misidentification problems of certain West Indian tree-ferns (Cyatheaceae), which have been resolved with the conclusion that none of these segregates is endangered since all are abundant locally. Since all tree-ferns are listed on the Convention on International Trade (CITES), more clarification of the status of these plants in other regions of the world is going to be needed, as the listing procedure under the Berne Criteria demands; biological, as well as trade, information.

Many type specimens of species known only from the type collection probably represent truly distinguishable species which still occur in nature and are vulnerable, but the resources of conservationists are insufficient to seek them all, even if the locality data for the places whence the specimens came are fairly detailed. A list of Colombian plants known only from the type collection was derived by Fernández-Pérez (1977) from several monographs in the "Flora Neotropica" series, indicating two such species in *Swartzia* (Leguminosae), one in Brunelliaceae, seven in Chrysobalanaceae, three in Dichapetalaceae, one in Caryocaraceae, and nineteen in the Pitcairnioideae of Bromeliaceae.

Unfortunately, there are some plants about which we have little hope of understanding, which have simply disappeared. In the case of the distinctive "Halakau Red" hibiscus not even a type specimen was preserved, for the plant was never given a scientific name. The plant was once in cultivation and believed to represent the only native red *Hibiscus* of the island of Hawaii, but it eventually died out without being propagated. Its taxonomic relations can only be guessed at, as it was believed that it possibly could have even been an imported *H. boryanus*.

In an essay on "Endemic taxa and the taxonomist", Richardson (1978) has indicated many of the practical and theoretical problems to be encountered during the taxonomist's quest to recognize taxa, i.e. to recognize discontinuous variation. During the areal progression of species in a monophyletic assemblage across the passage of time, the species undergo an origin, expansion, stabilization, diversification, migration and/or fragmentation, and then a period of contraction, relictualism, and finally extinction. The taxonomist is, of course, studying populations during a certain phase of the species' evolutionary journey, in which stage it may be either a neoendemic, holoendemic or palaeoendemic, depending on how far down the road to extinction it happens to be. All species begin as neoendemics and end as palaeoendemics (Richardson 1978).

For example, Morton (1976) has stated that:

> any statistics on endemism in the Canadian flora are virtually meaningless because most of our endemics are young endemics which have barely diverged sufficiently to be regarded as species. As a result there is little agreement amongst botanists whether to treat them as species or as infra-specific variation. However, this should in no way detract from the importance of these plants, and there are many of them, or of the habitats in which they occur and are evolving.

Further to this point is the observation by Argus and White (1978) that some plants considered to be endemic to Alberta province, such as *Carex athabascensis* and *Draba kananaskis*, are members of taxonomically difficult groups and their taxonomic status may be questioned.

The original stock of a plant may have mutated and its progeny have survived variously in response to competition for ecological niches, or changes in the composition of the community causing great stress, or pest pressure, or through adaptive radiation; it may have arisen through quantum speciation; and it somewhere may have an allopatric or sympatric counterpart species. Amidst this welter of phenomena the taxonomist must delineate the taxa. We may receive an idea of this milieu from the remarks of Polunin (1970), who noted, with reference to botanical conservation in the Arctic, that there are in all ten species of arctic-endemic vascular plants which need to be preserved. However, these ten consist of *Poa nascopieana* and *Carex jacobi-peteri* which are little known and somewhat doubtful species; *Ranunculus spitsbergenensis* and *Saxifraga nathorstii* which are apparently of hybrid origin but nonetheless highly characteristic and restricted in range; an unconfirmed record of another species; four species-segregates which were upheld while excluding taxa that were "sunk" years before; and another taxon which had recently been authoritatively set up as a species.

The very problems of rarity itself are among those which we are seeking to solve by means of conserving examples of rare endemics in nature parks and preserves for future investigation.

REFERENCES

Argus, G. (1978). "List of Canadian Flora Affected by CITES". CITES Reports No. 4. Canadian Wildlife Service, Ottawa.

Argus, G. W. and White, D. J. (1978). "The Rare Vascular Plants of Alberta", Syllogeus Series No. 17. National Museums of Canada, Ottawa.

Ayensu, E. S. (1981). Assessment of threatened plant species in the United States. *In* "The Biological Aspect of Rare Plant Conservation" (H. Synge, ed.) pp. 19–58. John Wiley, London, New York.

Ayensu, E. S. and DeFilipps, R. A. (1978). "Endangered and Threatened Plants of the United States". Smithsonian Institution and World Wildlife Fund – US, Washington, D.C.

Case, F. W. and Case, R. B. (1974). *Sarracenia alabamensis*, a newly recognised species from central Alabama. *Rhodora*, **76,** 650–665.

Case, F. W. and Case, R. B. (1976). The *Sarracenia rubra* complex. *Rhodora*, **78,** 270–325.

D'Arcy, W. G. (1977). Endangered landscapes in Panama and Central America: the threat to plant species. *In* "Extinction is Forever" (G. T. Prance and T. S. Elias, eds) pp. 89–104. New York Botanical Garden, New York.

Fernández-Pérez, A. (1977). The preparation of the endangered species list of Colombia. *In* "Extinction is Forever" (G. T. Prance and T. S. Elias, eds) pp. 117–127. New York Botanical Garden, New York.

Hardin, J. W. and Committee (1977). Vascular plants. *In* "Endangered and Threatened Plants and Animals of North Carolina" (J. E. Cooper, S. S. Robinson and J. B. Funderburg, eds) pp. 56–142. North Carolina State Museum of Natural History, Raleigh, N.C.

Howard, R. A. (1977). Conservation and the endangered species of plants in the Carribbean Islands. *In* "Extinction is Forever" (G. T. Prance and T. S. Elias, eds) pp. 105–114. New York Botanical Garden, New York.

Morton, J. K. (1976). Recent changes in the Canadian flora. *In* "Man's Impact on the Canadian Flora" (J. K. Morton, ed.). Supplement to the Canadian Botanical Association Bulletin **9(1),** 13–16.

Polunin, N. (1970). Botanical conservation in the Arctic. *Biological Conservation* **2(3),** 197–205.

Richardson, I. B. K. R. (1978). Endemic taxa and the taxonomist. *In* "Essays in Plant Taxonomy" (H. E. Street, ed.) pp. 245–262. Academic Press, London, New York.

Schnell, D. E. (1977). Infraspecific variation in *Sarracenia rubra* Walt.: some observations. *Castanea*, **42,** 149–170.

Small, E. (1979). "The Species Problem in *Cannabis*: Science and Semantics"; Vol. 1 "Science"; Vol. 2 "Semantics". Corpus Information Services, Toronto.

Small, E. and Cronquist, A. (1976). A practical and natural taxonomy for *Cannabis*. *Taxon* **25(4),** 405–435.

Spellenberg, R. W. (1970). *Panlcum shastense* (Graminae), a sterile Hybrid between *P. pacificum* and *P. scribnerianum*. *Brittonia* **22(2),** 154–162.

US Department of the Interior, Fish and Wildlife Service (1980). Endangered and threatened wildlife and plants: review of plant taxa for listing as endangered or threatened species. *Federal Register*, **45(242),** 82479-82569 (15 December 1980).

Subject Index

Editor-in-Chief, Special Volume Series

D. L. HAWKSWORTH PhD DSc FLS FIBiol

Systematics Association Publications

1. BIBLIOGRAPHY OF KEY WORKS FOR THE IDENTICATION OF THE BRITISH FAUNA AND FLORA *3rd edition* (1967)
   Edited by G. J. Kerrich, R. D. Meikle and N. Tebble Out of print
2. FUNCTION AND TAXONOMIC IMPORTANCE (1959)
   Edited by A. J. Cain
3. THE SPECIES CONCEPT IN PALAEONTOLOGY (1956)
   Edited by P. C. Sylvester-Bradley
4. TAXONOMY AND GEOGRAPHY (1962)
   Edited by D. Nichols
5. SPECIATION IN THE SEA (1963)
   Edited by J. P. Harding and N. Tebble
6. PHENETIC AND PHYLOGENETIC CLASSIFICATION (1964)
   Edited by V. H. Heywood and J. McNeil Out of print
7. ASPECTS OF TETHYAN BIOGEOGRAPHY (1967)
   Edited by C. G. Adams and D. V. Ager
8. THE SOIL ECOSYSTEM (1969)
   Edited by H. Sheals
9. ORGANISMS AND CONTINENTS THROUGH TIME (1973)†
   Edited by N. F. Hughes

Published by the Association

Systematics Association Special Volumes

1. THE NEW SYSTEMATICS (1940)
   Edited by Julian Huxley (Reprinted 1971)
2. CHEMOTAXONOMY AND SEROTAXONOMY (1968)*
   Edited by J. G. Hawkes
3. DATA PROCESSING IN BIOLOGY AND GEOLOGY (1971)*
   Edited by J. L. Cutbill
4. SCANNING ELECTRON MICROSCOPY (1971)*
   Edited by V. H. Heywood
5. TAXONOMY AND ECOLOGY (1973)*
   Edited by V. H. Heywood
6. THE CHANGING FLORA AND FAUNA OF BRITAIN (1974)*
   Edited by D. L. Hawksworth
7. BIOLOGICAL IDENTIFICATION WITH COMPUTERS (1975)*
   Edited by R. J. Pankhurst

* Published by Academic Press for the Systematics Association

†Published by the Palaeontological Association in conjunction with the Systematics Association

8. LICHENOLOGY: PROGRESS AND PROBLEMS (1976)★
Edited by D. H. Brown, D. L. Hawksworth and R. H. Bailey
9. KEY WORKS TO THE FAUNA AND FLORA OF THE BRITISH ISLES AND NORTHWESTERN EUROPE (1978)★
Edited by G. J. Kerrich, D. L. Hawksworth and R. W. Sims
10. MODERN APPROACHES TO THE TAXONOMY OF RED AND BROWN ALGAE (1978)★
Edited by D. E. G. Irvine and J. H. Price
11. BIOLOGY AND SYSTEMATICS OF COLONIAL ORGANISMS (1979)★
Edited by G. Larwood and B. R. Rosen
12. THE ORIGIN OF MAJOR INVERTEBRATE GROUPS (1979)★
Edited by M. R. House
13. ADVANCES IN BRYOZOOLOGY (1979)★
Edited by G. P. Larwood and M. B. Abbot
14. BRYOPHYTE SYSTEMATICS (1979)★
Edited by G. C. S. Clarke and J. G. Duckett
15. THE TERRESTRIAL ENVIRONMENT AND THE ORIGIN OF LAND VERTEBRATES (1980)★
Edited by A. L. Panchen
16. CHEMOSYSTEMATICS: PRINCIPLES AND PRACTICE (1980)★
Edited by F. A. Bisby, J. G. Vaughan and C. A. Wright
17. THE SHORE ENVIRONMENT: METHODS AND ECOSYSTEMS (2 Volumes) (1980)★
Edited by J. H. Price, D. E. G. Irvine and W. F. Farnham
18. THE AMMONOIDEA (1981)★
Edited by M. R. House and J. R. Senior
19. BIOSYSTEMATICS OF SOCIAL INSECTS (1981)★
Edited by E. Howse and J.-L. Clément
20. GENOME EVOLUTION (1982)★
Edited by G. A. Dover and R. B. Flavell
21. PROBLEMS OF PHYLOGENETIC RECONSTRUCTION (1982)★
Edited by K. A. Joysey and A. E. Friday
22. CONCEPTS IN NEMATODE SYSTEMATICS (1983)★
Edited by A. R. Stone, H. M. Platt and L. F. Khalil
23. EVOLUTION, TIME AND SPACE: THE EMERGENCE OF THE BIOSPHERE (1983)★
Edited by R. W. Sims, J. H. Price and P. E. S. Whalley
24. PROTEIN POLYMORPHISM: ADAPTIVE AND TAXONOMIC SIGNIFICANCE (1983)★
Edited by G. S. Oxford and D. Rollinson
25. CURRENT CONCEPTS IN PLANT TAXONOMY (1984)★
Edited by V. H. Heywood and D. M. Moore
26. DATABASES IN SYSTEMATICS (1984)★
Edited by R. Allkin and F. A. Bisby

★ Published by Academic Press for the Systematics Association